On Normalized Integral Table Algebras
(Fusion Rings)

T0212015

For further volumes:
www.springer.com/series/6253

Algebra and Applications

Volume 16

Algebra and Applications aims to publish well written and carefully refereed monographs with up-to-date information about progress in all fields of algebra, its classical impact on commutative and noncommutative algebraic and differential geometry, K-theory and algebraic topology, as well as applications in related domains, such as number theory, homotopy and (co)homology theory, physics and discrete mathematics.

Particular emphasis will be put on state-of-the-art topics such as rings of differential operators, Lie algebras and super-algebras, group rings and algebras, C^*-algebras, Kac-Moody theory, arithmetic algebraic geometry, Hopf algebras and quantum groups, as well as their applications. In addition, Algebra and Applications will also publish monographs dedicated to computational aspects of these topics as well as algebraic and geometric methods in computer science.

Zvi Arad · Xu Bangteng · Guiyun Chen ·
Effi Cohen · Arisha Haj Ihia Hussam ·
Mikhail Muzychuk

On Normalized Integral Table Algebras (Fusion Rings)

Generated by a Faithful Non-real Element of Degree 3

Springer

Zvi Arad
Department of Mathematics
Bar Ilan University
Ramat Gan, 52900
Israel
and
Netanya Academic College
1 University Street
Netanya
Israel
aradtzvi@netanya.ac.il

Xu Bangteng
Department of Mathematics and Statistics
Eastern Kentucky University
Richmond, KY, 40475
USA
bangteng.xu@eku.edu

Guiyun Chen
Department of Mathematics
Southwest University
Chongqing, 400715
People's Republic of China
gychen@swu.edu.cn

Effi Cohen
Department of Mathematics
Bar Ilan University
Ramat Gan, 52900
Israel
Cohene3@walla.com

Arisha Haj Ihia Hussam
Department of Mathematics
Alqasemi Academic College of Education
Baqa El-Gharbieh, 30100
Israel
Hussam10@walla.co.il

Mikhail Muzychuk
Netanya Academic College
Kibbutz Galuyot 16
Netanya, 42 365
Israel
muzy@netanya.ac.il

ISSN 1572-5553   e-ISSN 2192-2950
ISBN 978-1-4471-2703-1   ISBN 978-0-85729-850-8 (eBook)
DOI 10.1007/978-0-85729-850-8
Springer London Dordrecht Heidelberg New York

British Library Cataloguing in Publication Data
A catalogue record for this book is available from the British Library

Mathematics Subject Classification: 05E10, 05E30, 13A99, 20C15, 20N20, 81T40

Cover design: VTeX UAB, Lithuania

Printed on acid-free paper

Springer is part of Springer Science+Business Media (www.springer.com)

Preface

Arad and Chen proved in [AC] and [AC1] that a Normalized Integral Table Algebra (fusion ring) (A, B) generated by a non-real faithful element $b_3 \in B$ of degree three the non-identity elements of which have minimal degree 3 satisfies the condition $b_3 \bar{b}_3 = 1 + b_8$ where $b_8 \in B$ is an element of degree 8. They also showed that the general case naturally splits into four main sub-cases:

(1) $(A, B) \cong_x (CH(PSL(2, 7)), Irr(PSL(2, 7)))$;
(2) $b_3^2 = b_4 + b_5$ where $b_4, b_5 \in B$ are elements of degrees 4 and 5;
(3) $b_3^2 = \bar{b}_3 + b_6$ where $b_6 \in B$ is a non-real element of degree 6;
(4) $b_3^2 = c_3 + b_6$ where $c_3, b_6 \in B$ are elements of degrees 3 and 6, $c_3 \neq b_3, \bar{b}_3$.

The cases (1), (3) and (4) are considered in Chap. 2. Chapter 3 deals with the case (2). Chapters 4 and 5 analyze the most complicated case—the third one. We developed new original methods for enumerating NITAs in the title. Using the developed technique we settled the above cases almost completely.

Acknowledgements

Zvi Arad is sincerely grateful to Eastern Kentucky University (EKU) for inviting him to head the Wilson Endowment Chair in the Department of Mathematics and Statistics in Spring 2009. The warm hospitality and friendly atmosphere there enabled him to conduct the joint research of Chap. 3 in the spring semester of 2009.

Effie Cohen submitted Chaps. 4 and 5 of this book in partial fulfillment of the requirements for the Ph.D. degree in the Department of Mathematics, Bar-Ilan University, Ramat Gan, Israel, under the supervision of Prof. Zvi Arad and Prof. Malka Schaps.

The authors thank the anonymous referees for their helpful suggestions and comments.

The author thanks Mrs. Miriam Beller for her great efforts to print the first version of the book and for her helpful remarks.

Contents

Chapter 1
Introduction

The study of the properties of products of conjugacy classes of finite groups is an old branch of finite group theory. This topic was extensively studied in the 1980's. The book [AH2], Products of Conjugacy Classes in Groups, edited by Z. Arad and M. Herzog, provides a comprehensive picture of the results obtained during this period.

It was realized by several researchers that this investigation could be extended to products of irreducible characters. We refer the reader to the papers [AH1, BC]. This led to the notion of Table Algebra, introduced by Z. Arad and H. Blau in [AB], in order to study in a uniform way the decomposition of products of conjugacy classes $Cla(G)$ and irreducible characters $Irr(G)$ of finite groups G. Since then the theory of Table Algebras has been extensively developed. Today the theory of Table Algebras is an important branch of modern algebra.

This book gives a classification of Normalized Integral Table Algebras (Fusion Rings) generated by a faithful non-real element of degree 3. It contains an Introduction and 4 chapters. Zvi Arad is a co-author of each chapter of the book and planned the outline of attack to the solution of the classification of the family of Table Algebras. The co-authors of the 5 chapters are as follows: The Introduction (Chap. 1) was written by Zvi Arad and Effi Cohen. Chapter 2 is a joint research of Zvi Arad, Guiyun Chen and Arisha Haj Ihia Hussam. Chapter 3 is a joint research of Zvi Arad and Xu Bangteng. Chapter 4 is a joint research of Zvi Arad and Effi Cohen. Chapter 5 is a joint research of Zvi Arad, Effi Cohen and Mikhail Muzychuk.

We state the definition of Table Algebra and give two important examples of Table Algebras.

Definition 1.1 Let $B := \{b_1 = 1, b_2, \ldots, b_k\}$ be a distinguished basis of a finite dimensional, associative and commutative algebra A over the complex field $\mathbb{C}$ with identity element 1_A. Then (A, B) is a Table Algebra (B is a table basis and $|B|$ is the dimension of the Table Algebra (A, B)) if the following hold:

TA1. For all i, j, m, we have $b_i b_j = \sum_{m=1}^{k} \lambda_{ijm} b_m$ with λ_{ijm} a nonnegative real number.

Z. Arad et al., *On Normalized Integral Table Algebras (Fusion Rings)*,
Algebra and Applications 16, DOI 10.1007/978-0-85729-850-8_1,
© Springer-Verlag London Limited 2011

TA2. There is an algebra automorphism (denoted by –) of A whose order divides 2, such that $b_i \in B$ implies that $\bar{b}_i \in B$ (then $\bar{i}$ is defined by $b_{\bar{i}} = \bar{b}_i$ and $b_i \in B$ is called real if $b_{\bar{i}} = b_i$).

TA3. There is a coefficient function $g : B \times B \longrightarrow \mathbb{R}^+$ such that $\lambda_{ijm} = g(b_i, b_m)\lambda_{\bar{j}mi}$ where λ_{ijm}, $\lambda_{\bar{j}mi}$ are defined in TA1 and TA2 for all i, j, m.

Remark For all i, j, m as in TA1, we define $Irr(b_i b_j)$ to be the set $\{b_m | \lambda_{ijm} \neq 0\}$.

The methods of research that were developed and the results of Table Algebras provide, in many cases, surprising and innovative results that were not known before. The results can be used in finite group theory and graph theory.

Definition 1.2 Let (A, B) be a Table Algebra. A subset $D \subseteq B$ is called a table subset of B if $D \neq \emptyset$ and $Irr(b_i b_j) \subseteq D$ for all $b_i, b_j \in D$. A subalgebra of A generated by some table subset of B is called a table subalgebra of (A, B). We say that (A, B) is simple if the only table subsets of B are B and $\{1\}$.

Definition 1.3 Let (A, B) be a Table Algebra and fix $c \in B$. The set generated by c is defined as $B_c := \{\bigcup_{i=1}^{\infty} Irr(c^i)\}$. An element $b \in B$ will be called faithful if $B_b = B$. Note that B_c is a table subset of B.

Definition 1.4 If (A, B) is a Table Algebra, then, by [AB], there exists a unique algebra homomorphism $| \, | : A \longrightarrow \mathbb{R}$ such that $|b| = |\bar{b}| > 0$, $\forall b \in B$. We call $| \, |$ the degree homomorphism. The positive real numbers $\{|b|\}_{b \in B}$ are called the degrees of (A, B). An element $b_i \in B$ is called standard if $|b_i| = \lambda_{i\bar{i}1}$. A Table Algebra (A, B) is called standard if all elements in B are standard. If for all $b_i \in B$, $\lambda_{i\bar{i}1} = 1$, then (A, B) is a Normalized Table Algebra. A Table Algebra (A, B) is called homogenous of degree λ if $|x| = \lambda$ for all $x \in B \setminus \{1\}$. A Table Algebra is called Integral if for all $b \in B$ $|b| \in \mathbb{N}$ and for all i, j, m, $\lambda_{ijm} \in \mathbb{N} \cup \{0\}$. The number $o(B) := \sum_{i=1}^{k} \frac{|b_i|^2}{\lambda_{i\bar{i}1}}$ is called the order of B. In particular if (A, B) is Normalized, then $o(B) = \sum_{b \in B} |b|^2$.

Definition 1.5 Two Table Algebras (A, B) and (A', B') are called isomorphic (denoted $B \cong B'$) when there exists an algebra isomorphism $\psi : A \longrightarrow A'$ such that $\psi(B)$ is a rescaling of B'; the algebras are called strictly isomorphic (denoted $B \cong_x B'$) when $\psi(B) = B'$. Therefore, $B \cong_x B'$ means that B and B' yield the same structure constants.

Table Algebras, as defined, may be considered a special class of C-algebras, introduced by Y. Kawada [K] and G. Hoheisel [H]. More precisely, a Table Algebra is a C-algebra where the structure constants are nonnegative. Each finite group yields two natural Table Algebras: the Table Algebra of conjugacy classes and the Table Algebra of irreducible characters.

Example 1.1 Let G be a finite group. Let $Ch(G)$ denote the set of all complex valued class functions on G, a commutative algebra under pointwise addition and multiplication. Let $Irr(G)$ be the set of irreducible characters of G, a basis for $Ch(G)$. Then $(Ch(G), Irr(G))$ satisfies all the axioms TA1–TA3 where—extends linearly from $\chi_i \longrightarrow \bar{\chi}_i$ (complex conjugate characters) and $|\chi_i| = \chi_i(1)$ for all $\chi_i \in Irr(G)$. Here each λ_{ijm} is a nonnegative integer and each $g(\chi_i, \chi_j) = 1$. (Clearly, Example 1.1 is a Normalized Integral Table Algebra.)

Example 1.2 Let G be a finite group and let $Z(\mathbb{C}G)$ be the center of the complex group algebra. If $C_1 = 1$, $C_2, \ldots, C_k$ are the conjugacy classes of G. Let $b_i = \sum_{g \in C_i} g$ for each C_i, and let $Cla(G) = \{b_1, \ldots, b_k\}$. Then $(Z(\mathbb{C}G), Cla(G))$ satisfies TA1–TA3, where—extends linearly from $b_i \longrightarrow \bar{b}_i = \sum_{g \in C_i} g^{-1}$, and $|b_i| = |C_i|$. Here also each λ_{ijm} is a nonnegative integer, but $g(b_i, b_m) = |C_i|/|C_m|$. (Clearly, Example 1.2 is a Standard Table Algebra.)

"Groups occur as the algebraic abstraction of sets of symmetries, closed under composition and inversion, of geometric spaces. Table algebras may be regarded as playing a role analogous to that of groups for a generalization of geometric spaces; namely, for highly symmetric combinatorial configurations. Perhaps the most central instance of this perspective is the view of a Table Algebra as an abstraction of the adjacency algebra of an association scheme." See [B2].

We can find the axioms of the Table Algebra (A, B) in many different types of algebras which were the subjects of intensive research over the last hundred years. In the excellent survey article [B2] of Harvey Blau, there is a comprehensive picture of Table Algebras and an explanation of the strong connection between Table Algebras to different types of algebras. The following is a partial list of those algebras: Schur rings, C-algebras, association schemes (Banai Itô), hypergroups (Wildberger), posets, Bose-Mesner algebras, Hopf algebras, distance regular graphs, products of conjugacy classes, products of ordinary characters, fusion rings (a concept from physics), fusion rule algebras, pseudo groups (Brauer), coherent algebras (Higman), generalized circulants of finite groups (Kazhdan-Lusztig).

Normalized Integral Table Algebras are also known as integral commutative fusion rings, or integral fusion rule algebras. These are an important part of the theory of fusion categories, which is now a very active research area. It is related to conformal field theory and quantum physics, as well as to Hopf algebras, and is pursued independently of Table Algebras.

Table Algebras generated by elements of small degree were extensively investigated, as follows:

Homogenous Integral Table Algebras of degrees $1, 2, 3, 4$; see [AB, A, B, ABFMMX] and [AEM].

Integral Table Algebras (denoted ITA) generated by a faithful element of degree 2 without non-identity basis elements of degree 1; see [B].

ITA generated by a faithful non-real element of degree 3, see [B1].

Standard Integral Table Algebras (denoted SITA) generated by a faithful non-real element of degree 3; see [AAFM].

SITA generated by a faithful non-real element of degree 4; see [AM].

Normalized Integral Table Algebras (denoted NITA) generated by a faithful element of degree two without non-identity elements of degree 1; see [FK].

NITA generated by a faithful element of degree 3 without non-identity basis elements of degree 1 or 2; see [AC] and [AC1].

In [AC] and [AC1], the classification has not been completed. However, Arad and Chen have proved the following Theorem.

We use the following notation. If (A, B) is a Table Algebra then $b_n, c_n \in B$ denote elements of B of degree n.

Theorem *Let* (A, B) *be NITA generated by a non-real element* $b_3 \in B$ *of degree* 3 *and without non-identity basis elements of degree* 1 *or* 2. *Then* $b_3\bar{b}_3 = 1 + b_8$, $b_8 \in B$ *and one of the following holds:*

1. $(A, B) \cong_x (CH(PSL(2, 7)), Irr(PSL(2, 7)))$.
2. $b_3^2 = b_4 + b_5$ *where* $b_4, b_5 \in B$.
3. $b_3^2 = \bar{b}_3 + b_6$ *where* $b_6 \in B$ *is a non-real element.*
4. $b_3^2 = c_3 + b_6$, *where* $c_3, b_6 \in B$ *and* $c_3 \neq b_3, \bar{b}_3$.

In this book we investigate Cases 2–4 of the above theorem, and we have made progress in order to complete the classification. To this end, we use the following definitions:

Definition 1.6 Let $B := \{b_1 = 1, b_2, \ldots, b_l, \ldots\}$ be a distinguished basis of a countable dimensional, associative and commutative algebra A over the complex field $\mathbb{C}$ with identity element 1_A. If $a \in A$ and $b_i \in B$, let $k(b_i, a)$ be the coefficient of b_i in a. Then (A, B) is a Countable Table Algebra (B is a table basis and $|B|$ is the dimension of the Table Algebra (A, B)) if the following hold:

TA1. For all $i, j, m \in \mathbb{N}$ $b_i b_j = \sum_{m=1}^{k} b_{ijm} b_m$ where $k \in \mathbb{N}$ with b_{ijm} a nonnegative real number.

TA2. There is an algebra automorphism (denoted by –) of A whose order divides 2, such that $b_i \in B$ implies that $\bar{b}_i \in B$ (then $\bar{i}$ is defined by $b_{\bar{i}} = \bar{b}_i$ and $b_i \in B$ is called real if $b_{\bar{i}} = b_i$).

TA3. There is a function $g : B \times B \longrightarrow \mathbb{R}^+$ such that $b_{ijm} = g(b_i, b_m) b_{\bar{j}mi}$ where $b_{ijm}, b_{\bar{j}mi}$ are defined in TA1 and TA2 for all i, j, m.

TA4. There exists a unique algebra homomorphism $| | : A \longrightarrow \mathbb{C}$ such that $|b| = |\bar{b}| > 0$, $b \in B$. We call it the degree homomorphism. The positive real numbers $\{|b|\}_{b \in B}$ are called the degrees of (A, B).

If for all $b_i \in B$, $b_{i\bar{i}1} = 1$, then (A, B) is a Normalized Countable Table Algebra (denoted C-NTA) and if in addition for all $b \in B$ $|b| \in \mathbb{N}$ and for all i, j, m, $b_{ijm} \in \mathbb{N} \cup \{0\}$, then (A, B) is a Normalized Integral Countable Table Algebra (denote by C-NITA).

Note 1.1 If the dimension of a Countable Table Algebra is finite then by [AB, Lemma 2.9], axiom TA4 is a consequence of axioms TA1–TA3.

In Chaps. 2–4 we assume that the Table Algebras are NITA of finite dimension. If we assume instead that the Table Algebras are C-NITA, then the same proofs of Chaps. 2–4 give us C-NITA of finite dimension, i.e., Table Algebras.

Definition 1.7 Let $D = \{(k, n) | n \in \mathbb{N}, k$ is an odd number with $-1 \leq k \leq 2n + 1\}$. Define $f : D \to \mathbb{N} \cup \{0\}$ by

$$f(k, n) = \frac{(n + 1)(k + 1)(2n - k + 1)}{8}.$$

By definition, $f(-1, n) = 0$, $f(2n + 1, n) = 0$, $f(1, 1) = 1$ and $f(1, 2) = 3$.

Definition 1.8 Let (A, B) be a C-NITA with a faithful non-real element b_3 such that $b_3^2 = \bar{b}_3 + b_6$ where $b_6 \in B$. For $t \in \mathbb{N}$ with $t \geq 2$, or $t = \infty$, define an *array* C_t as

$$C_t = \{b_{(2j+1,i)}\}_{i=1; j=-1}^{t+1; i},$$

where $b_{(-1,i)} = b_{(2i+1,i)} = 0$ for $1 \leq i \leq t + 1$, $b_{(k,i)} \in B$ for all $1 \leq i \leq t + 1$ and odd k with $1 \leq k \leq 2i - 1$, and the following properties hold:

1. $b_{(1,1)} = 1$, $b_{(1,2)} = b_3$ and $b_{(k,i)} = \bar{b}_{(2i-k,i)}$ for all $1 \leq i \leq t + 1$ and odd k with $1 \leq k \leq 2i - 1$.
2. $b_3 b_{(k,i)} = b_{(k-2,i-1)} + b_{(k+2,i)} + b_{(k,i+1)}$ for all $2 \leq i \leq t$ and odd k with $1 \leq k \leq 2i - 1$.

Remark If an array C_t exists for $t > 2$, then properties 1 and 2 imply that C_{t-1} exists, with $C_{t-1} \subseteq C_t$ in the obvious sense. It is a straightforward consequence of properties 1 and 2 and induction on i that

$$|b_{(k,i)}| = f(k, i), \quad \text{for all } 1 \leq i \leq t + 1 \text{ and odd } k \text{ with } 1 \leq k \leq 2i + 1.$$

Definition 1.9 Let (A, B) be a C-NITA as in Definition 1.8. If for some $n \in \mathbb{N}$ there exists an array C_n for (A, B), but there does not exist an array C_{n+1}, then n is called the *stopping number* of (A, B). All such Table Algebras with stopping number $n \in \mathbb{N}$ are denoted (A, B, C_n). If there are arrays C_n for all $n \in \mathbb{N}$, $n \geq 2$, then the array C_∞ exists. In this case, we say that the stopping number for (A, B) is ∞, and denote such a Table Algebra as (A, B, C_∞). In either case, we say that (A, B) is a C_n-Table Algebra.

Remarks

a) We will show in Theorem 5.1 that there exists no C-NITA (A, B, C_n) such that $\infty > n \geq 43$. The classification of C-NITA (A, B, C_n) such that $4 \leq n \leq 42$ is still open. If there exists a C-NITA (A, B, C_n), it could be an infinite dimensional C-NITA. For a given $n \in \mathbb{N}$, there may exist several (perhaps infinitely many) nonisomorphic (A, B, C_n) of either finite or infinite dimension.

b) A C-NITA (A, B) is called a C_∞-Table Algebra (denoted by (A, B, C_∞)) if C_∞ exists and 1, 2 of Definition 1.8 hold. In this case b_3 faithful implies that $B = C_\infty \setminus \{0\}$. We will show in Chap. 5, Theorem 5.2 that (A, B, C_∞) is a uniquely determined C-NITA. Note that (A, B, C_∞) is an infinite dimensional C-NITA as defined in Definitions 1.8 and 1.9.

Now we state the main theorem of this book.

Main Theorem *Let (A, B) be a C-NITA generated by a non-real element $b_3 \in B$ of degree 3 and without non-identity basis elements of degree 1 and 2. Then $b_3\bar{b}_3 = 1 + b_8$, $b_8 \in B$, and (A, B) is of one of the following types:*

(1) *$(A, B) \cong_x (A, B, C_2)$ which is the unique C_2-Table Algebra and is strictly isomorphic to $(CH\, PSL(2, 7), Irr\, PSL(2, 7))$ of dimension 6.*

(2) *$(A, B) \cong_x (A, B, C_3)$ which is the unique C_3-Table Algebra and is strictly isomorphic to $(CH(3 \cdot A_6), Irr(3 \cdot A_6))$ of dimension 17.*

(3) *There exists a unique C_∞-Table Algebra which is strictly isomorphic to C-NITA generated by the non-real polynomial representation of dimension 3 of $SL(3, \mathbb{C})$.*

(4) *There exists no C-NITA (A, B, C_n) such that $43 \leq n$ and the classification of the Table Algebra (A, B, C_n) such that $4 \leq n \leq 42$ is still open.*

(5) *There exist $c_3, b_6 \in B$, $c_3 \neq b_3$ or $\bar{b}_3$, such that $b_3^2 = c_3 + b_6$ and either $(b_3b_8, b_3b_8) = 3$ or 4. If $(b_3b_8, b_3b_8) = 3$ and c_3 is non-real, then $(A, B) \cong_x (A(3 \cdot A_6 \cdot 2), B_{32})$ of dimension 32. (See Theorem 2.9 of Chap. 2 for the definition of this specific NITA.) If $(b_3b_8, b_3b_8) = 3$ and c_3 is real, then $(A, B) \cong_x (A(7 \cdot 5 \cdot 10), B_{22})$ of dimension 22. (See Theorem 2.10 of Chap. 2 for the definition of this specific NITA.)*

The Case $(b_3b_8, b_3b_8) = 4$ is still open.

For the complete classification, we will have to solve the open problems in Cases 4 and 5 of the Main Theorem.

References

[A]   Arad, Z.: Homogeneous integral table algebras of degrees two, three and four with a faithful element. In: Group St. Andrews 1997 in Bath, I. London Math. Soc. Lecture Notes Series, vol. 260, pp. 20–29 (1999)

[AAFM]   Arad, Z., Arisha, H., Fisman, E., Muzychuk, M.: Integral standard table algebras with a faithful nonreal element of degree 3. J. Algebra **231**(2), 473–483 (2000)

[AB]   Arad, Z., Blau, H.I.: On table algebras and applications to finite group theory. J. Algebra **138**, 137–185 (1991)

[ABFMMX]   Arad, Z., Blau, H.I., Fisman, E., Miloslavsky, V., Muzychuk, M., Xu, B.: Homogeneous Integral Table Algebra of Degree Three, A Trilogy, Memoirs of the AMS, vol. 144 (2000), no. 684

[AC]   Arad, Z., Chen, G.: On normalized table algebras generated by a faithful non-real element of degree 3. In: Shum, K.P. (ed.) Advances in Algebra, Proc. of the ICN Satellite Conf. in Algebra and Related Topics, pp. 13–37. World Scientific, Singapore (Dec. 2003)

[AC1]   Arad, Z., Chen, G.: On four normalized table algebras generated by a faithful nonreal element of degree 3. J. Algebra **283**, 457–484 (2005)

[AEM]   Arad, Z., Erez, Y., Muzychuk, M.: On homogeneous standard integral table algebras of degree 4. Comm. Algebra **34**, 463–519 (2006)

[AH1]   Arad, Z., Herzog, M. (eds.): Survey on Products of Conjugacy Classes in Groups. Mathematical Lecture Notes Series, vol. 2, pp. 2.01–2.04 (1984)

[AH2]   Arad, Z., Herzog, M. (eds.): Products of Conjugacy Classes in Groups. LNM, vol. 1112. Springer Verlag, Berlin (1985)

[AM]   Arad, Z., Muzychuk, M. (eds.): Standard Integral Table Algebras Generated by a Non-real Element of Small Degree. Lecture Notes in Mathematics (2002)

[B]   Blau, H.I.: Homogenous integral table algebras of degree two. Algebra Colloq. **4**(4), 393–408 (1997)

[B1]   Blau, H.I.: Integral table algebras and Bose-Mesner algebras with a faithful nonreal element of degree 3. J. Algebra **231**(2), 484–545 (2000)

[B2]   Blau, H.I.: Table algebras. Eur. J. Comb. **30**, 1426–1455 (2009)

[BC]   Blau, H.I., Chillag, D.: On powers of characters and powers of conjugacy classes of a finite group. Proc. Am. Math. Soc. **298**, 7–10 (1986)

[FK]   Fröhlich, J., Kerler, T.: Quantum Groups, Quantom Categories and Quantum Field Theory, Lecture Notes in Mathematics, vol. 1542 (1993)

[H]   Hoheisel, G.: Uber Charaktere. Monatsch. f. Math. Phys. **48**, 448–456 (1939)

[K]   Kawada, Y.: Uber den Dualitatssatz der Charaktere nichtcommutativer Gruppen. Proc. Phys. Math. Soc. Jpn. **24**(3), 97–109 (1942)

Chapter 2
Splitting of the Main Problem into Four Sub-cases

2.1 Introduction

Each of Chaps. 2–5 of this book is self-contained. Therefore the definitions in the different parts of the book may sometimes appear with different signs.

The concept of Table Algebra was introduced by Z. Arad and H. Blau in [AB] in order to study in a uniform way properties of products of conjugacy classes and of irreducible characters of a finite group.

Definition 2.1 A Table Algebra (A, B) is a finite dimensional, commutative algebra A with identity element 1 over the complex numbers C, and a distinguished base $B = \{b_1 = 1, b_2, \ldots, b_k\}$ such that the following properties hold (where (b_i, a) denotes the coefficient of b_i in $a \in A$, a written as a linear combination of B; and where R^* denotes $R^+ \cup \{0\}$, the set of non-negative real numbers):

(I) For all i, j, m, $b_i b_j = \sum_m \delta_{ijm} b_m$ with δ_{ijm} a nonnegative real number.
(II) There is an algebra automorphism (denoted by $^-$) of A, whose order divides 2, such that $b_i \in B$ implies that $\bar{b}_i \in B$. (Then $\bar{i}$ is defined by $\bar{b}_i = b_{\bar{i}}$.)
(III) Hypothesis (II) holds and there is a function $g : B \times B \to R^+$ (the positive reals) such that

$$(b_m, b_i b_j) = g(b_i, b_m) \cdot (b_i, \bar{b}_j b_m),$$

where $g(b_i, b_m)$ is independent of j for all i, j, m.

B is called the table basis of (A, B). Clearly $1 \in B$, we always use b_1 to denote base element 1, and $B^\sharp$ to denote $B \setminus \{b_1\}$. The elements of B are called the irreducible components of (A, B), and nonzero nonnegative linear combinations of elements of B with coefficients in R^+ are called components. If $a = \sum_{m=1}^k \lambda_m b_m$ is a component ($\lambda_m \in R^*$) then $Supp\{a\} = \{b_m | \lambda_m \neq 0\}$ is called the set of irreducible constituents of a. An element $a \in A$ is called a real element if $a = \bar{a}$.

Two Table Algebras (A, B) and (A', B') are called isomorphic (denoted $B \cong B'$) when there exists an algebra isomorphism $\psi : A \to A'$ such that $\psi(B)$ is a rescaling of B'; and the algebras are called exactly isomorphic (denoted $B \cong_x B'$) when $\psi(B) = B'$. So $B \cong_x B'$ means that B and B' yield the same structure constants.

Z. Arad et al., *On Normalized Integral Table Algebras (Fusion Rings)*, Algebra and Applications 16, DOI 10.1007/978-0-85729-850-8_2, © Springer-Verlag London Limited 2011

Proposition 2.2 of [AB] shows that if (A, B) is a Table Algebra, then there exists a basis B', which consists of suitable positive real scalar multiples b'_i of the elements b_i of B, $g(b'_i, b'_j) = 1$ for any $b'_i, b'_j \in B'$. Such a basis B' is called a normalized basis. Now $Supp(b'_{i_1} b'_{i_2} \cdots b'_{i_t})$ consists of the corresponding scalar multiples of $Supp(b_{i_1} b_{i_2} \cdots b_{i_t})$, for any sequence $i_1, i_2, \ldots, i_t$, of indices. So in the proof of any proposition which identifies the irreducible constituents of a product of basis elements, we may assume that B is normalized.

Suppose that B is normalized. It follows from (III) and [AB, Sect. 2] that A has a positive definite Hermitian form, with B as an orthonormal basis, and such that

$$(a, bc) = (a\bar{b}, c)$$

for all $a, b, c \in \mathcal{R}B$.

It follows from Sect. 2 in [AB] that there exists an algebra homomorphism f from A to C such that $f(B) \subseteq R^+$. For an element $b \in B$, $f(b)$ is called the degree of b.

Each finite group G yields two natural Table Algebras: the Table Algebra of conjugacy classes (denoted by $(ZC(G), Cl(G))$) and the Table Algebra of generalized characters (denoted by $(Ch(G), Irr(G))$).

Both Table Algebras arising from group theory have an additional property: their structure constants and degrees are non-negative integers. Such algebras were defined in [B1] as Integral Table Algebras (denoted ITA).

Each of the elements of a Table Algebra is contained in a unique table subalgebra which may be considered as a table subalgebra generated by this element. So it is natural to start the study of Integral Table Algebras from those which are generated by a single element. Normalized Integral Table Algebras (denoted NITA) generated by an element of degree 2 were completely classified by H. Blau in [B].

The finite linear groups in dimension $n \leq 7$ have been completely classified. See Feit [F]. Representation theory and properties of finite groups are heavily used for the proofs. For $n = 2, 3, 4$ see Blichfeldt [Bl]. For $n = 5$ see Brauer [Br]. For $n = 6$ see Lindsey [L]. For $n = 7$ see Wales [W]. In order to generalize these results to Normalized Integral Table Algebras, the authors began to classify Normalized Integral Table Algebras (A, B) generated by a faithful nonreal element of degree 3 and without nonidentity irreducible elements of degree 1 or 2 in [CA], and arrived at the following theorem:

Theorem 2.1 *Let (A, B) be NITA generated by a nonreal element $b_3 \in B$ of degree 3 and without non-identity basis element of degree 1 or 2. Then $b_3\bar{b}_3 = 1 + b_8$ and one of the following holds*:

(1) $(A, B) \cong_x (Ch(PSL(2, 7)), Irr(PSL(2, 7)))$;
(2) $b_3^2 = b_4 + b_5$, *where $b_4 \in B$ and $b_5 \in B$*;
(3) $b_3^2 = \bar{b}_3 + b_6$, *where $b_6 \in B$ is a nonreal element*;
(4) $b_3^2 = c_3 + b_6$, *where $c_3, b_6 \in B$ and $c_3 \neq b_3, \bar{b}_3$.*

The purpose of this chapter is to investigate NITA generated by a nonreal element b_3 that satisfies one of the conditions (2), (3) and (4) in Theorem 2.1. The following theorem is proved:

Main Theorem 1 *Let (A, B) be a NITA generated by a nonreal element $b_3 \in B$ of degree 3 and without non-identity basis element of degree 1 or 2. Then $b_3\bar{b}_3 = 1 + b_8$, $b_8 \in B$ and one of the following holds:*

(1) *There exists a real element $b_6 \in B$ such that $b_3^2 = \bar{b}_3 + b_6$ and $(A, B) \cong_x$ $(Ch(PSL(2, 7)), Irr(PSL(2, 7)))$.*

(2) *There exists $b_4, b_5 \in B$ such that $b_3^2 = b_4 + b_5$, $(b_3b_8, b_3b_8) = 3$ and $(b_4^2, b_4^2) = 3$.*

(3) *There exists $b_6, b_{10}, b_{15} \in B$, where b_6 is nonreal such that*

$$b_3^2 = \bar{b}_3 + b_6, \qquad \bar{b}_3 b_6 = b_3 + b_{15}, \qquad b_3 b_6 = b_8 + b_{10}$$

and $(b_3b_8, b_3b_8) = 3$. Moreover if b_{10} is real, then $(A, B) \cong_x (CH(3 \cdot A_6), Irr(3 \cdot A_6))$.

(4) *There exists $c_3, b_6 \in B$, $c_3 \neq b_3, \bar{b}_3$, such that $b_3^2 = c_3 + b_6$ and either $(b_3b_8, b_3b_8) = 3$ or 4.*

When $(b_3b_8, b_3b_8) = 3$. If c_3 is nonreal, then (A, B) is exactly isomorphic to $(A(3 \cdot A_6 \cdot 2), B_{32})$ (see Theorem 2.9). If c_3 is real, then (A, B) is exactly isomorphic to the NITA $(A(7 \cdot 5 \cdot 10), B_{22})$ of dimension 22 (see Theorem 2.10).

Proof The theorem follows from Theorem 2.1 and Theorems 2.2, 2.7, 2.9 and 2.10 below.   □

2.2 Two NITA Generated by a Non-real Element of Degree 3 not Derived from a Group and Lemmas

Here we give two examples of NITA not induced from group theory, which has either 32 or 22 basis elements. The NITA of dimension 32 contains a subalgebra which is strictly isomorphic to the algebra of dimension 17 of characters of the group $3 \cdot A_6$, we denote it as $(A(3 \cdot A_6 \cdot 2), B_{32})$, where its basis is B_{32}.

For B_{32} there are three table subsets: $\{1\} \subseteq C \subseteq D \subseteq B$, where

$$C = \{1, b_8, x_{10}, b_5, c_5, c_8, x_9\},$$

$$D = C \cup \{c_3, \bar{c}_3, d_3, \bar{d}_3, c_9, \bar{c}_9, b_6, \bar{b}_6, y_5, \bar{y}_{15}\},$$

$$B_{32} = D \cup \{b_3, \bar{b}_3, r_3, x_6, \bar{x}_6, s_6, b_9, \bar{b}_9, x_{15}, \bar{x}_{15}, t_{15}, d_9, y_3, z_3, \bar{z}_3\}.$$

From the equations below one can check that B_{32} has the following table subsets: $\{1\} \subseteq C \subseteq E = C \cup \{r_3, s_6, t_{15}, d_9, y_3\} \subseteq B_{32}$. The table subsets E and D are two maximal table subsets of B_{32}.

The structure algebra constants of C:

$$\begin{aligned}
b_5^2 &= 1 + x_9 + x_{10} + b_5,\\
b_5c_5 &= c_8 + b_8 + x_9,\\
b_5c_8 &= c_5 + c_8 + b_8 + x_9 + x_{10},\\
b_5b_8 &= c_5 + c_8 + b_8 + x_9 + x_{10},\\
b_5x_9 &= c_8 + c_5 + b_8 + b_5 + x_9 + x_{10},\\
b_5x_{10} &= c_8 + b_8 + x_9 + b_5 + 2x_{10},
\end{aligned}$$

$$c_5^2 = 1 + x_9 + x_{10} + c_5,$$
$$c_5 c_8 = b_5 + c_8 + b_8 + x_9 + x_{10},$$
$$c_5 b_8 = b_5 + c_8 + b_8 + x_9 + x_{10},$$
$$c_5 x_9 = b_8 + c_8 + b_5 + c_5 + x_9 + x_{10},$$
$$\underline{c_5 x_{10} = b_8 + c_8 + x_9 + c_5 + 2x_{10},}$$

$$b_8^2 = 1 + b_5 + c_5 + c_8 + 2b_8 + x_9 + 2x_{10},$$
$$b_8 x_9 = b_5 + c_5 + 2c_8 + b_8 + 2x_9 + 2x_{10},$$
$$\underline{b_8 x_{10} = b_5 + c_5 + 2c_8 + 2b_8 + 2x_9 + 2x_{10},}$$

$$c_8^2 = 1 + b_5 + c_5 + 2c_8 + b_8 + x_9 + 2x_{10},$$
$$c_8 b_8 = b_5 + c_5 + c_8 + b_8 + 2x_9 + 2x_{10},$$
$$c_8 x_9 = b_5 + c_5 + c_8 + 2b_8 + 2x_9 + 2x_{10},$$
$$\underline{c_8 x_{10} = b_5 + c_5 + 2c_8 + 2b_8 + 2x_9 + 2x_{10},}$$

$$x_9^2 = 1 + b_5 + c_5 + 2c_8 + 2b_8 + 2x_9 + 2x_{10},$$
$$x_9 x_{10} = b_5 + c_5 + 2c_8 + 2b_8 + 2x_9 + 3x_{10},$$
$$\underline{x_{10}^2 = 1 + 2b_5 + 2c_5 + 2c_8 + 2b_8 + 3x_9 + 2x_{10}.}$$

The structure algebra constants of D (since $C \subseteq D$ we will not repeat the equations of C as shown above):

$$c_3 \bar{c}_3 = 1 + b_8,$$
$$c_3^2 = \bar{c}_3 + \bar{b}_6,$$
$$c_3 d_3 = x_9,$$
$$c_3 \bar{d}_3 = \bar{c}_9,$$
$$c_3 b_6 = \bar{c}_3 + \bar{y}_{15},$$
$$c_3 \bar{b}_6 = b_8 + x_{10},$$
$$c_3 c_9 = d_3 + \bar{c}_9 + \bar{y}_{15},$$
$$c_3 \bar{c}_9 = c_8 + x_9 + x_{10},$$
$$c_3 y_{15} = \bar{b}_6 + 2\bar{y}_{15} + \bar{c}_9,$$
$$c_3 \bar{y}_{15} = b_5 + c_5 + c_8 + b_8 + x_9 + x_{10},$$
$$c_3 b_8 = c_3 + b_6 + y_{15},$$
$$c_3 x_{10} = b_6 + c_9 + y_{15},$$
$$c_3 b_5 = y_{15},$$
$$c_3 c_5 = y_{15},$$
$$c_3 c_8 = y_{15} + c_9,$$
$$c_3 x_9 = y_{15} + c_9 + \bar{d}_3,$$

$$d_3\bar{d}_3 = 1 + c_8,$$
$$d_3^2 = \bar{d}_3 + b_6,$$
$$d_3 b_6 = c_8 + x_{10},$$
$$d_3 \bar{b}_6 = \bar{d}_3 + y_{15},$$
$$d_3 c_9 = b_8 + x_9 + x_{10},$$
$$d_3 \bar{c}_9 = c_3 + y_{15} + c_9,$$
$$d_3 y_{15} = b_5 + c_5 + c_8 + b_8 + x_9 + x_{10},$$
$$d_3 \bar{y}_{15} = 2y_{15} + b_6 + c_9,$$
$$d_3 b_8 = \bar{c}_9 + \bar{y}_{15},$$
$$d_3 x_{10} = \bar{y}_{15} + \bar{b}_6 + \bar{c}_9,$$
$$d_3 b_5 = \bar{y}_{15},$$
$$d_3 c_5 = \bar{y}_{15},$$
$$d_3 c_8 = \bar{y}_{15} + b_6 + d_3,$$
$$d_3 x_9 = \bar{y}_{15} + \bar{c}_3 + \bar{c}_9,$$

$$b_6 \bar{b}_6 = 1 + b_5 + c_5 + c_8 + b_8 + x_9,$$
$$b_6^2 = 2\bar{b}_6 + \bar{y}_{15} + \bar{c}_9,$$
$$b_6 c_9 = 2\bar{y}_{15} + 2\bar{c}_9 + \bar{b}_6,$$
$$b_6 \bar{c}_9 = b_5 + c_5 + c_8 + b_8 + 2x_9 + x_{10},$$
$$b_6 \bar{y}_{15} = b_5 + c_5 + 2c_8 + 2b_8 + 2x_9 + 3x_{10},$$
$$b_6 y_{15} = 4\bar{y}_{15} + \bar{b}_6 + \bar{c}_3 + d_3 + 2\bar{c}_9,$$
$$b_6 b_5 = b_6 + y_{15} + c_9,$$
$$b_6 c_5 = y_{15} + b_6 + c_9,$$
$$b_6 c_8 = \bar{d}_3 + b_6 + 2y_{15} + c_9,$$
$$b_6 b_8 = c_3 + b_6 + c_9 + 2y_{15},$$
$$b_6 x_9 = 2y_{15} + b_6 + 2c_9,$$
$$b_6 x_{10} = 3y_{15} + c_3 + \bar{d}_3 + c_9,$$

$$y_{15} \bar{y}_{15} = 1 + 3b_5 + 3c_5 + 5c_8 + 5b_8 + 6x_9 + 6x_{10},$$
$$y_{15}^2 = 9\bar{y}_{15} + 4\bar{b}_6 + 2\bar{c}_3 + 2d_3 + 6\bar{c}_9,$$
$$y_{15} b_5 = 3y_{15} + b_6 + c_3 + \bar{d}_3 + 2c_9,$$
$$y_{15} c_5 = 3y_{15} + b_6 + c_3 + \bar{d}_3 + 2c_9,$$
$$y_{15} c_8 = 5y_{15} + c_3 + \bar{d}_3 + 2b_6 + 3c_9,$$
$$y_{15} b_8 = 5y_{15} + c_3 + \bar{d}_3 + 2b_6 + 3c_9,$$
$$y_{15} x_9 = 6y_{15} + 2b_6 + \bar{d}_3 + c_3 + 3c_9,$$
$$y_{15} x_{10} = 6y_{15} + 3b_6 + c_3 + \bar{d}_3 + 4c_9,$$

$$c_9\bar{c}_9 = 1 + 2c_8 + 2b_8 + 2x_9 + 2x_{10} + c_5 + b_5,$$
$$c_9^2 = 3\bar{y}_{15} + d_3 + \bar{c}_3 + 2\bar{b}_6 + 2\bar{c}_9,$$
$$c_9 y_{15} = 6\bar{y}_{15} + 2\bar{b}_6 + \bar{c}_3 + d_3 + 3\bar{c}_9,$$
$$c_9 \bar{y}_{15} = 2b_5 + 2c_5 + 3c_8 + 3b_8 + 3x_9 + 4x_{10},$$
$$c_9 b_5 = 2y_{15} + b_6 + c_9,$$
$$c_9 c_5 = 2y_{15} + b_6 + c_9,$$
$$c_9 b_8 = 3y_{15} + \bar{d}_3 + b_6 + 2c_9,$$
$$c_9 c_8 = 3y_{15} + b_6 + c_3 + 2c_9,$$
$$c_9 x_9 = c_3 + 2c_9 + 3y_{15} + \bar{d}_3 + 2b_6,$$
$$c_9 x_{10} = c_3 + 2c_9 + 4y_{15} + \bar{d}_3 + b_6.$$

From the equations one can see that C and D are table subsets of B_{32}. Furthermore, the NITA subalgebra $(\langle D \rangle, D)$ of dimension 17 is strictly isomorphic to $(CH(3 \cdot A_6), Irr(3 \cdot A_6))$ while our NITA of dimension 32 is not induced from a finite group G.

The following equations describe other products of basis elements in B_{32}:

$$b_3\bar{b}_3 = 1 + b_8,$$
$$b_3^2 = c_3 + b_6,$$
$$b_3 r_3 = \bar{c}_3 + \bar{b}_6,$$
$$b_3 x_6 = c_3 + y_{15},$$
$$b_3 \bar{x}_6 = b_8 + x_{10},$$
$$b_3 s_6 = \bar{c}_3 + \bar{y}_{15},$$
$$b_3 b_9 = y_{15} + c_9 + \bar{d}_3,$$
$$b_3 \bar{b}_9 = x_9 + c_8 + x_{10},$$
$$b_3 x_{15} = b_6 + 2y_{15} + c_9,$$
$$b_3 \bar{x}_{15} = b_5 + c_5 + b_8 + c_8 + x_9 + x_{10},$$
$$b_3 t_{15} = b_6 + \bar{c}_9 + 2\bar{y}_{15},$$
$$b_3 d_9 = \bar{c}_9 + \bar{y}_{15} + d_3,$$
$$b_3 y_3 = \bar{c}_9,$$
$$b_3 z_3 = x_9,$$
$$b_3 \bar{z}_3 = c_9,$$
$$b_3 b_8 = b_3 + x_6 + x_{15},$$
$$b_3 x_{10} = x_6 + x_{15} + b_9,$$
$$b_3 b_5 = x_{15},$$
$$b_3 c_5 = x_{15},$$
$$b_3 c_8 = x_{15} + b_9,$$
$$b_3 x_9 = \bar{z}_3 + b_9 + x_{15},$$
$$b_3 c_3 = r_3 + s_6,$$
$$b_3 \bar{c}_3 = b_3 + \bar{x}_6,$$
$$b_3 d_3 = \bar{b}_9,$$
$$b_3 \bar{d}_3 = d_9,$$
$$b_3 c_9 = t_{15} + d_9 + y_3,$$
$$b_3 \bar{c}_9 = \bar{b}_9 + \bar{x}_{15} + z_3,$$
$$b_3 b_6 = r_3 + t_{15},$$
$$b_3 \bar{b}_6 = b_3 + \bar{x}_{15},$$
$$b_3 y_{15} = s_6 + 2t_{15} + d_9,$$
$$b_3 \bar{y}_{15} = \bar{x}_6 + 2\bar{x}_{15} + b_9,$$

$$r_3^2 = 1 + b_8,$$
$$r_3 x_6 = \bar{c}_3 + \bar{y}_{15},$$
$$r_3 s_6 = b_8 + x_{10},$$
$$r_3 b_9 = d_3 + \bar{c}_9 + \bar{y}_{15},$$
$$r_3 x_{15} = \bar{b}_6 + \bar{c}_9 + 2\bar{y}_{15},$$
$$r_3 t_{15} = b_5 + c_5 + b_8 + c_8 + x_9 + x_{10},$$
$$r_3 d_9 = c_8 + x_9 + x_{10},$$
$$r_3 y_3 = x_9,$$
$$r_3 z_3 = c_9,$$
$$r_3 b_8 = s_6 + r_3 + t_{15},$$
$$r_3 x_{10} = s_6 + t_{15} + d_9,$$
$$r_3 b_5 = t_{15},$$
$$r_3 c_5 = t_{15},$$
$$r_3 c_8 = t_{15} + d_9,$$
$$r_3 x_9 = t_{15} + d_9 + y_3,$$
$$r_3 c_3 = \bar{b}_3 + \bar{x}_6,$$
$$r_3 d_3 = b_9,$$
$$r_3 c_9 = \bar{b}_9 + \bar{x}_{15} + z_3,$$
$$r_3 b_6 = \bar{x}_{15} + \bar{b}_3,$$
$$r_3 y_{15} = \bar{x}_6 + 2\bar{x}_{15} + \bar{b}_9,$$

$$x_6^2 = 2b_6 + y_{15} + c_9,$$
$$x_6 \bar{x}_6 = 1 + b_8 + b_5 + c_5 + c_8 + x_9,$$
$$x_6 s_6 = 2\bar{b}_6 + \bar{c}_9 + \bar{y}_{15},$$
$$x_6 b_9 = \bar{b}_6 + 2\bar{c}_9 + 2\bar{y}_{15},$$
$$x_6 \bar{b}_9 = b_5 + c_5 + x_{10} + b_8 + c_8 + 2x_9,$$
$$x_6 x_{15} = 4y_{15} + b_6 + 2c_9 + d_3 + c_3,$$
$$x_6 \bar{x}_{15} = b_5 + c_5 + 2b_8 + 3x_{10} + 2c_8 + 2x_9,$$
$$x_6 t_{15} = \bar{c}_3 + d_3 + 2\bar{c}_9 + 4\bar{y}_{15} + \bar{b}_6,$$
$$x_6 d_9 = \bar{b}_6 + 2\bar{c}_9 + 2\bar{y}_{15},$$
$$x_6 y_3 = d_3 + \bar{y}_{15},$$
$$x_6 z_3 = x_{10} + c_8,$$
$$x_6 \bar{z}_3 = \bar{d}_3 + y_{15},$$
$$x_6 b_8 = b_3 + x_6 + 2x_{15} + b_9,$$
$$x_6 x_{10} = b_9 + b_3 + \bar{z}_3 + 3x_{15},$$
$$x_6 b_5 = x_6 + b_9 + x_{15},$$
$$x_6 c_5 = x_6 + b_9 + x_{15},$$
$$x_6 c_8 = x_6 + b_9 + 2x_{15} + \bar{z}_3,$$
$$x_6 x_9 = 2b_9 + x_6 + 2x_{15},$$
$$x_6 c_3 = r_3 + t_{15},$$
$$x_6 \bar{c}_3 = b_3 + \bar{x}_{15},$$
$$x_6 d_3 = z_3 + \bar{x}_{15},$$
$$x_6 \bar{d}_3 = y_3 + t_{15},$$
$$x_6 c_9 = 2t_{15} + 2d_9 + s_6,$$
$$x_6 \bar{c}_9 = 2\bar{x}_{15} + \bar{x}_6 + 2\bar{b}_9,$$
$$x_6 b_6 = 2s_6 + t_{15} + d_9,$$
$$x_6 \bar{b}_6 = 2\bar{x}_6 + \bar{b}_9 + \bar{x}_{15},$$
$$x_6 y_{15} = r_3 + 4t_{15} + s_6 + 2d_9 + y_3,$$
$$x_6 \bar{y}_{15} = b_3 + z_3 + \bar{x}_6 + 4\bar{x}_{15} + 2\bar{b}_9,$$

$$s_6^2 = 1 + b_5 + c_5 + b_8 + c_8 + x_9,$$
$$s_6 b_9 = \bar{b}_6 + 2\bar{c}_9 + 2\bar{y}_{15},$$
$$s_6 x_{15} = \bar{c}_3 + d_3 + \bar{b}_6 + 2\bar{c}_9 + 4\bar{y}_{15},$$
$$s_6 t_{15} = b_5 + c_5 + 2b_8 + 2c_8 + 2x_9 + 3x_{10},$$
$$s_6 d_9 = b_5 + c_5 + 2x_9 + b_8 + c_8 + x_{10},$$
$$s_6 y_3 = c_8 + x_{10},$$
$$s_6 z_3 = \bar{d}_3 + y_{15},$$
$$s_6 b_8 = r_3 + s_6 + 2t_{15} + d_9,$$
$$s_6 x_{10} = r_3 + y_3 + d_9 + 3t_{15},$$
$$s_6 b_5 = s_6 + d_9 + t_{15},$$
$$s_6 c_5 = s_6 + t_{15} + d_9,$$
$$s_6 c_8 = y_3 + s_6 + d_9 + 2t_{15},$$
$$s_6 x_9 = s_6 + 2d_9 + 2t_{15},$$
$$s_6 c_3 = \bar{b}_3 + \bar{x}_{15},$$
$$s_6 d_3 = \bar{z}_3 + x_{15},$$
$$s_6 c_9 = \bar{x}_6 + 2\bar{b}_9 + 2\bar{x}_{15},$$
$$s_6 b_6 = 2\bar{x}_6 + \bar{b}_9 + \bar{x}_{15},$$
$$s_6 y_{15} = \bar{b}_3 + z_3 + 4\bar{x}_{15} + \bar{x}_6 + 2\bar{b}_9,$$

$$b_9 \bar{b}_9 = 1 + b_5 + c_5 + 2b_8 + 2c_8 + 2x_9 + 2x_{10},$$
$$b_9^2 = \bar{d}_3 + c_3 + 2b_6 + 2c_9 + 3y_{15},$$
$$b_9 x_{15} = 6y_{15} + 3c_9 + \bar{d}_3 + 2b_6 + c_3,$$
$$b_9 \bar{x}_{15} = 2b_5 + 2c_5 + 3c_8 + 3b_8 + 4x_{10} + 3x_9,$$
$$b_9 t_{15} = d_3 + \bar{c}_3 + 6\bar{y}_{15} + 3\bar{c}_9 + 2\bar{b}_6,$$
$$b_9 d_9 = \bar{c}_3 + d_3 + 2\bar{b}_6 + 2\bar{c}_9 + 3\bar{y}_{15},$$
$$b_9 y_3 = \bar{c}_3 + \bar{y}_{15} + \bar{c}_9,$$
$$b_9 z_3 = b_8 + x_9 + x_{10},$$
$$b_9 \bar{z}_3 = c_9 + c_3 + y_{15},$$
$$b_9 b_8 = \bar{z}_3 + x_6 + 2b_9 + 3x_{15},$$
$$b_9 x_{10} = \bar{z}_3 + b_3 + 2b_9 + x_6 + 4x_{15},$$
$$b_9 b_5 = x_6 + b_9 + 2x_{15},$$
$$b_9 c_5 = b_9 + 2x_{15} + x_6,$$
$$b_9 c_8 = r_3 + t_{15} + 2b_9 + x_6 + 2x_{15},$$

$$b_9 x_9 = 2b_9 + 3x_{15} + \bar{z}_3 + b_3 + 2x_6,$$
$$b_9 c_3 = t_{15} + d_9 + \bar{y}_3,$$
$$b_9 \bar{c}_3 = z_3 + \bar{b}_9 + \bar{x}_{15},$$
$$b_9 d_3 = \bar{b}_3 + \bar{b}_9 + \bar{x}_{15},$$
$$b_9 \bar{d}_3 = r_3 + t_{15} + d_9,$$
$$b_9 c_9 = r_3 + y_3 + 2s_6 + 2d_9 + 3t_{15},$$
$$b_9 \bar{c}_9 = \bar{b}_3 + z_3 + 2\bar{x}_6 + 3\bar{x}_{15} + 2\bar{b}_9,$$
$$b_9 b_6 = s_6 + 2d_9 + 2t_{15},$$
$$b_9 \bar{b}_6 = 2\bar{b}_9 + 2\bar{x}_{15} + \bar{x}_6,$$
$$b_9 \bar{y}_{15} = 6\bar{x}_{15} + \bar{b}_3 + z_3 + 2\bar{x}_6 + 3\bar{b}_9,$$
$$b_9 y_{15} = r_3 + y_3 + 2s_6 + 6t_{15} + 3d_9,$$

$$x_{15}^2 = 2c_3 + 2\bar{d}_3 + 4b_6 + 6c_9 + 9y_{15},$$
$$x_{15}\bar{x}_{15} = 1 + 6x_9 + 3b_5 + 3c_5 + 5b_8 + 5c_8 + 6x_{10},$$
$$x_{15}t_{15} = 4\bar{b}_6 + 6\bar{c}_9 + 9\bar{y}_{15} + 2\bar{c}_3 + 2d_3,$$
$$x_{15}d_9 = \bar{c}_3 + d_3 + 2\bar{b}_6 + 3\bar{c}_9 + 6\bar{y}_{15},$$
$$x_{15}y_3 = \bar{b}_6 + \bar{c}_9 + 2\bar{y}_{15},$$
$$x_{15}\bar{z}_3 = b_6 + c_9 + 2y_{15},$$
$$x_{15}z_3 = b_5 + c_5 + b_8 + c_8 + x_9 + x_{10},$$
$$x_{15}b_8 = b_3 + \bar{z}_3 + 2x_6 + 5x_{15} + 3b_9,$$
$$x_{15}x_{10} = b_3 + \bar{z}_3 + 3x_6 + 4b_9 + 6x_{15},$$
$$x_{15}b_5 = \bar{z}_3 + b_3 + x_6 + 2b_9 + 3x_{15},$$
$$x_{15}c_5 = b_3 + \bar{z}_3 + x_6 + 2b_9 + 3x_{15},$$
$$x_{15}c_8 = b_3 + \bar{z}_3 + 2x_6 + 3b_9 + 5x_{15},$$
$$x_{15}x_9 = b_3 + \bar{z}_3 + 2x_6 + 3b_9 + 6x_{15},$$
$$x_{15}c_3 = 2t_{15} + s_6 + d_9,$$
$$x_{15}\bar{c}_3 = 2\bar{x}_{15} + \bar{x}_6 + \bar{b}_9,$$
$$x_{15}d_3 = \bar{x}_6 + 2\bar{x}_{15} + \bar{b}_9,$$
$$x_{15}\bar{d}_3 = s_6 + 2t_{15} + d_9,$$
$$x_{15}c_9 = r_3 + y_3 + 2s_6 + 6t_{15} + 3d_9,$$
$$x_{15}\bar{c}_9 = \bar{b}_3 + z_3 + 3\bar{b}_9 + 2\bar{x}_6 + 6\bar{x}_{15},$$
$$x_{15}b_6 = r_3 + s_6 + y_3 + 2d_9 + 4t_{15},$$
$$x_{15}\bar{b}_6 = \bar{b}_3 + z_3 + 2\bar{b}_9 + 4\bar{x}_{15} + \bar{x}_6,$$
$$x_{15}y_{15} = 2y_3 + 2r_3 + 4s_6 + 6d_9 + 9t_{15},$$
$$x_{15}\bar{y}_{15} = 2z_3 + 9\bar{x}_{15} + 2\bar{b}_3 + 4\bar{x}_6 + 6\bar{b}_9,$$

$$t_{15}^2 = 1 + 5b_8 + 5c_8 + 3b_5 + 3c_5 + 6x_{10} + 6x_9,$$
$$t_{15}d_9 = 2b_5 + 2c_5 + 3b_8 + 3c_8 + 3x_9 + 4x_{10},$$
$$t_{15}y_3 = b_5 + c_5 + b_8 + c_8 + x_9 + x_{10},$$
$$t_{15}z_3 = b_6 + c_9 + 2y_{15},$$
$$t_{15}b_8 = r_3 + y_3 + 5t_{15} + 2s_6 + 3d_9,$$
$$t_{15}x_{10} = r_3 + y_3 + 4d_9 + 3s_6 + 6t_{15},$$
$$t_{15}b_5 = r_3 + y_3 + s_6 + 2d_9 + 3t_{15},$$
$$t_{15}c_5 = r_3 + y_3 + s_6 + 2d_9 + 3t_{15},$$
$$t_{15}c_8 = r_3 + y_3 + 2s_6 + 3d_9 + 5t_{15},$$
$$t_{15}x_9 = r_3 + y_3 + 3d_9 + 2s_6 + 6t_{15},$$
$$t_{15}c_3 = \bar{x}_6 + \bar{b}_9 + 2\bar{x}_{15},$$
$$t_{15}d_3 = x_6 + b_9 + 2x_{15},$$
$$t_{15}c_9 = \bar{b}_3 + z_3 + 3\bar{b}_9 + 2\bar{x}_6 + 6\bar{x}_{15},$$
$$t_{15}b_6 = \bar{b}_3 + z_3 + \bar{x}_6 + 2\bar{b}_9 + 4\bar{x}_{15},$$
$$t_{15}y_{15} = 2\bar{b}_3 + 2z_3 + 4\bar{x}_6 + 6\bar{b}_9 + 9\bar{x}_{15},$$

$$d_9^2 = 1 + b_5 + c_5 + 2b_8 + 2c_8 + 2x_9 + 2x_{10},$$
$$d_9 y_3 = x_9 + b_8 + x_{10},$$
$$d_9 z_3 = c_3 + c_9 + y_{15},$$
$$d_9 b_8 = y_3 + s_6 + 2d_9 + 3t_{15},$$
$$d_9 x_{10} = y_3 + r_3 + s_6 + 2d_9 + 4t_{15},$$
$$d_9 b_5 = s_6 + d_9 + 2t_{15},$$
$$d_9 c_5 = s_6 + d_9 + 2t_{15},$$
$$d_9 c_8 = r_3 + s_6 + 2d_9 + 3t_{15},$$
$$d_9 x_9 = r_3 + y_3 + 2s_6 + 2d_9 + 3t_{15},$$
$$d_9 c_3 = \bar{b}_9 + z_3 + \bar{x}_{15},$$
$$d_9 d_3 = b_3 + b_9 + x_{15},$$
$$d_9 c_9 = \bar{b}_3 + z_3 + 2\bar{x}_6 + 2\bar{b}_9 + 3\bar{x}_{15},$$
$$d_9 b_6 = \bar{x}_6 + 3\bar{b}_9 + 2\bar{x}_{15},$$
$$d_9 y_{15} = \bar{b}_3 + z_3 + 2\bar{x}_6 + 3\bar{b}_9 + 6\bar{x}_{15},$$

$$y_3^2 = 1 + c_8,$$
$$y_3 z_3 = \bar{d}_3 + b_6,$$
$$y_3 b_8 = d_9 + t_{15},$$
$$y_3 x_{10} = s_6 + t_{15} + d_9,$$
$$y_3 b_5 = t_{15},$$
$$y_3 c_5 = t_{15},$$
$$y_3 c_8 = s_6 + t_{15} + y_3,$$
$$y_3 x_9 = t_{15} + d_9 + r_3,$$
$$y_3 c_3 = \bar{b}_9,$$
$$y_3 d_3 = x_6 + \bar{z}_3,$$
$$y_3 c_9 = \bar{b}_3 + \bar{b}_9 + \bar{x}_{15},$$
$$y_3 b_6 = \bar{x}_{15} + z_3,$$
$$y_3 y_{15} = \bar{x}_6 + \bar{b}_9 + 2\bar{x}_{15},$$

$$z_3 \bar{z}_3 = 1 + c_8,$$
$$z_3^2 = d_3 + \bar{b}_6,$$
$$z_3 b_8 = \bar{b}_9 + \bar{x}_{15},$$
$$z_3 x_{10} = \bar{x}_6 + \bar{b}_9 + \bar{x}_{15},$$
$$z_3 b_5 = \bar{x}_{15},$$
$$z_3 c_5 = \bar{x}_{15},$$
$$z_3 c_8 = \bar{x}_6 + \bar{x}_{15} + z_3,$$
$$z_3 x_9 = \bar{b}_3 + \bar{b}_9 + \bar{x}_{15},$$
$$z_3 c_3 = b_9,$$
$$z_3 \bar{c}_3 = d_9,$$
$$z_3 d_3 = s_6 + y_3,$$
$$z_3 \bar{d}_3 = \bar{z}_3 + x_6,$$
$$z_3 c_9 = b_3 + b_9 + x_{15},$$
$$z_3 \bar{c}_9 = r_3 + d_9 + t_{15},$$
$$z_3 b_6 = \bar{z}_3 + x_{15},$$
$$z_3 \bar{b}_6 = y_3 + t_{15},$$
$$z_3 y_{15} = x_6 + b_9 + 2x_{15},$$
$$z_3 \bar{y}_{15} = s_6 + 2t_{15} + d_9.$$

Now let us see the following example of NITA of dimension 22, which satisfies $b_3^2 = r_3 + s_6$, r_3 is real, denoted as $(A(7 \cdot 5 \cdot 10), B_{22})$, where $B_{22} = \{1, b_8, x_{10}, b_5, c_5, c_8, x_9, r_3, y_3, s_6, t_{15}, d_9, b_3, \bar{b}_3, t_6, \bar{t}_6, b_{15}, \bar{b}_{15}, y_9, \bar{y}_9, x_3, \bar{x}_3\}$. The interesting thing is that this example also contains table subsets:

$$\{1\} \subseteq C \subseteq E \subseteq B_{22},$$

where C and E are as defined in the previous example of dimension 32. In particular both examples of dimensions 22 and 32 each contain the same sub-table algebra of dimension 12 generated by the basis E. Also E is a maximal table subset of B_{22}. The equations of products of the basis elements of C and E are as in the example of $A(3 \cdot A_6 \cdot 2, 32)$ described above

$$
\begin{aligned}
b_3\bar{b}_3 &= 1 + b_8, \\
b_3^2 &= r_3 + s_6, \\
b_3 t_6 &= b_8 + x_{10}, \\
b_3 \bar{t}_6 &= r_3 + t_{15}, \\
b_3 b_{15} &= 2t_{15} + s_6 + d_9, \\
b_3 \bar{b}_{15} &= b_8 + x_{10} + b_5 + c_5 + c_8 + x_9, \\
b_3 y_9 &= t_{15} + d_9 + y_3, \\
b_3 \bar{y}_9 &= c_8 + x_9 + x_{10}, \\
b_3 x_3 &= d_9, \\
b_3 \bar{x}_3 &= x_9, \\
b_3 b_8 &= b_3 + \bar{t}_6 + b_{15}, \\
b_3 x_{10} &= b_{15} + \bar{t}_6 + y_9, \\
b_3 b_5 &= b_{15}, \\
b_3 c_5 &= b_{15}, \\
b_3 c_8 &= b_{15} + y_9, \\
b_3 x_9 &= b_{15} + y_9 + x_3, \\
b_3 r_3 &= \bar{b}_3 + t_6, \\
b_3 y_3 &= \bar{y}_9, \\
b_3 s_6 &= \bar{b}_3 + \bar{b}_{15}, \\
b_3 t_{15} &= 2\bar{b}_{15} + t_6 + \bar{y}_9, \\
b_3 d_9 &= \bar{b}_{15} + \bar{y}_9 + \bar{x}_3,
\end{aligned}
$$

$$t_6 \bar{t}_6 = 1 + b_5 + c_5 + b_8 + c_8 + x_9,$$
$$t_6^2 = 2s_6 + t_{15} + d_9,$$
$$t_6 b_{15} = 2b_8 + 2c_8 + b_5 + c_5 + 3x_{10} + 2x_9,$$
$$t_6 \bar{b}_{15} = c_3 + b_6 + 4t_{15} + y_3 + 2d_9,$$
$$t_6 y_9 = b_8 + b_5 + c_5 + c_8 + 2x_9 + x_{10},$$
$$t_6 \bar{y}_9 = s_6 + 2d_9 + 2t_{15},$$
$$t_6 x_3 = x_{10} + c_8,$$
$$t_6 \bar{x}_3 = t_{15} + y_3,$$
$$t_6 b_8 = \bar{b}_3 + 2\bar{b}_{15} + t_6 + \bar{y}_9,$$
$$t_6 x_{10} = 3\bar{b}_{15} + \bar{y}_9 + \bar{b}_3 + \bar{x}_3,$$
$$t_6 b_5 = \bar{b}_{15} + t_6 + \bar{y}_9,$$
$$t_6 c_5 = \bar{b}_{15} + t_6 + \bar{y}_9,$$
$$t_6 c_8 = 2\bar{b}_{15} + t_6 + \bar{y}_9 + \bar{x}_3,$$
$$t_6 x_9 = 2\bar{b}_{15} + t_6 + 2\bar{y}_9,$$
$$t_6 r_3 = b_{15} + b_3,$$
$$t_6 y_3 = b_{15} + x_3,$$
$$t_6 s_6 = 2\bar{t}_6 + b_{15} + y_9,$$
$$t_6 t_{15} = 2y_9 + x_3 + b_3 + \bar{t}_6 + 4b_{15},$$
$$t_6 d_9 = 2y_9 + 2b_{15} + \bar{t}_6,$$

$$b_{15} \bar{b}_{15} = 1 + 3b_5 + 3c_5 + 5b_8 + 5c_8 + 6x_{10} + 6x_9,$$
$$b_{15}^2 = 2r_3 + 2y_3 + 4s_6 + 6d_9 + 9t_{15},$$
$$b_{15} y_9 = r_3 + y_3 + 2s_6 + 3d_9 + 6t_{15},$$
$$b_{15} \bar{y}_9 = 2b_5 + 2c_5 + 3b_8 + 3c_8 + 3x_9 + 4x_{10},$$
$$b_{15} x_3 = s_6 + 2t_{15} + d_9,$$
$$b_{15} \bar{x}_3 = c_5 + c_8 + x_9 + b_8 + x_{10} + b_5,$$
$$b_{15} b_8 = 5b_{15} + 3y_9 + x_3 + b_3 + 2\bar{t}_6,$$
$$b_{15} x_{10} = 6b_{15} + 4y_9 + b_3 + x_3 + 3\bar{t}_6,$$
$$b_{15} b_5 = b_3 + x_3 + \bar{t}_6 + 2y_9 + 3b_{15},$$
$$b_{15} c_5 = 3b_{15} + 2y_9 + b_3 + x_3 + \bar{t}_6,$$
$$b_{15} c_8 = 5b_{15} + 3y_9 + x_3 + b_3 + 2\bar{t}_6,$$
$$b_{15} x_9 = 6b_{15} + 3y_9 + x_3 + b_3 + 2\bar{t}_6,$$
$$b_{15} r_3 = 2\bar{b}_{15} + t_6 + \bar{y}_9,$$
$$b_{15} y_3 = 2\bar{b}_{15} + \bar{y}_9 + t_6,$$
$$b_{15} s_6 = 4\bar{b}_{15} + 2\bar{y}_9 + \bar{b}_3 + \bar{x}_3 + t_6,$$
$$b_{15} t_{15} = 9\bar{b}_{15} + 6\bar{y}_9 + 2\bar{b}_3 + 2\bar{x}_3 + 4t_6,$$
$$b_{15} d_9 = 6\bar{b}_{15} + 3\bar{y}_9 + \bar{b}_3 + \bar{x}_3 + 2t_6,$$

$$y_9 \bar{y}_9 = 1 + b_5 + c_5 + 2b_8 + 2c_8 + 2x_9 + 2x_{10},$$
$$y_9^2 = r_3 + y_3 + 2s_6 + 2d_9 + 3t_{15},$$
$$y_9 x_3 = r_3 + d_9 + t_{15},$$
$$y_9 \bar{x}_3 = b_8 + x_{10} + x_9,$$
$$y_9 b_8 = \bar{t}_6 + 2y_9 + 3b_{15} + x_3,$$
$$y_9 x_{10} = \bar{t}_6 + 2y_9 + b_3 + x_3 + 4b_{15},$$
$$y_9 b_5 = \bar{t}_6 + y_9 + 2b_{15},$$
$$y_9 c_5 = \bar{t}_6 + y_9 + 2b_{15},$$
$$y_9 c_8 = 3b_{15} + b_3 + 2y_9 + \bar{t}_6,$$
$$y_9 x_9 = 2\bar{t}_6 + 2y_9 + 3b_{15} + x_3 + b_3,$$
$$y_9 r_3 = \bar{b}_{15} + \bar{y}_9 + \bar{x}_3,$$
$$y_9 y_3 = \bar{b}_{15} + \bar{b}_3 + \bar{y}_9,$$
$$y_9 s_6 = 2\bar{b}_{15} + t_6 + 2\bar{y}_9,$$
$$y_9 t_{15} = 6\bar{b}_{15} + 2t_6 + \bar{b}_3 + 3\bar{y}_9 + \bar{x}_3,$$
$$y_9 d_9 = 3\bar{b}_{15} + 2t_6 + \bar{b}_3 + 2\bar{y}_9 + \bar{x}_3,$$

$$x_3 \bar{x}_3 = 1 + c_8,$$
$$x_3^2 = s_6 + y_3,$$
$$x_3 b_8 = b_{15} + y_9,$$
$$x_3 x_{10} = b_{15} + \bar{t}_6 + y_9,$$
$$x_3 b_5 = b_{15},$$
$$x_3 c_5 = b_{15},$$
$$x_3 c_8 = x_3 + b_{15} + \bar{t}_6,$$
$$x_3 x_9 = b_3 + y_9 + b_{15},$$
$$x_3 r_3 = \bar{y}_9,$$
$$x_3 y_3 = \bar{x}_3 + t_6,$$
$$x_3 s_6 = \bar{b}_{15} + \bar{x}_3,$$
$$x_3 t_{15} = 2\bar{b}_{15} + t_6 + \bar{y}_9,$$
$$x_3 d_9 = \bar{b}_{15} + \bar{b}_3 + \bar{y}_9.$$

If we define a Table Algebra (A, B) for $R, S \subseteq B$,

$$R * S = \{\cup \, Supp(R \cdot S) | r \in R, s \in S\},$$

then: $Cb_3 * C\bar{b}_3 = Cr_3^2 = Cc_3 * C\bar{c}_3 = C$, C is an identity. $(Cb_3)^2 = C\bar{c}_3$, $Cb_3 * Cc_3 = Cr_3$, $Cb_3 * C\bar{c}_3 = C\bar{b}_3$ and $Cb_3 Cr_3 = C\bar{c}_3$. For the definitions of abelian table algebra, quotient of table algebra and its basis and the sub-table algebra generated by an element $b \in B$ denoted by $\langle b \rangle = B_b$, see [AB] and [B1]. The definitions of a linear and faithful element $b \in B$ can be also found there.

One can derive from Table 2.1 that $B_{32} = \bigcup_{i=1}^{10} Bb_3^i = \bigcup_{i=0}^{\infty} Bb_3^i$ and b_3 is a faithful element of B_{32} that generates B_{32}. Also the quotient table algebra with basis B_{32}/C is strictly isomorphic to the Table Algebra induced by the cyclic group Z_6 of order 6.

One can derive from Table 2.2 that $B_{22} = \bigcup_{i=1}^{7} Bb_3^i = \bigcup_{i=0}^{\infty} Bb_3^i$ and b_3 is a faithful element of B_{22} generates B_{22}. Also the quotient table algebra with basis

Table 2.1 NITA $(A(3 \cdot A_6 \cdot 2), B_{32})$

The powers of b_3	Constituents of the powers	Basis of quotient of B_{32}/C
b_3^2	c_3, b_6	$C = \{1, b_8, x_{10}, b_5, c_5, c_8, x_9\}$
b_3^3	r_3, s_6, t_{15}	$Cb_3 = \{b_3, x_6, x_{15}, b_9, \bar{z}_3\}$
b_3^4	$\bar{c}_3, \bar{b}_6, \bar{y}_{15}, \bar{c}_9$	$C\bar{b}_3 = \{\bar{b}_3, \bar{x}_6, \bar{x}_{15}, \bar{b}_9, z_3\}$
b_3^5	$\bar{b}_3, \bar{x}_6, \bar{x}_{15}, \bar{b}_9, z_3$	$Cr_3 = \{r_3, t_{15}, d_9, y_3, s_6\}$
b_3^6	$1, b_8, x_{10}, b_5, c_5, c_8, x_9$	$Cc_3 = \{c_3, b_6, y_{15}, c_9, \bar{d}_3\}$
b_3^7	$b_3, x_6, x_{15}, b_9, \bar{z}_3$	$C\bar{c}_3 = \{\bar{c}_3, \bar{b}_6, \bar{y}_{15}, \bar{c}_9, d_3\}$
b_3^8	$c_3, b_6, y_{15}, c_9, \bar{d}_3$	
b_3^9	$r_3, s_6, t_{15}, d_9, y_3$	
b_3^{10}	$\bar{c}_3, \bar{b}_6, \bar{y}_{15}, \bar{c}_9, d_3$	

Table 2.2 NITA $(A(7 \cdot 5 \cdot 10), B_{22})$

The powers of b_3	New constituents of the powers	Basis of quotient of B_{32}/C
b_3^2	r_3, s_6	$C = \{1, b_8, x_{10}, b_5, c_5, c_8, c_9\}$
b_3^3	$\bar{b}_3, t_6, \bar{b}_{15}$	$C\bar{b}_3 = \{\bar{b}_3, t_6, \bar{b}_{15}, \bar{y}_9, \bar{x}_3\}$
b_3^4	$1, b_8, x_{10}, b_5, c_5, c_8, x_9$	$Cr_3 = \{r_3, s_6, t_{15}, d_9, y_3\}$
b_3^5	$b_3, \bar{t}_6, b_{15}, y_9, x_3$	$Cb_3 = \{b_3, \bar{t}_6, b_{15}, y_9, x_3\}$
b_3^6	$r_3, s_6, t_{15}, d_9, y_{15}, y_3$	
b_3^7	$\bar{b}_3, t_6, \bar{b}_{15}, \bar{y}_9, \bar{x}_3$	

B_{22}/C is strictly isomorphic to the Table Algebra induced by the cyclic group Z_n of order 4.

In order to prove the Main Theorem 2.1, the following fact is frequently used:

Lemma 2.1 *Let* $e_m, f_m, u_n, v_n \in B$ *such that* $e_m \bar{e}_m = f_m \bar{f}_m$. *Then* $(e_m u_n, e_m u_n) = (\bar{e}_m u_n, \bar{e}_m u_n) = (f_m v_n, f_m v_n) = (\bar{f}_m v_n, \bar{f}_m v_n)$.

Proof (1) follows from $(e_m u_n, e_m u_n) = (e_m \bar{e}_m, u_n \bar{u}_n) = (f_m \bar{f}_m, u_n \bar{u}_n) = (f_m u_n, f_m u_n)$.

Note that $(t_3 \bar{t}_3, b_8), (s_4 \bar{s}_4, b_8) \le 1$, and we obtain (2).   □

Lemma 2.2 *Let* $b_3, t_3, s_4 \in B$, $b_3 \bar{b}_3 = 1 + b_8$. *Then* $(b_3 t_3, b_3 t_3), (b_3 s_4, b_3 s_4) \le 2$. *If* $(b_3 t_3, b_3 t_3) = 2$, *then* $t_3 \bar{t}_3 = 1 + b_8$.

Proof If $(b_3 x_4, b_3 x_4) \ge 3$, then $(x_4 \bar{x}_4, b_8) \ge 2$ for $b_3 \bar{b}_3 = 1 + b_8$, a contradiction. (1) follows.

If $(b_3 c_3, b_3 c_3) = 2$, then $(b_3 \bar{b}_3, c_3 \bar{c}_3) = 2$. (2) follows from $b_3 \bar{b}_3 = 1 + b_8$.   □

2.3   NITA Generated by b_3 and Satisfying $b_3^2 = b_4 + b_5$

In this section, we shall investigate NITA generated by b_3 and satisfying $b_3^2 = b_4 + b_5$, and obtain the following theorem:

Theorem 2.2   *Let (A, B) be a NITA generated by a nonreal element $b_3 \in B$ of degree 3 and without non-identity basis element of degree 1 or 2. Then $b_3\bar{b}_3 = 1 + b_8$, $b_8 \in B$. Assume that $b_3^2 = b_4 + b_5$, $b_4 \in B$ and $b_5 \in B$. Then $(\bar{b}_5 b_5, b_8) = 1$, $(b_4^2, b_4^2) = 3$ and $(b_3 b_8, b_3 b_8) = 3$.*

Now we start to investigate NITA satisfying

$$b_3 \bar{b}_3 = 1 + b_8, \tag{2.1}$$

$$b_3^2 = b_4 + b_5. \tag{2.2}$$

Proof   Based on the above equations, one may set

$$\bar{b}_3 b_4 = b_3 + b_9, \quad \text{some } b_9 \in N^* B. \tag{2.3}$$

Since $(b_4 \bar{b}_4, b_8) \le 1$, we have that $(\bar{b}_3 b_4, \bar{b}_3 b_4) = 2$, $b_9 \in B$ and

$$b_4 \bar{b}_4 = 1 + b_8 + c_7, \quad \text{where } c_7 \in N^* B. \tag{2.4}$$

$\square$

Collecting the above equations, checking the associative laws for basis elements and making the necessary calculations and assumptions, we have the following lemma.

Lemma 2.3   *The following equations always hold:*

$$\bar{b}_3 b_4 = b_3 + b_9,$$
$$b_4 \bar{b}_4 = 1 + b_8 + c_7, \quad c_7 \in N^* B,$$

$\bar{b}_3 b_5 = b_3 + x_{12}, \quad x_{12} \in N^* B \qquad\qquad\quad$ *assumption,*

$b_3 b_8 = b_3 + b_9 + x_{12} \qquad\qquad\qquad\qquad$ *calculating $b_3^2 \bar{b}_3 = b_3(b_3 \bar{b}_3)$,*

$b_3 c_7 = b_9 + f_{12}, \quad f_{12} \in N^* B \qquad\qquad$ *assumption,*

$b_4 \bar{b}_9 = b_3 + b_9 + x_{12} + f_{12} \qquad\qquad\quad$ *calculating $b_3(\bar{b}_4 b_4) = (b_3 \bar{b}_4) b_4$,*

$b_4 b_8 = b_5 + b_3 b_9 \qquad\qquad\qquad\qquad\quad$ *calculating $b_3(\bar{b}_3 b_4) = (b_3 \bar{b}_3) b_4$,*

$\bar{b}_4 b_5 = b_8 + z_{12}, \quad z_{12} \in N^* B \qquad\qquad$ *assumption,*

$\bar{b}_3 b_9 = c_7 + b_8 + \bar{z}_{12} \qquad\qquad\qquad\qquad$ *calculating $\bar{b}_3^2 b_4 = \bar{b}_3(\bar{b}_3 b_4)$,*

$b_8^2 = 1 + c_7 + b_8 + \bar{z}_{12} + \bar{b}_3 x_{12} \qquad\qquad$ *calculating $(b_3 \bar{b}_3) b_8 = (b_3 b_8) \bar{b}_3$,*

$b_8 c_7 = b_8 + \bar{z}_{12} + \bar{b}_3 f_{12} \qquad\qquad\qquad$ *calculating $(b_3 \bar{b}_3) c_7 = (b_3 b_7) \bar{b}_3$,*

$b_5 \bar{b}_9 = b_9 + b_3 z_{12} \qquad\qquad\qquad\qquad\quad$ *calculating $(\bar{b}_3 b_4) \bar{b}_5 = \bar{b}_3(b_4 \bar{b}_5)$,*

$b_9 \bar{b}_9 = 1 + b_8 - z_{12} + \bar{z}_{12} + \bar{b}_3 f_{12} + \bar{b}_3 x_{12}$ *calculating $(b_4 \bar{b}_4) b_8 = \bar{b}_4(b_4 b_8)$.*

Proof Based on the known equations, checking associative laws one by one and making necessary assumptions, one can easily obtain the equations above. The proof of the last equation follows from

$$(b_4\bar{b}_4)b_8 = b_8 + b_8^2 + b_8c_7$$
$$= b_8 + 1 + c_7 + b_8 + \bar{z}_{12} + \bar{b}_3x_{12} + b_8 + \bar{z}_{12} + \bar{b}_3f_{12},$$
$$\bar{b}_4(b_4b_8) = \bar{b}_4b_5 + (\bar{b}_4b_3)b_9$$
$$= \bar{b}_4b_5 + \bar{b}_3b_9 + b_9\bar{b}_9$$
$$= b_8 + z_{12} + c_7 + b_8 + \bar{z}_{12} + b_9\bar{b}_9.$$

$\square$

Obviously $(b_5\bar{b}_5, b_8) \leq 3$, c_7 is either $c_3 + c_4$ or irreducible.

If $c_7 \notin B$, then there exist $c_3, c_4 \in B$ such that $c_7 = c_3 + c_4$. First we have the following lemma:

Lemma 2.4 *There exists no NITA generated by b_3 and satisfying $b_3\bar{b}_3 = 1 + b_8$,* $b_3^2 = b_4 + b_5$ *and* $c_7 = c_3 + c_4$.

Proof By the assumption, we have that

$$b_4\bar{b}_4 = 1 + b_8 + c_3 + c_4. \tag{2.5}$$

Then

$$b_3^2\bar{b}_4 = b_4\bar{b}_4 + b_5\bar{b}_4$$
$$= 1 + c_3 + c_4 + b_8 + \bar{b}_4b_5,$$
$$b_3(b_3\bar{b}_4) = b_3\bar{b}_3 + b_3\bar{b}_9$$
$$= 1 + b_8 + b_3\bar{b}_9.$$

Hence $(c_3, b_3\bar{b}_9) = (c_4, b_3\bar{b}_9) = 1$. Therefore

$$b_3c_3 = b_9, \tag{2.6}$$
$$b_3c_4 = d_3 + b_9, \quad \text{some } d_3 \in B, \ d_3 \neq b_3. \tag{2.7}$$

Now one has $(b_3\bar{d}_3, c_4) = 1$. Then

$$b_3\bar{d}_3 = c_4 + d_5, \quad \text{some } d_5 \in B, \tag{2.8}$$

from which $(b_3\bar{b}_3, d_3\bar{d}_3) = 2$ follows. Hence

$$d_3\bar{d}_3 = 1 + b_8. \tag{2.9}$$

By (2.5), we have that $(b_4c_3, b_4) = 1$. Let

$$b_4c_3 = b_4 + y_8, \quad y_8 \in N^* B. \tag{2.10}$$

Since $(b_3\bar{b}_3)c_3 = c_3 + c_3b_8$, $\bar{b}_3(b_3c_3) = \bar{b}_3b_9$, one has that $\bar{b}_3b_9 = c_3 + c_3b_8$. But $(c_4, \bar{b}_3b_9) = 1$. Then $(c_4, c_3b_8) = 1$ and so $(c_4c_3, b_8) = 1$. Thus

$$c_3c_4 = x_4 + b_8, \quad \text{some } x_4 \in B. \tag{2.11}$$

By (2.10) we have that

$$2 \leq (b_4c_3, b_4c_3) = (b_4\bar{b}_4, c_3^2) = (1 + c_3 + c_4 + b_8, c_3^2),$$

which implies that c_3^2 equals one of

$$1 + b_8, \qquad 1 + c_3 + x_5, \qquad 1 + c_4 + y_4, \qquad 1 + 2c_4.$$

By (2.11), we know that c_3^2 cannot be $1 + c_4 + y_4$ or $1 + 2c_4$. And by (2.6), $c_3^2 \neq 1 + b_8$. Hence

$$c_3^2 = 1 + c_3 + x_5. \tag{2.12}$$

Therefore $(b_4c_3, b_4c_3) = (c_3^2, b_4\bar{b}_4) = 2$, which implies that $y_8 \in B$.

By Lemmas 2.3, (2.8) and (2.9), we have

$$b_3(b_3\bar{b}_3) = b_3 + b_3b_8 = 2b_3 + b_9 + x_{12},$$
$$\bar{b}_3(b_3^2) = \bar{b}_3b_4 + \bar{b}_3b_5 = b_3 + b_9 + \bar{b}_3b_5,$$
$$(b_3\bar{d}_3)d_3 = c_4d_3 + d_5d_3.$$

Hence $b_3b_8 = b_9 + \bar{b}_3b_5$, and $b_9 \in Supp\{c_4d_3 + d_5d_3\}$.

Assume that $b_9 \in Supp\{d_3d_5\}$. By (2.8), it follows that

$$d_3d_5 = b_9 + b_3 + l_3, \quad l_3 \in B,$$
$$\bar{b}_3b_5 = c_4d_3 + l_3,$$

so that $b_3b_8 = b_9 + l_3 + c_4d_3$, which implies that $b_8 \in Supp\{b_3\bar{l}_3\}$. Hence $l_3 = b_3$. So $(d_3d_5, b_3) = 3$, which implies that $(b_4\bar{d}_3, d_5) = 2$, a contradiction to (2.8). Thus $b_9 \in Supp\{c_4d_3\}$ and

$$c_4d_3 = b_3 + b_9. \tag{2.13}$$

Thus $(d_3\bar{d}_3, c_4^2) = 2$, we may assume that

$$c_4^2 = 1 + b_8 + f_7, \quad \text{some } f_7 \in N^* B. \tag{2.14}$$

Then $c_4 \notin Supp\{c_4x_5\}$. Otherwise, $c_4 \in Supp\{c_4x_5\}$, which implies that $x_5 \in Supp\{c_4^2\}$, and so $f_7 = x_5 + x_2$, a contradiction.

Since

$$c_3^2 c_4 = c_4 + c_3 c_4 + x_5 c_4$$
$$= c_4 + x_4 + b_8 + c_4 x_5,$$
$$c_3(c_3 c_4) = c_3 b_8 + c_3 x_4,$$

we have that

$$c_3 b_8 + c_3 x_4 = c_4 + x_4 + b_8 + c_4 x_5.$$

But the left hand side of the above equation contains two c_4 by (2.11). Hence $c_4 = x_4$ by $c_4 \notin Supp\{c_4 x_5\}$. Therefore

$$c_3 c_4 = b_8 + c_4, \tag{2.15}$$

which implies that $(c_4^2, c_3) = 1$ and

$$c_4^2 = 1 + b_8 + c_3 + e_4, \quad \text{some } e_4 \in B.$$

Now we have the following equations:

$$b_3 c_4^2 = b_3 + b_3 b_8 + b_3 c_3 + b_3 e_4$$
$$= 2b_3 + 2b_9 + x_{12} + b_3 e_4,$$
by (2.7) and (2.13) $c_4(b_3 c_4) = c_4 b_9 + c_4 d_3$
$$= c_4 b_9 + b_9 + b_3.$$

Then $c_4 b_9 = b_3 + b_9 + x_{12} + b_3 e_4$. Since $d_3 \in Supp\{c_4 b_9\}$, one has that $d_3 \in Supp\{x_{12}\}$ or $b_3 e_4$.

If $d_3 \in Supp\{x_{12}\}$, then $d_3 \in Supp\{b_3 b_8\}$, which implies that $b_3 \bar{d}_3 = 1 + b_8$. Thus $b_3 = d_3$, which implies that $b_3 \bar{b}_3 = c_4 + d_5$, a contradiction. Hence $d_3 \in Supp\{b_3 e_4\}$, $e_4 \in Supp\{b_3 \bar{d}_3\}$. Therefore by (2.8) $e_4 = c_4$ and

$$\text{by (2.7)}\quad c_4^2 = 1 + c_3 + c_4 + b_8 = b_4 \bar{b}_4,$$
$$c_4 b_9 = b_3 + d_3 + 2b_9 + x_{12}.$$

Furthermore

$$b_3 c_4^2 = b_3 + b_3 b_8 + b_3 c_4 + b_3 c_3$$
$$= b_3 + b_3 + b_9 + x_{12} + d_3 + b_9 + b_9,$$
$$(b_3 \bar{b}_4) b_4 = \bar{b}_3 b_4 + \bar{b}_9 b_4$$
$$= b_3 + b_9 + b_4 \bar{b}_9,$$
$$(b_3 c_4) c_4 = c_4 b_9 + d_3 c_4$$
$$= c_4 b_9 + b_3 + b_9.$$

Thus

$$b_4 \bar{b}_9 = b_3 + d_3 + 2b_9 + x_{12} = c_4 b_9.$$

Hence by (2.6) $(b_3 c_3)c_4 = b_9 c_4 = b_3 + d_3 + 2b_9 + x_{12}$. But by (2.7) $(b_3 c_4)c_3 = c_3 b_9 + c_3 d_3$.

Since $d_3 \notin Supp\{c_3 d_3\}$, we have that $d_3 \in Supp\{c_3 b_9\}$

$$c_3 d_3 = b_9,$$
$$c_3 b_9 = b_3 + d_3 + b_9 + x_{12}. \tag{2.16}$$

Therefore $(b_3 c_3, c_3 d_3) = 1$, so $(b_3 \bar{d}_3, c_3^2) = 1$. By (2.8) and (2.12), we have that $(b_3 \bar{d}_3, x_5) = 1$, which implies that $x_5 = d_5$. Consequently d_5 is real. Then

$$b_3 \bar{d}_3 = c_4 + d_5 = \bar{b}_3 d_3,$$
$$c_3^2 = 1 + c_3 + d_5.$$

Moreover $(b_3^2, d_3^2) = (b_3 \bar{d}_3, \bar{b}_3 d_3) = 2$. So

$$d_3^2 = 1 + 2b_4 \text{ or } b_4 + b_5.$$

Since (2.9) implies that d_3 is nonreal. So

$$d_3^2 = b_4 + b_5.$$

By (2.8), we may state that

$$\bar{b}_3 d_5 = \bar{d}_3 + u_{12}, \quad u_{12} \in N^* B.$$

Then

$$\bar{b}_3 c_3^2 = \bar{b}_3 + \bar{b}_3 c_3 + \bar{b}_3 d_5$$
$$= \bar{b}_3 + \bar{b}_9 + \bar{d}_3 + u_{12},$$
$$(\bar{b}_3 c_3)c_3 = c_3 \bar{b}_9.$$

Thus

$$c_3 b_9 = b_3 + d_3 + b_9 + \bar{u}_{12}.$$

Hence $x_{12} = \bar{u}_{12}$ by (2.16) and so $b_3 d_5 = d_3 + x_{12}$. Since $c_4 c_3^2 = c_4 + c_3 c_4 + d_5 c_4$ and $c_3(c_3 c_4) = c_3 c_4 + c_3 b_8$ by (2.15), we have that

$$c_3 b_8 = c_4 + c_4 d_5.$$

Now, by $c_4^2 = b_4 \bar{b}_4$, we have that

$$c_3 c_4^2 = c_3 + c_3^2 + c_3 c_4 + c_3 b_8$$
$$= c_3 + 1 + c_3 + d_5 + c_4 + b_8 + c_4 + c_4 d_5,$$

$$(c_3 b_4)\bar{b}_4 = b_4 \bar{b}_4 + y_8 \bar{b}_4$$
$$= 1 + c_3 + c_4 + b_8 + \bar{b}_4 y_8.$$

So

$$\bar{b}_4 y_8 = c_3 + c_4 + d_5 + c_4 d_5.$$

We know that y_8 is irreducible, then $(c_4 b_4, y_8) = 1$. But $(b_4 c_4, b_4 c_4) = 4$. Hence $(c_4 b_4 - y_8, c_4 b_4 - y_8) = 3$, which is impossible for $L_1(B) = \{1\}$ and $L_2(B) = \emptyset$. The lemma follows.   □

Now we begin to investigate NITA such that $c_7 \in B$ is based on the value of $(b_5 \bar{b}_5, b_8)$.

Lemma 2.5   *There exists no NITA such that $c_7 \in B$ and $(b_5 \bar{b}_5, b_8) = 3$.*

Proof It is easy to see that $(b_5 \bar{b}_5, b_8) = 3$ implies that $(\bar{b}_3 b_5, \bar{b}_3 b_5) = 4$. By Lemma 2.3, we may assume that

$$x_{12} = x + y + z, \quad x, y, z \in B,$$
$$\bar{b}_3 b_5 = b_3 + x + y + z,$$
$$b_3 b_8 = b_3 + b_9 + x + y + z.$$

From the last equation above, one has that b_8 is contained in $\bar{b}_3 x$, $\bar{b}_3 y$ and $\bar{b}_3 z$. Hence no one of x, y and z has degree 3. Therefore

$$x_{12} = x_4 + y_4 + z_4,$$
$$\bar{b}_3 x_4 = r_4 + b_8, \quad \text{some } r_4 \in B,$$
$$\bar{b}_3 y_4 = s_4 + b_8, \quad \text{some } s_4 \in B,$$
$$\bar{b}_3 z_4 = t_4 + b_8, \quad \text{some } t_4 \in B.$$

Furthermore, $b_3 x_4$, $b_3 y_4$ and $b_3 z_4$ have exactly two constituents. Let

$$b_3 x_4 = b_5 + r_7, \quad \text{some } r_7 \in B,$$
$$b_3 y_4 = b_5 + s_7, \quad \text{some } s_7 \in B,$$
$$b_3 z_4 = b_5 + t_7, \quad \text{some } t_7 \in B.$$

Since

$$\bar{b}_3 (b_3 x_4) = \bar{b}_3 b_5 + \bar{b}_3 r_7$$
$$= b_3 x_4 + y_4 + z_4 + \bar{b}_3 r_7,$$
$$b_3 (\bar{b}_3 x_4) = b_3 b_8 + b_3 r_4$$
$$= b_3 + b_9 + x_4 + y_4 + z_4 + b_3 r_4,$$

we have that $\bar{b}_3 r_7 = b_9 + b_3 r_4$. By the same reasoning we have that $\bar{b}_3 s_7 = b_9 + b_3 s_4$ and $\bar{b}_3 t_7 = b_9 + b_3 t_4$. Hence $r_7, s_7, t_7 \in Supp\{b_3 b_9\}$.

If r_7, s_7 and t_7 are distinct, then

$$b_3 b_9 = b_4 + r_7 + s_7 + t_7 + e_2, \quad \text{some } e_2 \in B,$$

a contradiction. Hence at least two of r_7, s_7 and t_7 are equal, without loss of generality, let $r_7 = s_7$. Then $(b_3 x_4, b_3 y_4) = (\bar{b}_3 x_4, \bar{b}_3 y_4) = 2$. Thus $r_4 = s_4$. By Lemma 2.3, we have

$$b_8^2 = 1 + 4b_8 + c_7 + z_{12} + 2r_4 + t_4.$$

So $\bar{z}_{12} = 2r_4 + t_4$. Lemma 2.3 implies that

$$\bar{b}_3 b_9 = c_7 + b_8 + 2r_4 + t_4.$$

Then $(\bar{b}_3 b_9, r_4) = (b_9, b_3 r_4) \geq 2$, a contradiction, from which the lemma follows. $\square$

Lemma 2.6   *There exists no NITA such that $c_7 \in B$ and $(b_5 \bar{b}_5, b_8) = 2$.*

Suppose that $(b_5 \bar{b}_5, b_8) = 2$. Let

$$b_5 \bar{b}_5 = 1 + 2b_8 + d_8, \quad \text{some } d_8 \in N^* B.$$

Since $(\bar{b}_3 b_5, \bar{b}_3 b_5) = (b_3 \bar{b}_3, b_5 \bar{b}_5) = 3$, by Lemma 2.3, we have that $x_{12} = x + y$ and

$$\bar{b}_3 b_5 = b_3 + x + y,$$
$$b_3 b_8 = b_3 + b_9 + x + y.$$

Thus $b_8 \in Supp\{\bar{b}_3 x\}$, $Supp\{\bar{b}_3 y\}$, which implies that degrees of x and y are ≥ 4. We have three possibilities: $(f(x), f(y)) = (4, 8), (5, 7), (6, 6)$.

We shall divide the proof into three propositions, Propositions 2.1, 2.2, 2.3 from which the lemma follows.

Proposition 2.1   *There exists no NITA such that $b_5 \bar{b}_5 = 1 + 2b_8 + d_8$ and $(f(x), f(y)) = (4, 8)$.*

Proof   Assume $(f(x), f(y)) = (4, 8)$. Then $x_{12} = x_4 + y_8$ and we may set

$$\bar{b}_3 x_4 = b_8 + x_4', \quad x_4' \in B,$$
$$\bar{b}_3 y_8 = b_8 + y_{16}, \quad y_{16} \in N^* B, \tag{2.17}$$

Then

$$b_8^2 = 1 + c_7 + 3b_8 + x_4' + \bar{z}_{12} + y_{16}.$$

Hence

$$(b_3\bar{b}_3)^2 = 2 + 5b_8 + c_7 + x_4' + \bar{z}_{12} + y_{16},$$

$$b_3^2\bar{b}_3^2 = b_4\bar{b}_4 + b_4\bar{b}_5 + \bar{b}_4b_5 + b_5\bar{b}_5$$

$$= 1 + b_8 + c_7 + b_4\bar{b}_5 + \bar{b}_4b_5 + 1 + 2b_8 + d_8,$$

so $z_{12} + d_8 = x_4' + y_{16}$.

Now we have two cases, $x_4' \in Supp\{d_8\}$ or $x_4' \notin Supp\{d_8\}$.

Case I Suppose that $x_4' \in Supp\{z_{12}\}$. Then there exists s_8 such that

$$z_{12} = x_4' + s_8,$$

$$y_{16} = d_8 + s_8.$$

Since $(\bar{b}_4b_5, \bar{b}_4b_5) = (b_4\bar{b}_4, b_5\bar{b}_5) = 3$, we have that $s_8 \in B$ and

$$\bar{b}_3b_9 = c_7 + b_8 + \bar{x}'_4 + \bar{s}_8.$$

Hence $b_9 \in Supp\{b_3x_4'\}$, which means that $(b_3\bar{x}_4', b_3\bar{x}_4') = 2$. So $(b_3x_4', b_3x_4') = 2$. Since $\bar{b}_3x_4 = b_8 + x_4'$, we may set $b_3\bar{x}_4' = a_3 + b_9$ and $b_3x_4' = x_4 + e_8$, for some $a_3, e_8 \in B$.

Suppose that $b_3\bar{a}_3 = x_4' + l_5$, for some $l_5 \in B$. Then

$$b_3(\bar{b}_3x_4') = b_3\bar{b}_9 + b_3\bar{a}_3$$

$$= c_7 + b_8 + x_4' + s_8 + x_4' + l_5,$$

$$\bar{b}_3(b_3x_4') = \bar{b}_3x_4 + \bar{b}_3e_8$$

$$= b_8 + x_4' + \bar{b}_3e_8.$$

Hence

$$\bar{b}_3e_8 = c_7 + s_8 + x_4' + l_5.$$

Therefore $e_8 \in Supp\{b_3c_7\}$. Let $b_3c_7 = b_9 + e_8 + e_4$, $e_4 \in B$. By $(b_3\bar{b}_3)\bar{a}_3 = \bar{b}_3(b_3\bar{a}_3)$, we obtain

$$\bar{a}_3b_8 = \bar{b}_9 + \bar{b}_3l_5.$$

Since $(\bar{a}_3b_8, \bar{a}_3b_8) = (a_3\bar{a}_3, b_8^2) = (b_3\bar{b}_3, b_8^2) = 4$, we have that $(\bar{b}_3l_5, \bar{b}_3l_5) = 3$. So $(b_3l_5, b_3l_5) = 3$. Let

$$b_3l_5 = e_8 + \alpha_3 + \beta_4,$$

where $\alpha_3, \beta_4 \in B$. Obviously $\alpha_3 \neq b_3$.

$$b_3(\bar{b}_3e_8) = b_3c_7 + b_3b_8 + b_3x_4' + b_3l_5$$

$$= 2b_9 + 3e_8 + e_4 + b_3 + 2x_4 + y_8 + \alpha_3 + \beta_4,$$

$$(b_3\bar{b}_3)e_8 = e_8 + e_8b_8.$$

Then

$$e_8 b_8 = 2b_9 + 2e_8 + e_4 + b_3 + 2x_4 + y_8 + \alpha_3 + \beta_4.$$

Hence $e_8 \in Supp\{b_3 b_8\}$, which implies that $e_8 = y_8$. Hence $b_3 c_7 = b_9 + y_8 + e_4$, so that $c_7 \in Supp\{\bar{b}_3 y_8\}$, which implies that $c_7 \in Supp\{y_{16}\} = \{d_8, s_8\}$, a contradiction.

Case II NITA satisfying $x_4' \in Supp\{d_8\}$.

If $x_4' \in d_8$, then $d_8 = x_4' + \bar{x}_4'$ (x_4' maybe real). Then

$$b_5 \bar{b}_5 = 1 + 2b_8 + x_4' + \bar{x}_4'.$$

Hence $y_{16} = \bar{x}_4' + z_{12}$ and

$$b_8^2 = 1 + c_7 + 3b_8 + x_4' + \bar{x}_4' + z_{12} + \bar{z}_{12}, \qquad (2.18)$$

$$\bar{b}_3 y_8 = b_8 + \bar{x}_4' + z_{12}, \qquad (2.19)$$

$$b_9 \bar{b}_9 = 1 + 3b_8 + x_4' + z_{12} + \bar{x}_4' + \bar{b}_3 f_{12}, \qquad (2.20)$$

by (2.17). Therefore $(y_8, b_3 \bar{x}_4') = 1$. Let

$$b_3 \bar{x}_4' = y_8 + t_4, \, b_3 x_4' = x_4 + j_8, \qquad \text{where } t_4, j_8 \in B. \qquad (2.21)$$

Since

$$b_3 (b_5 \bar{b}_5) = b_3 + 2b_3 b_8 + b_3 x_4' + b_3 \bar{x}_4'$$
$$= 3b_3 + 3b_9 + 2x_4 + 3y_8 + t_4 + x_4 + j_8,$$
$$b_5 (b_3 \bar{b}_5) = b_5 \bar{b}_3 + b_5 \bar{x}_4 + b_5 \bar{y}_8$$
$$= b_3 + x_4 + y_8 + b_5 \bar{x}_4 + b_5 \bar{y}_8,$$

we have that

$$\bar{x}_4 b_5 + \bar{y}_8 b_5 = 2b_3 + 2b_9 + 2x_4 + 2y_8 + t_4 + j_8.$$

Since

$$x_4' (\bar{b}_3 b_4) = b_3 x_4' + b_9 x_4'$$
$$= x_4 + j_8 + b_9 x_4',$$
$$(\bar{b}_3 x_4') b_4 = \bar{y}_8 b_4 + \bar{t}_4 b_4,$$

we have that $x_4 \in Supp\{b_4 \bar{y}_8\}$ or $Supp\{\bar{t}_4 b_4\}$.

We assert that $x_4 \in Supp\{b_4 \bar{t}_4\}$. Otherwise $x_4 \in Supp\{b_4 \bar{y}_8\}$, and $y_8 \in Supp\{b_4 \bar{x}_4\}$. But $b_9 \in Supp\{b_4 \bar{x}_4\}$ by Lemma 2.3, a contradiction. Hence $x_4 \in Supp\{b_4 \bar{t}_4\}$, which implies that $t_4 \in Supp\{b_4 \bar{x}_4\}$. Therefore

$$b_4 \bar{x}_4 = t_4 + b_9 + r_3, \qquad r_3 \in B. \qquad (2.22)$$

Since

$$\bar{b}_3(b_4b_8) = \bar{b}_3b_5 + \bar{b}_3(b_3b_9)$$
$$= b_3 + x_4 + y_8 + b_9 + b_3(\bar{b}_3b_9)$$
$$= b_3 + x_4 + y_8 + b_9 + b_3c_7 + b_3b_8 + b_3\bar{z}_{12}$$
$$= b_3 + x_4 + y_8 + b_9 + b_9 + f_{12} + b_3 + x_4 + y_8 + b_9 + b_3\bar{z}_{12},$$
$$b_4(\bar{b}_3b_8) = b_4\bar{b}_3 + b_4\bar{x}_4 + b_4\bar{y}_8 + b_4\bar{b}_9$$
$$= b_3 + b_9 + t_4 + b_9 + r_3 + b_4\bar{y}_8 + b_3 + b_9 + x_4 + y_8 + f_{12},$$
$$(\bar{b}_3b_4)b_8 = b_3b_8 + b_8b_9$$
$$= b_3 + x_4 + y_8 + b_9 + b_8b_9,$$

we have that

$$x_4'b_8 = b_8 + z_{12} + b_3\bar{t}_4,$$
$$b_8b_9 = b_3 + x_4 + y_8 + b_9 + y_{12} + b_3\bar{z}_{12},$$
$$b_4\bar{y}_8 + b_9 + t_4 + r_3 = x_4 + y_8 + b_3\bar{z}_{12}.$$

We continue the proof based on $t_4 = x_4$ or $t_4 \neq x_4$.

Subcase I There exists no NITA such that $t_4 = x_4$.

If $t_4 = x_4$, then $(x_4, b_3\bar{x}_4') = 1$ by $b_3\bar{x}_4' = y_8 + t_4$, which implies that $x_4' \in Supp\{b_3\bar{x}_4\} = \{b_8, \bar{x}_4'\}$. And $\bar{b}_3t_4 = \bar{b}_3x_4 = b_8 + x_4'$. Hence x_4' is real and

$$b_5\bar{b}_5 = 1 + 2b_8 + 2x_4', \tag{2.23}$$
$$b_3x_4' = x_4 + y_8. \tag{2.24}$$

Thus

$$\bar{b}_3(b_3x_4') = \bar{b}_3y_8 + \bar{b}_3x_4$$
$$= b_8 + x_4' + \bar{z}_{12} + b_8 + x_4',$$
$$(b_3\bar{b}_3)x_4' = x_4' + x_4'b_8.$$

So

$$x_4'b_8 = x_4' + 2b_8 + z_{12}.$$

Since

$$b_3(b_5\bar{b}_5) = b_3 + 2b_3b_8 + 2b_3x_4'$$
$$= b_3 + 2b_3 + 2b_9 + 2x_4 + 2y_8 + 2x_4 + 2y_8,$$
$$b_5(b_3\bar{b}_5) = \bar{b}_3b_5 + \bar{x}_4b_5 + \bar{y}_8b_5$$
$$= b_3 + x_4 + y_8 + \bar{x}_4b_5 + \bar{y}_8b_5,$$

we have that $\bar{x}_4 b_5 + \bar{y}_8 b_5 = 2b_3 + 2b_9 + 3x_4 + 3y_8$. Since $b_3 \in Supp\{\bar{x}_4 b_5\}$ and $b_3 \in Supp\{\bar{y}_8 b_5\}$, there are exactly two possibilities:

$$\bar{x}_4 b_5 = b_3 + b_9 + y_8, \qquad \bar{y}_8 b_5 = b_3 + b_9 + 3x_4 + 2y_8,$$
$$\bar{x}_4 b_5 = b_3 + b_9 + 2x_4, \qquad \bar{y}_8 b_5 = b_3 + b_9 + x_4 + 3y_8.$$

If $\bar{x}_4 b_5 = b_3 + b_9 + y_8$ and $\bar{y}_8 b_5 = b_3 + b_9 + 3x_4 + 2y_8$, then $(\bar{y}_8 b_5, x_4) = 3$, which implies that $(b_5 \bar{x}_4, y_8) = 3$, a contradiction.

If $\bar{x}_4 b_5 = b_3 + b_9 + 2x_4$ and $\bar{y}_8 b_5 = b_3 + b_9 + x_4 + 3y_8$, then $(\bar{x}_4 b_5, x_4) = 3$. So $(x_4^2, b_5) = 2$. It follows that $(x_4^2, x_4^2) \geq 5$, which is impossible for either $x_4 \bar{x}_4 = 1 + b_8 + r_7, r_7 \in B$ or $x_4 \bar{x}_4 = 1 + b_8 + r_4 + r_3, r_3, r_4 \in B$. Hence Subcase I follows.

Subcase II There exists no NITA such that $t_4 \neq x_4$.

If $t_4 \neq x_4$, by $b_4 \bar{y}_8 + b_9 + t_4 + r_3 = x_4 + y_8 + b_3 \bar{z}_{12}$, we have that $r_3 + t_4 + b_9$ is a part of $b_3 \bar{z}_{12}$, $x_4 + y_8$ is a part of $b_4 \bar{y}_8$ and $b_3 + x_4 + y_8 + b_9 + y_{12} + r_3 + t_4 + b_9$ is a part of $b_8 b_9$.

If $b_9 \in Supp\{y_{12}\}$, then there exists $h_3 \in B$ such that $y_{12} = h_3 + b_9$. So $b_3 c_7 = h_3 + 2b_9$, $b_3 \bar{h}_3$ contains c_7, which is impossible for $L_2(B) = \emptyset$. Hence $b_9 \notin Supp\{y_{12}\}$.

Since $(b_9 b_9, b_8) = 3$, so $(b_8 b_9, b_9) = 3$, which implies that $(b_3 \bar{z}_{12}, b_9) = 2$ and there exists $x_{11} \in N^* B$ of degrees 11 such that

$$b_3 \bar{z}_{12} = 2b_9 + t_4 + r_3 + x_{11},$$
$$b_4 \bar{y}_8 = x_4 + y_8 + b_9 + x_{11}.$$

The following leads to a contradiction based on the z_{12} representation. Since $(z_{12}, z_{12}) = 2$, we have that $z_{12} = \alpha + \beta$, where $(f(\alpha), f(\beta)) = (3, 9), (4, 8), (5, 7)$ or $(6, 6)$.

Step 1 There exists no NITA such that $z_{12} = \alpha_3 + \beta_9$.
  If $z_{12} = \alpha_3 + \beta_9$, then $b_3(\bar{\alpha}_3 + \bar{\beta}_9) = r_3 + t_4 + b_9 + x_{11}$, $b_3 \bar{\alpha}_3 = b_9$ and

$$\bar{b}_3(b_3 \bar{\alpha}_3) = c_7 + b_8 + \bar{\alpha}_3 + \bar{\beta}_9,$$
$$(\bar{b}_3 b_3)\bar{\alpha}_3 = \bar{\alpha}_3 + \bar{\alpha}_3 b_8.$$

Thus

$$\alpha_3 b_8 = c_7 + b_8 + \beta_9,$$

which implies that $(\alpha_3 b_8, \alpha_3 b_8) = 3$. Hence $(\alpha_3 \bar{\alpha}_3, b_8^2) = 3$. It follows that $\alpha_3 \bar{\alpha}_3 = 1 + x_4' + \bar{x}_4'$ by the expression b_8^2 (see (2.18)), which contradicts Theorem 2.1.

Step 2 There exists no NITA such that $z_{12} = \alpha_4 + \beta_8$.
  If $z_{12} = \alpha_4 + \beta_8$, then $b_3(\bar{\alpha}_4 + \bar{\beta}_8) = r_3 + t_4 + 2b_9 + x_{11}$. Since $b_9 \in Supp\{b_3 \bar{\alpha}_4\}$, we have that

$$b_3 \bar{\alpha}_4 = r_3 + b_9, \qquad b_3 \bar{\beta}_8 = t_4 + b_9 + x_{11}.$$

Let $\bar{b}_3 r_3 = \bar{\alpha}_4 + e_5$. Then

$$\bar{b}_3(b_3\bar{\alpha}_4) = \bar{b}_3 b_9 + \bar{b}_3 r_3$$
$$= c_7 + b_8 + \bar{\alpha}_4 + \bar{\beta}_8 + \bar{\alpha}_4 + e_5,$$
$$\bar{\alpha}_4(b_3\bar{b}_3) = \bar{\alpha}_4 + \bar{\alpha}_4 b_8,$$

which means that $\alpha_4 b_8 = c_7 + b_8 + \beta_8 + e_5$. Since $\bar{b}_3 y_8 = \alpha_4 + \beta_8 + \bar{x}'_4 + b_8$, we have that $b_3 \alpha_4 = y_8 + \gamma_4$, $\gamma_4 \in B$.

Since

$$\bar{b}_3(b_3\alpha_4) = \bar{b}_3 y_8 + \bar{b}_3 \gamma_4$$
$$= b_8 + \alpha_4 + \beta_8 + \bar{x}'_4 + b_3\gamma_4,$$
$$(\bar{b}_3 b_3)\alpha_4 = \alpha_4 + \alpha_4 b_8$$
$$= 2\alpha_4 + c_7 + b_8 + \beta_8 + e_5,$$

we have that $\bar{x}'_4 + b_3\gamma_4 = \alpha_4 + c_7 + e_5$, which implies that $\bar{x}'_4 = \alpha_4$. But by (2.21) $b_3 x'_4 = j_8 + x_4$, $j_8 \in B$ and $b_3\bar{\alpha}_4 = b_9 + r_3$, a contradiction.

Step 3 There exists no NITA such that $z_{12} = \alpha_5 + \beta_7$.
   If $z_{12} = \alpha_5 + \beta_7$, then

$$b_3(\bar{\alpha}_5 + \bar{\beta}_7) = 2b_9 + r_3 + t_4 + 2b_9 + x_{11}.$$

We need a sum of degree 6 to make up $b_3\bar{\alpha}_5$ for $b_9 \in Supp\{b_3\bar{\alpha}_5\}$.
   If $r_3 \in Supp\{b_3\bar{\beta}_7\}$, then $\bar{b}_3 r_3$ contains an element of degree 2, a contradiction. Hence $r_3 \in Supp\{b_3\bar{\alpha}_5\}$. We may set

$$b_3\bar{\alpha}_5 = b_9 + r_3 + \delta_3, \quad \delta_3 \in Supp\{x_{11}\}.$$

Furthermore $\bar{b}_3 r_3 = \bar{\alpha}_5 + d_4$, for some $d_4 \in B$. It follows that $r_3\bar{r}_3 = 1 + b_8$. Now we have that $(\alpha_5\bar{\alpha}_5, b_8) = 2$ and

$$2b_3 + x_4 + y_8 + b_9 = b_3(b_3\bar{b}_3) = r_3(b_3\bar{r}_3) = r_3\alpha_5 + r_3\bar{d}_4,$$

from which the following holds:

$$r_3\alpha_5 + r_3\bar{d}_4 = 2b_3 + x_4 + y_8 + b_9.$$

Comparing the two sides, the following equations hold:

$$r_3\bar{d}_4 = b_3 + b_9,$$
$$r_3\alpha_5 = b_3 + x_4 + y_8.$$

Therefore

$$\bar{b}_3(r_3\alpha_5) = b_3\bar{b}_3 + x_4\bar{b}_3 + y_8\bar{b}_3$$
$$= 1 + b_8 + b_8 + x_4' + b_8 + \alpha_5 + \beta_7 + \bar{x}_4',$$
$$\alpha_5(\bar{b}_3r_3) = \alpha_5\bar{\alpha}_5 + d_4\alpha_5.$$

Since $(\alpha_5\bar{\alpha}_5, b_8) = 2$, we can only have that

$$\alpha_5\bar{\alpha}_5 = 1 + 2b_8 + x_4' + \bar{x}_4',$$
$$d_4\alpha_5 = b_8 + \alpha_5 + \beta_7.$$

Hence $d_4 = x_4'$ or $\bar{x}_4'$. So $\bar{b}_3r_3 = \alpha_5 + x_4'$ or $\alpha_5 + \bar{x}_4'$.

If $\bar{b}_3r_3 = \alpha_5 + \bar{x}_4'$, then $r_3 \in Supp\{b_3\bar{x}_4'\} = \{y_8, t_4\}$ by (2.21), a contradiction. Hence $\bar{b}_3r_3 = \alpha_5 + x_4'$, which implies that $r_3 \in Supp\{b_3x_4'\} = \{x_4, j_8\}$, a contradiction.

Step 4 There exists no NITA such that $z_{12} = \alpha_6 + \beta_6$.

If $z_{12} = \alpha_6 + \beta_6$, then

$$b_3(\bar{\alpha}_6 + \bar{\beta}_6) = 2b_9 + t_4 + r_3 + x_{11}.$$

Hence $\alpha_6 \neq \beta_6$. Without loss of generality, let $r_3 \in Supp\{b_3\alpha_6\}$. We may set $\bar{b}_3r_3 = \bar{\alpha}_6 + d_3$, which implies that $r_3\bar{r}_3 = 1 + b_8$. Therefore we have the following equation:

$$\alpha_6r_3 + \bar{d}_3r_3 = (b_3\bar{r}_3)r_3$$
$$= (b_3\bar{b}_3)b_3$$
$$= 2b_3 + b_9 + x_4 + y_8.$$

It is easy to see that we cannot sum up $\bar{d}_3r_3$, a contradiction. Proposition 2.1 follows. □

Proposition 2.2 *There exists no NITA such that $(f(x), f(y)) = (5, 7)$.*

Proof If $(f(x), f(y)) = (5, 7)$, that is to say, $x_{12} = x_5 + y_7$, and

$$\bar{b}_3b_5 = b_3 + x_5 + y_7,$$
$$b_3b_8 = b_3 + b_9 + x_5 + y_7.$$

So we may set

$$\bar{b}_3x_5 = b_8 + e_7, \quad e_7 \in N^*B,$$
$$b_3x_5 = b_5 + e_{10}, \quad e_{10} \in N^*B,$$
$$\bar{b}_3y_7 = b_8 + y_{13}, \quad y_{13} \in N^*B,$$
$$b_3y_7 = b_5 + y_{16}, \quad y_{16} \in N^*B.$$

Hence

$$\bar{b}_3(b_3b_8) = \bar{b}_3b_3 + \bar{b}_3b_9 + \bar{b}_3x_5 + \bar{b}_3y_7$$
$$= 1 + b_8 + c_7 + b_8 + \bar{z}_{12} + b_8 + e_7 + b_8 + y_{13},$$
$$(b_3\bar{b}_3)b_8 = b_8 + b_8^2,$$
$$(\bar{b}_3b_5)b_3 = b_3^2 + b_3x_5 + b_3y_7$$
$$= b_4 + 3b_5 + e_{10} + y_{16},$$
$$(b_3\bar{b}_3)b_5 = b_5 + b_5b_8,$$

and one has that

$$b_8^2 = 1 + c_7 + e_7 + 3b_8 + \bar{z}_{12} + y_{13},$$
$$b_5b_8 = b_4 + 2b_5 + e_{10} + y_{16}.$$

On the other hand,

$$(b_3\bar{b}_3)^2 = 1 + 2b_8 + 1 + c_7 + e_7 + 3b_8 + \bar{z}_{12} + y_{13},$$
$$b_3^2\bar{b}_3^2 = 1 + b_8 + c_7 + b_8 + z_{12} + b_8 + \bar{z}_{12} + 1 + 2b_8 + d_8,$$

which implies that $z_{12} + d_8 = e_7 + y_{13}$.

Step 1 There exists no NITA such that $e_7 \notin Supp\{z_{12}\}$.

Suppose that $e_7 \notin Supp\{z_{12}\}$. Since $e_7 \notin Supp\{d_8\}$, there exist $\alpha_3 \in Supp\{z_{12}\}$, $g_4 \in Supp\{d_8\}$ or $\alpha_4 \in Supp\{z_{12}\}$, $g_3 \in Supp\{d_8\}$ such that

$$e_7 = \alpha_3 + g_4 \text{ or } \alpha_4 + g_3.$$

Substep 1 There exists no NITA such that $e_7 = \alpha_3 + g_4$.

If $e_7 = \alpha_3 + g_4$, then there exist $z_9 \in N^*B$ and $h_4 \in B$ such that $z_{12} = \alpha_3 + z_9$ and $d_8 = g_4 + h_4$. Furthermore,

$$\bar{b}_4b_5 = \alpha_3 + z_9 + b_8,$$
$$\bar{b}_3b_9 = c_7 + b_8 + \bar{\alpha}_3 + \bar{z}_9,$$
$$\bar{b}_3x_5 = b_8 + \alpha_3 + g_4,$$

Hence

$$b_3\alpha_3 = x_5 + i_4, \quad i_4 \in B,$$
$$b_3\bar{\alpha}_3 = b_9,$$

which is impossible for $(b_3\bar{\alpha}_3, b_3\bar{\alpha}_3) = 1$ and $(b_3\alpha_3, b_3\alpha_3) = 2$.

Substep 2 There exists no NITA such that $e_7 = \alpha_4 + g_3$.

If $e_7 = \alpha_4 + g_3$, then there exist $z_8 \in Supp\{N^*B\}$ and $h_5 \in B$ such that $\bar{z}_{12} = \alpha_4 + z_8$, $d_8 = g_3 + h_5$. Hence

$$\bar{b}_3 x_5 = g_3 + \alpha_4 + b_8,$$

$$b_5 \bar{b}_5 = 1 + 2b_8 + g_3 + h_5,$$

$$\bar{b}_3 y_7 = b_8 + z_8 + h_5.$$

So there exists $l_4 \in B$ and $n_8 \in N^* B$ such that $b_3 g_3 = x_5 + l_4$ and $b_3 h_5 = y_7 + n_8$. Hence $g_3 \bar{g}_3 = 1 + b_8$ and $\bar{b}_3 l_4 = g_3 + m_9$. Here $m_9 \in B$, otherwise $\bar{l}_4 l_4$ will contains two b_8, which is a contradiction.

Since

$$b_3 (b_3 \bar{b}_3) = 2b_3 + b_9 + x_5 + y_7,$$

$$(b_3 g_3) \bar{g}_3 = \bar{g}_3 x_5 + \bar{g}_3 l_4,$$

it follows that $\bar{g}_3 l_4 = b_3 + b_9$ and

$$\bar{g}_3 x_5 = b_3 + x_5 + y_7 = \bar{b}_3 b_5. \tag{2.25}$$

We assert that $Supp\{n_8\} \cap \{b_3, x_5, y_7, l_4\} = \emptyset$. Obviously $y_7 \notin Supp\{n_8\}$. By the expressions of $b_3 h_5$ and $b_3 \bar{b}_3$, we have that $b_3 \notin Supp\{n_8\}$. If $x_5 \in Supp\{n_8\}$, then $h_5 \in Supp\{\bar{b}_3 x_5\}$, a contradiction. If $l_4 \in Supp\{n_8\}$, then $h_5 \in Supp\{\bar{b}_3 l_4\}$, a contradiction. Since

$$b_3 (b_5 \bar{b}_5) = b_3 + 2b_3 b_8 + b_3 g_3 + b_3 h_5$$

$$= 3b_3 + 2b_9 + 2x_5 + 2y_7 + x_5 + l_4 + y_7 + n_8,$$

$$(b_3 \bar{b}_5) b_5 = b_5 (\bar{b}_3 + \bar{x}_5 + \bar{y}_7)$$

$$= b_3 + x_5 + y_7 + \bar{x}_5 b_5 + \bar{y}_7 b_5,$$

it holds that

$$b_5 \bar{x}_5 + b_5 \bar{y}_7 = 2b_3 + 2b_9 + 2x_5 + 2y_7 + l_4 + n_8. \tag{2.26}$$

Now we assert that $(b_5 \bar{y}_7, b_9) = 1$. In fact, it follows from that $b_5 \bar{b}_9 = b_9 + b_3 \alpha_4 + b_3 z_8$ by Lemma 2.3 and $(b_3 \alpha_4, y_7) = 0$, $(b_3 z_8, y_7) = 1$. Therefore $(b_5 \bar{x}_5, b_3) = (b_5 \bar{x}_5, b_9) = 1$ and $(b_5 \bar{y}_7, b_3) = (b_5 \bar{y}_7, b_9) = 1$.

By (2.25), $(\bar{g}_3 x_5, \bar{b}_3 b_5) = 3$, so $(b_3 \bar{g}_3, b_5 \bar{x}_5) = 3$. By (2.26), $b_5 \bar{x}_5$ cannot contain three multiples of a constituent of degree 3 or two multiples of a constituent of degree 6. Then the common constituents of $b_3 \bar{g}_3$ and $b_5 \bar{x}_5$ are degree 5 or 4, since $b_3 \bar{g}_3$ can now only have constituents of degree 5 and 4. From $(b_3 \bar{g}_3, b_5 \bar{x}_5) = 3$ and (2.26), we have that $b_5 \bar{x}_5$ must contain $2x_5$ and some element v_4 of degree 4 from $l_4 + n_8$. This leads to

$$b_5 \bar{x}_5 = b_3 + b_9 + 2x_5 + v_4,$$

a contradiction for the left hand side having degree 26.

Step 2 There exists no NITA such that $e_7 \in Supp\{z_{12}\}$.

If $e_7 \in Supp\{z_{12}\}$, then there exists $w_5 \in B$ such that $z_{12} = e_7 + w_5$. So $y_{13} = d_8 + w_5$ and

$$\bar{b}_3 x_5 = b_8 + e_7, \tag{2.27}$$

$$\bar{b}_3 y_7 = b_8 + d_8 + w_5, \tag{2.28}$$

$$\bar{b}_4 b_5 = w_5 + e_7 + b_8, \tag{2.29}$$

$$\bar{b}_3 b_9 = \bar{w}_5 + c_7 + \bar{e}_7 + b_8. \tag{2.30}$$

Since $(\bar{b}_4 b_5, \bar{b}_4 b_5) = 3$, we have that $e_7, w_5 \in B$. By Lemma 2.3

$$b_5 \bar{b}_9 = b_9 + b_3 e_7 + b_3 w_5. \tag{2.31}$$

Since $y_7 \in Supp\{b_3 w_5\}$, $b_9 \in Supp\{b_3 \bar{w}_5\}$, and $x_5 \in Supp\{b_3 e_7\}$, we may set

$$b_3 w_5 = y_7 + v_8, \quad v_8 \in N^* B, \tag{2.32}$$

$$b_3 \bar{w}_5 = b_9 + d_6, \quad d_8 \in N^* B, \tag{2.33}$$

$$b_3 c_7 = b_9 + f_{12}, \quad f_{12} \in N^* B, \tag{2.34}$$

$$b_3 e_7 = x_5 + h_{16}, \quad h_{16} \in N^* B, \tag{2.35}$$

$$b_3 \bar{e}_7 = b_9 + g_{12}, \quad g_{12} \in N^* B. \tag{2.36}$$

Case 1 There exists no NITA such that either v_8 or d_6 is reducible.

It is easy to see that when either v_8 or d_6 is reducible, this will imply that the other one is also reducible. Let $d_6 = d_3 + g_3$. Then

$$b_3 \bar{w}_5 = d_3 + g_3 + b_9.$$

It is easy to show that $d_3 \neq g_3$. Hence we may set

$$b_3 \bar{d}_3 = w_5 + \delta_4,$$

$$b_3 \bar{g}_3 = w_5 + \psi_4,$$

which implies that $d_3 \bar{d}_3 = g_3 \bar{g}_3 = 1 + b_8$. If $\delta_4 = \psi_4$, then $b_3 \bar{d}_3 = b_3 \bar{g}_3$, $(b_3 \bar{d}_3, b_3 \bar{g}_3) = 2$, which implies that $d_3 \bar{g}_3 = 1 + b_8$. Thus $g_3 = d_3$ and $b_3 \bar{w}_5 = b_9 + 2g_3$, so that $(b_3 \bar{g}_3, w_5) = 2$, a contradiction. Hence $\delta_4 \neq \psi_4$. We calculate $b_3^2 \bar{b}_3 = b_3 d_3 \bar{d}_3 = b_3 g_3 \bar{g}_3$. We have that

$$(b_3 \bar{b}_3)b_3 = 2b_3 + b_9 + x_5 + y_7,$$

$$(b_3 \bar{d}_3)d_3 = w_5 d_3 + \delta_4 d_3,$$

$$(b_3 \bar{g}_3)g_3 = w_5 g_3 + \psi_4 g_3.$$

It is easy to see that

$$\delta_4 d_3 = \psi_4 g_3 = b_3 + b_9,$$

$$w_5 d_3 = w_5 g_3 = b_3 + x_5 + y_7.$$

On the other hand, we have that

$$w_5 \bar{w}_5 = 1 + 2b_8 + u_8,$$

$$\bar{b}_3(g_3 \psi_4) = \bar{b}_3(d_3 \delta_4) = \bar{b}_3 b_3 + \bar{b}_3 b_9$$

$$= 1 + b_8 + c_7 + \bar{e}_7 + \bar{w}_5 + b_8,$$

$$(\bar{b}_3 g_3)\psi_4 = \bar{w}_5 \psi_4 + \bar{\psi}_4 \psi_4,$$

$$(\bar{b}_3 d_3)\delta_4 = \bar{w}_5 \delta_4 + \bar{\delta}_4 \delta_4,$$

so that

$$\bar{w}_5 \psi_4 = \bar{w}_5 \delta_4 = \bar{e}_7 + \bar{w}_5 + b_8,$$

$$\psi_4 \bar{\psi}_4 = \delta_4 \bar{\delta}_4 = 1 + b_8 + c_7.$$

Hence $\psi_4, \delta_4 \in Supp\{w_5 \bar{w}_5\}$. Therefore

$$w_5 \bar{w}_5 = 1 + 2b_8 + \delta_4 + \psi_4.$$

But $(g_3 \bar{d}_3, w_5 \bar{w}_5) = (g_3 w_5, d_3 w_5) = 3$. We have that $1 \in Supp\{g_3 \bar{d}_3\}$, which implies that $g_3 = d_3$, a contradiction.

Case 2 There exists no NITA such that v_8 and d_6 are irreducible.

Suppose v_8 and d_6 are irreducible. We assert that f_{12} is irreducible. Otherwise, let $f_{12} = r + s$. Suppose that $c_7 \in Supp\{\bar{b}_3 r\}$, which implies that $f(r) \geq 4$.
By Lemma 2.3, one has that

$$b_8 c_7 = e_7 + w_5 + b_8 + b_3 \bar{f}_{12},$$

$$\bar{b}_9 b_9 = 1 + d_8 + 3b_8 + e_7 + w_5 + b_3 \bar{f}_{12}.$$

Subcase 1 w_5 is not real.

If w_5 is real, then $y_7 + v_8 = b_3 w_5 = b_3 \bar{w}_5 = b_9 + d_6$ by (2.32) and (2.33), a contradiction.

Subcase 2 $f_{12} \notin B$.

By the expression $b_8 c_7$, we have that $\bar{w}_5 \in Supp\{b_3 \bar{f}_{12}\}$. If f_{12} is irreducible, then $f_{12} \in Supp\{b_3 w_5\} = \{y_7, v_8\}$ by (2.32), a contradiction. So f_{12} is reducible.

Subcase 3 e_7 is not real.

If e_7 is real, then there exists an element $u_7 \in N^* B$ by (2.27) and (2.30), such that

$$b_3 e_7 = x_5 + b_9 + u_7.$$

Hence $(e_7^2, b_8) \geq 2$, so $(e_7 b_8, e_7) \geq 2$. By the expressions of $b_8 c_7$ and b_8^2, we have that $(e_7 b_8, c_7) = 1$, $(e_7 b_8, b_8) \geq 2$. Thus there exists an element $x_{19} \in N^* B$ of degree 19 such that

$$e_7 b_8 = 2e_7 + 2b_8 + c_7 + x_{19}.$$

On the other hand, from (2.27) and (2.30), we have that

$$\bar{b}_3 (b_3 e_7) = \bar{b}_3 x_5 + \bar{b}_3 b_9 + \bar{b}_3 u_7$$

$$= b_8 + e_7 + c_7 + e_7 + \bar{w}_5 + b_8 + \bar{b}_3 u_7,$$

$$(\bar{b}_3 b_3) e_7 = e_7 + b_8 e_7$$

$$= 3e_7 + 2b_8 + c_7 + x_{19}.$$

Then $\bar{w}_5 + \bar{b}_3 u_7 = e_7 + x_{19}$. Since the right side is real, we have that $w_5 \in Supp\{\bar{b}_3 u_7\}$ by Substep 1, which implies that there exists a constituent t of u_7 such that $w_5 \in Supp\{\bar{b}_3 t\}$. Then $t \in Supp\{b_3 w_5\}$, so $t = y_7$ or v_8. It follows that $u_7 = y_7$. By (2.28) implies that

$$w_5 + \bar{w}_5 + b_8 + d_8 = \bar{w}_5 + \bar{b}_3 y_7 = e_7 + x_{19},$$

a contradiction.

Subcase 4 Case 2 follows.

By Subcases 1, 2, 3, we have that e_7 and w_5 are not real, and f_{12} is reducible. Note that $b_8 c_7$ is real, by the expression $b_8 c_7$, we have that $b_3 \bar{f}_{12}$ contains $\bar{w}_5$ and $\bar{e}_7$. Hence there exists a constituent r of f_{12} such that $b_3 \bar{r}$ contains $\bar{w}_5$, which implies that $r \in Supp\{b_3 w_5\}$. Therefore $r = y_7$ or v_8.

If $r = y_7$, then $b_3 \bar{y}_7 = b_8 + d_8 + \bar{w}_5$ by (2.28). But (2.34) means that $b_3 \bar{y}_7$ contains c_7, a contradiction.

If $r = v_8$, then there exists s_4 such that $f_{12} = v_8 + s_4$. Since $c_7 \in Supp\{b_3 \bar{s}_4\}$ by (2.34), we have that $\bar{e}_7 \notin Supp\{b_3 \bar{s}_4\}$. So $\bar{e}_7 \in Supp\{b_3 \bar{v}_8\}$. Hence there exists a basis element a_5 of degree 5 such that

$$b_3 \bar{v}_8 = \bar{w}_5 + \bar{e}_7 + c_7 + a_5, \tag{2.37}$$

$$b_3 \bar{s}_4 = c_7 + c_5, \tag{2.38}$$

$$b_3 c_7 = v_8 + s_4 + b_9. \tag{2.39}$$

Checking $b_3 (b_5 \bar{b}_5) = b_5 (b_3 \bar{b}_5)$, this implies that

$$b_5 \bar{x}_5 + b_5 \bar{y}_7 = 2b_3 + 2b_9 + x_5 + y_7 + b_3 d_8. \tag{2.40}$$

We assert that $d_8 \in B$. In fact, if $d_8 \notin B$, then $d_8 = 2x_4$ or $x_4 + y_4$.

If $d_8 = 2x_4$, then $\bar{b}_3 y_7 = b_8 + 2x_4 + w_5$ by (2.28). Hence $(b_3 x_4, y_7) = 2$, which is impossible.

If $d_8 = x_4 + y_4$, then $\bar{b}_3 y_7 = b_8 + x_4 + y_4 + w_5$ by (2.28). Let $b_3 x_4 = y_7 + p_5$ and $b_3 y_4 = y_7 + q_5$. By (2.40), it holds that

$$b_5 \bar{x}_5 + b_5 \bar{y}_7 = 2b_3 + 2b_9 + x_5 + 3y_7 + p_5 + q_5.$$

But $(b_5 \bar{x}_5, b_3) = (b_5 \bar{x}_5, b_9) = 1$. So it is necessary to find elements from the right hand side of the above equation such that their sum having degree 13 yields $b_5 \bar{x}_5$. This is impossible. Therefore $d_8 \in B$.

Since

$$\bar{b}_3 (b_3 w_5) = \bar{b}_3 y_7 + \bar{b}_3 v_8$$
$$= b_8 + d_8 + w_5 + c_7 + w_5 + e_7 + \bar{a}_5,$$
$$b_3 (\bar{b}_3 w_5) = b_3 \bar{b}_9 + b_3 \bar{d}_6$$
$$= w_5 + c_7 + e_7 + b_8 + b_3 \bar{d}_6,$$

then

$$b_3 \bar{d}_6 = d_8 + w_5 + \bar{a}_5. \tag{2.41}$$

Now we assert that c_5 and e_5 are nonreal. Otherwise, $d_6, v_8, \in Supp\{b_3 a_5\}$ by (2.37) and (2.41). So $1 \in Supp\{b_3 a_5\}$, a contradiction. Hence c_5 and a_5 are nonreal, and $c_5 = \bar{a}_5$ by the expression b_8^2. Hence $b_3 \bar{d}_6 = d_8 + w_5 + c_5$. Associated with (2.38), we have that $(b_3 \bar{c}_5, d_6) = (b_3 \bar{c}_5, s_4) = 1$, which implies that $(b_3 \bar{c}_5, b_3 \bar{c}_5) = 3$. Furthermore $(b_3 c_5, b_3 c_5) = 3$. Let

$$b_3 c_5 = v_8 + \delta_3 + \lambda_4.$$

Thus we may set $\bar{b}_3 \delta_3 = c_5 + \Delta_4$. Then $\bar{\delta}_3 \delta_3 = 1 + b_8$, and

$$b_3 (\delta_3 \bar{\delta}_3) = b_3 + b_3 b_8$$
$$= 2b_3 + b_9 + x_5 + y_7,$$
$$(b_3 \bar{\delta}_3) \delta_3 = \bar{c}_5 \delta_3 + \bar{\Delta}_4 \delta_3.$$

We have that

$$\bar{\Delta}_4 \delta_3 = b_3 + b_9 \quad \text{and} \quad \bar{c}_5 \delta_3 = b_3 + x_5 + y_7. \tag{2.42}$$

Furthermore

$$\bar{b}_3 (\bar{\Delta}_4 \delta_3) = \bar{b}_3 b_3 + \bar{b}_3 b_9$$
$$= 1 + 2b_8 + c_7 + \bar{w}_5 + \bar{e}_7,$$
$$(\bar{b}_3 \delta_3) \bar{\Delta}_4 = c_5 \bar{\Delta}_4 + \Delta_4 \bar{\Delta}_4.$$

It follows that

$$\Delta_4\bar{\Delta}_4 = 1 + b_8 + c_7 \quad \text{and} \quad c_5\bar{\Delta}_4 = b_8 + \bar{w}_5 + \bar{e}_7 = b_4\bar{b}_5.$$

Hence $(b_3 c_5)\bar{\Delta}_4 = b_3(c_5\bar{\Delta}_4) = b_3(b_4\bar{b}_5) = b_4(b_3\bar{b}_5)$. But

$$b_4(b_3\bar{b}_5) = b_4\bar{b}_3 + b_4\bar{x}_5 + b_4\bar{y}_7 = b_3 + b_9 + b_4\bar{x}_5 + b_4\bar{y}_7,$$

$$(b_3 c_5)\bar{\Delta}_4 = v_8\bar{\Delta}_4 + \delta_3\bar{\Delta}_4 + \lambda_4\bar{\Delta}_4.$$

So b_3 is a constituent of one of $v_8\bar{\Delta}_4$, $\delta_3\bar{\Delta}_4$ and $\lambda_4\bar{\Delta}_4$. Then one of v_8, δ_3 and λ_4 is a constituent of $b_3\Delta_4$. On the other hand $\delta_3 \in Supp\{b_3\Delta_4\}$ by (2.42). But $b_3\Delta_4$ contains at most two constituents, a contradiction. Subcase 4 follows, which concludes Step 2. Consequently the proposition follows. □

Proposition 2.3 *There exists no NITA such that* $(f(x), f(y)) = (6, 6)$.

Proof If $x_{12} = x_6 + y_6$, then

$$\bar{b}_3 b_5 = b_3 + x_6 + y_6,$$

$$\bar{b}_3 b_8 = b_3 + b_9 + x_6 + y_6.$$

Assume that

$$\bar{b}_3 x_6 = b_8 + h_{10}, \quad \text{for some } h_{10} \in N^*B,$$

$$\bar{b}_3 y_6 = b_8 + i_{10}, \quad \text{for some } i_{10} \in N^*B,$$

$$\bar{b}_3 x_6 = b_5 + g_{13}, \quad \text{for some } g_{13} \in N^*B,$$

$$\bar{b}_3 y_6 = b_5 + f_{13}, \quad \text{for some } f_{13} \in N^*B.$$

Therefore,

$$b_8^2 = 1 + c_7 + 3b_8 + h_{10} + i_{10} + \bar{z}_{12}.$$

By the above equations and Lemma 2.3, one has that

$$(\bar{b}_3 b_3)^2 = 2 + c_7 + 5b_8 + h_{10} + i_{10} + \bar{z}_{12},$$

$$b_3^2 b_3^2 = \bar{b}_4 b_4 + \bar{b}_4 b_5 + \bar{b}_5 b_4 + \bar{b}_5 b_5$$

$$= 2 + 5b_8 + c_7 + \bar{z}_{12} + z_{12} + d_8.$$

Hence

$$d_8 + z_{12} = h_{10} + i_{10}.$$

We assert that d_8 cannot be totally contained in one of h_{10} and i_{10}. Otherwise, without loss of generality, assume d_8 is contained in h_{10}. Then h_{10} contains an element of degree 2, a contradiction. Thus d_8 must be reducible and cannot be contained in one of h_{10} and i_{10}. For d_8, there are four possibilities:

$$d_8 = c_3 + d_5, \quad c_4 + d_4, \quad c_4 + \bar{c}_4, \quad \text{and} \quad 2c_4.$$

Step 1 There exists no NITA such that $d_8 = c_3 + d_5$.

If $d_8 = c_3 + d_5$, then $c_3 + d_5 + z_{12} = h_{10} + i_{10}$. Without loss of generality, let

$$h_{10} = c_3 + h_7, \qquad i_{10} = d_5 + i_5.$$

Then $z_{12} = h_7 + i_5$ and

$$\bar{b}_3 x_6 = c_3 + h_7 + b_8, \tag{2.43}$$
$$\bar{b}_3 y_6 = d_5 + i_5 + b_8, \tag{2.44}$$
$$\bar{b}_4 b_5 = b_8 + i_5 + h_7, \tag{2.45}$$
$$\bar{b}_3 b_9 = c_7 + b_8 + \bar{i}_5 + \bar{h}_7, \tag{2.46}$$
$$b_8^2 = 1 + c_7 + 3b_8 + c_3 + h_7 + d_5 + i_5 + \bar{h}_7 + \bar{i}_5. \tag{2.47}$$

Hence c_3 is real. By (2.45) and $(\bar{b}_4 b_5, \bar{b}_4 b_5) = 3$, we have that $h_7 \in B$. Set $b_3 c_3 = x_6 + u_3$, $u_3 \in B$ by Lemma 2.2. Thus $(b_3 c_3, b_3 c_3) = 2$ so that

$$c_3^2 = 1 + b_8.$$

Moreover,

$$(b_3 c_3)c_3 = x_6 c_3 + u_3 c_3,$$
$$b_3 c_3^2 = b_3^2 \bar{b}_3 = b_3 + b_3 b_8$$
$$= 2b_3 + b_9 + x_6 + y_6.$$

Hence

$$c_3 u_3 = b_3 + x_6 \text{ or } b_3 + y_6,$$
$$c_3 x_6 = b_3 + b_9 + y_6 \text{ or } b_3 + b_9 + x_6.$$

If $c_3 u_3 = b_3 + x_6$, $c_3 x_6 = b_3 + b_9 + y_6$, then $c_3 \in Supp\{\bar{u}_3 x_6\}$. But

$$(b_3 \bar{b}_3)(c_3^2) = 2 + 5b_8 + c_7 + c_3 + d_5 + h_7 + \bar{h}_7 + i_5 + \bar{i}_5,$$
$$(b_3 c_3)(\bar{b}_3 c_3) = x_6 \bar{x}_6 + u_3 \bar{x}_6 + \bar{u}_3 x_6 + 1 + b_8.$$

Note that $u_3 \bar{x}_6 + \bar{u}_3 x_6$ contains two c_3, and so we come to a contradiction. Therefore $c_3 u_3 = b_3 + y_6$, $c_3 x_6 = b_3 + b_9 + x_6$. Let $\bar{b}_3 u_3 = c_3 + w_6$. Then $u_3 \bar{u}_3 = 1 + b_8$ by Lemma 2.2 and

$$(c_3b_3)\bar{b}_3 = \bar{b}_3x_6 + \bar{b}_3u_3$$
$$= b_8 + c_3 + h_7 + c_3 + w_6.$$

Since $\varsigma_3(b_3\bar{b}_3)$ is real, it follows that h_7 and w_6 are real. Since

$$(\bar{b}_3u_3)c_3 = c_3^2 + c_3w_6$$
$$= 1 + b_8 + c_3w_6,$$
$$(\bar{b}_3c_3)u_3 = \bar{x}_6u_3 + \bar{u}_3u_3$$
$$= 1 + b_8 + \bar{x}_6u_3,$$
$$\bar{b}_3(c_3u_3) = \bar{b}_3b_3 + \bar{b}_3y_6$$
$$= 1 + b_8 + b_8 + d_5 + i_5,$$

we have that

$$c_3w_6 = \bar{x}_6u_3 = d_5 + i_5 + b_8.$$

Since c_3, w_6 are real, i_5 is also real by the above equation. By (2.43), (2.44) and (2.46), one has that

$$b_3i_5 = y_6 + b_9,$$
$$b_3h_7 = b_9 + x_6 + l_6, \quad \text{for some } l_6 \in N^*B.$$

Considering $c_3^2u_3\bar{u}_3$, we have that

$$(c_3u_3)(c_3\bar{u}_3) = b_3\bar{b}_3 + \bar{b}_3y_6 + b_3\bar{y}_6 + y_6\bar{y}_6$$
$$= 1 + b_8 + 2b_8 + 2d_5 + 2i_5 + y_6\bar{y}_6,$$
$$c_3^2(u_3\bar{u}_3) = 2 + 5b_8 + c_7 + c_3 + 2i_5 + 2h_7 + d_5.$$

Comparing the number of elements of degree 5, we come to a contradiction. Step 1 follows.

Step 2 There exists no NITA such that $d_8 = 2c_4$.

If $d_8 = 2c_4$, then $2c_4 + z_{12} = h_{10} + i_{10}$. There exist h_6 and i_6 such that $z_{12} = h_6 + i_6$. Since $(\bar{b}_4b_5, \bar{b}_4b_5) \leq 4$, we obtain that $h_6, i_6 \in B$. Furthermore

$$\bar{b}_3x_6 = b_8 + c_4 + h_6, \qquad \bar{b}_3y_6 = b_8 + c_4 + i_6.$$

Thus $b_3c_4 = x_6 + y_6$. Then we can set

$$c_4^2 = 1 + b_8 + g_7, \quad g_7 \in N^*B.$$

Since $b_5\bar{b}_5 = 1 + 2b_8 + 2c_4$, $(c_4b_5, b_5) = 2$. Furthermore, $(c_4b_5, c_4b_5) \geq 5$. Hence

$$c_4^2 = 1 + b_8 + c_4 + e_3$$

and $(c_4b_5, c_4b_5) = 5$. There exists $t_{10} \in B$ such that

$$c_4b_5 = 2b_5 + t_{10}.$$

Since

$$b_3(b_5\bar{b}_5) = b_3 + 2b_3b_8 + 2b_3c_4$$
$$= 3b_3 + 2b_9 + 4x_6 + 4y_6,$$
$$b_5(b_3\bar{b}_5) = b_5(\bar{b}_3 + \bar{x}_6 + \bar{y}_6)$$
$$= b_3 + x_6 + y_6 + \bar{x}_6b_5 + \bar{y}_6b_5,$$

we have that $b_5\bar{x}_6 + b_5\bar{y}_6 = 2b_3 + 2b_9 + 3x_6 + 3y_6$. On the other hand,

$$\bar{b}_3(c_4b_5) = 2\bar{b}_3b_5 + \bar{b}_3t_{10} = 2b_3 + 2x_6 + 2y_6 + \bar{b}_3t_{10},$$
$$(\bar{b}_3c_4)b_5 = \bar{x}_6b_5 + \bar{y}_6b_5 = 2b_3 + 2b_9 + 3x_6 + 3y_6.$$

Hence $\bar{b}_3t_{10} = 2b_9 + x_6 + y_6$, from which it follows that $(b_3b_9, t_{10}) = 2$. Therefore $(b_3b_9, b_3b_9) \geq 5$. But $(\bar{b}_3b_9, \bar{b}_3b_9) = 4$ by Lemma 2.3 and $h_6, i_6 \in B$, a contradiction.

Step 3 There exists no NITA such that $d_8 = c_4 + d_4$.

If $d_8 = c_4 + d_4$, then there exist $h_6, i_6 \in N^*B$ such that $h_{10} = c_4 + h_6$, $i_{10} = d_4 + i_6$, $z_{12} = h_6 + i_6$ and

$$\bar{b}_3x_6 = c_4 + h_6 + b_8, \tag{2.48}$$
$$\bar{b}_3y_6 = d_4 + i_6 + b_8, \tag{2.49}$$
$$\bar{b}_4b_5 = b_8 + i_6 + h_6, \tag{2.50}$$
$$\bar{b}_3b_9 = c_7 + b_8 + \bar{i}_6 + \bar{h}_6, \tag{2.51}$$
$$b_8^2 = 1 + c_7 + 3b_8 + c_4 + h_6 + d_4 + i_6 + \bar{h}_6 + \bar{i}_6. \tag{2.52}$$

By (2.50) and $(\bar{b}_4b_5, \bar{b}_4b_5) = 3$, we have that $i_6, h_6 \in B$. Now we may set

$$b_3c_4 = x_6 + z_6, \quad z_6 \in B, \tag{2.53}$$
$$b_3d_4 = y_6 + p_6, \quad p_6 \in B. \tag{2.54}$$

Substep 1 h_6 and i_6 are nonreal.

If h_6 is real, then there exist e_3 and f_3 such that

$$\bar{b}_3h_6 = \bar{b}_9 + \bar{x}_6 + e_3,$$
$$b_3e_3 = f_3 + h_6.$$

Thus $e_3\bar{e}_3 = 1 + b_8$ by Lemma 2.2. Since

$$b_3(e_3\bar{e}_3) = 2b_3 + b_9 + x_6 + y_6,$$
$$(b_3e_3)\bar{e}_3 = h_6\bar{e}_3 + f_3\bar{e}_3,$$

hence

$$f_3\bar{e}_3 = b_3 + x_6 \text{ or } b_3 + y_6,$$
$$h_6\bar{e}_3 = b_3 + b_9 + y_6 \text{ or } b_3 + b_9 + x_6.$$

Therefore

$$\bar{b}_3(f_3\bar{e}_3) = \bar{b}_3 b_3 + \bar{b}_3 x_6 \text{ or } \bar{b}_3 b_3 + \bar{b}_3 y_6$$
$$= 1 + 2b_8 + c_4 + h_6 \text{ or } 1 + 2b_8 + d_4 + i_6,$$
$$f_3(\bar{b}_3\bar{e}_3) = f_3(h_6 + \bar{f}_3)$$
$$= 1 + b_8 + f_3 h_6.$$

Thus

$$f_3 h_6 = b_8 + c_4 + h_6 \text{ or } b_8 + d_4 + i_6.$$

On the other hand, one has the following equations:

$$(b_3\bar{b}_3)h_6 = h_6 + h_6 b_8,$$
$$(b_3 h_6)\bar{b}_3 = (\bar{e}_3 + x_6 + b_9)\bar{b}_3$$
$$= \bar{b}_3 b_9 + \bar{b}_3 x_6 + \bar{b}_3 \bar{e}_3$$
$$= c_7 + 2b_8 + 3h_6 + \bar{i}_6 + c_4 + \bar{f}_3.$$

Then

$$h_6 b_8 = c_7 + c_4 + 2b_8 + 2h_6 + \bar{i}_6 + \bar{f}_3, \tag{2.55}$$

which implies that i_6 and f_3 are real.

Now let us calculate $(b_3\bar{b}_3)(e_3\bar{e}_3)$:

$$(b_3\bar{b}_3)(e_3\bar{e}_3) = 2 + 5b_8 + c_7 + c_4 + d_4 + 2h_6 + 2i_6,$$
$$(b_3 e_3)(\bar{b}_3\bar{e}_3) = (h_6 + f_3)^2$$
$$= h_6^2 + 2f_3 h_6 + f_3^2$$
$$= 1 + h_6^2 + 3b_8 + 2c_4 + 2h_6 \text{ or } 1 + h_6^2 + 3b_8 + 2d_4 + 2i_6.$$

Comparing the number of elements of degree 4, we have that $c_4 = d_4$, a contradiction.

Symmetrically, we can prove that i_6 is nonreal.

Substep 2 There exists no NITA such that $d_8 = c_4 + d_4$ and h_6, i_6 are nonreal.

Since $(b_3\bar{b}_3)c_4 = c_4 + c_4 b_8$ and $\bar{b}_3(b_3 c_4) = \bar{b}_3 x_6 + \bar{b}_3 z_6 = c_4 + h_6 + b_8 + \bar{b}_3 z_6$, we have that

$$c_4 b_8 = b_8 + h_6 + \bar{b}_3 z_6,$$

Since h_6 is nonreal and $c_4 b_8$ is real, $\bar{h}_6 \in Supp\{\bar{b}_3 z_6\}$. By (2.53), there exists $r_8 \in N^* B$ such that

$$\bar{b}_3 z_6 = c_4 + \bar{h}_6 + r_8.$$

By (2.51), there exists α_3 such that

$$b_3 \bar{h}_6 = b_9 + z_6 + \alpha_3.$$

Thus there exists γ_3 such that $b_3 \bar{\alpha}_3 = h_6 + \gamma_3$. Hence $\alpha_3 \bar{\alpha}_3 = 1 + b_8$ by Lemma 2.2.

Since

$$b_3(\alpha_3 \bar{\alpha}_3) = 2b_3 + b_9 + x_6 + y_6,$$
$$(b_3 \alpha_3)\bar{\alpha}_3 = h_6 \bar{\alpha}_3 + \gamma_3 \bar{\alpha}_3,$$

we have that

$$\gamma_3 \bar{\alpha}_3 = b_3 + x_6 \text{ or } b_3 + y_6,$$
$$h_6 \bar{\alpha}_3 = b_3 + b_9 + y_6 \text{ or } b_3 + b_9 + x_6.$$

Hence

$$\bar{b}_3(\gamma_3 \bar{\alpha}_3) = \bar{b}_3 b_3 + \bar{b}_3 x_6 \text{ or } \bar{b}_3 b_3 + \bar{b}_3 y_6$$
$$= 1 + 2b_8 + c_4 + h_6 \text{ or } 1 + 2b_8 + d_4 + i_6,$$
$$\gamma_3(\bar{b}_3 \bar{\alpha}_3) = \gamma_3 \bar{h}_6 + \gamma_3 \bar{\gamma}_3$$
$$= 1 + b_8 + \gamma_3 \bar{h}_6.$$

Therefore

$$\gamma_3 \bar{h}_6 = b_8 + c_4 + h_6 \text{ or } b_8 + d_4 + i_6.$$

Since

$$(b_3 \bar{b}_3)(\alpha_3 \bar{\alpha}_3) = 2 + 5b_8 + c_7 + c_4 + d_4 + 2h_6 + 2i_6,$$
$$(b_3 \alpha_3)(\bar{b}_3 \bar{\alpha}_3) = (h_6 + \gamma_3)(\bar{h}_6 + \bar{\gamma}_3)$$
$$= h_6 \bar{h}_6 + \gamma_3 \bar{h}_6 + \bar{\gamma}_3 h_6 + \gamma_3 \bar{\gamma}_3$$
$$= 1 + 3b_8 + 2c_4 + h_6 + \bar{h}_6 + h_6 \bar{h}_6$$
$$\text{or } 1 + 3b_8 + 2d_4 + i_6 + \bar{i}_6 + h_6 \bar{h}_6,$$

comparing the number of degree 4, we come to $c_4 = d_4$, a contradiction. So Substep 2 follows.

Step 4 There exists no NITA such that $d_8 = c_4 + \bar{c}_4$.

If $d_8 = c_4 + \bar{c}_4$, then $b_5\bar{b}_5 = 1 + 2b_8 + c_4 + \bar{c}_4$. And (2.48), (2.49), (2.50), (2.51), (2.52), (2.53) and (2.54) still hold with $d_4 = \bar{c}_4$. Thus

$$\bar{b}_3(b_3c_4) = \bar{b}_3x_6 + \bar{b}_3z_6,$$
$$= c_4 + h_6 + b_8 + \bar{b}_3z_6,$$
$$b_3(\bar{b}_3\bar{c}_4) = b_3\bar{y}_6 + b_3\bar{p}_6$$
$$= c_4 + \bar{i}_6 + b_8 + b_3\bar{p}_6.$$

Then

$$h_6 + \bar{b}_3z_6 = \bar{i}_6 + b_3\bar{p}_6. \tag{2.56}$$

Substep 1 There exists no NITA such that $h_6 = \bar{i}_6$.

If $h_6 = \bar{i}_6$, then it follows from (2.48) and (2.51) that there is $\alpha_3 \in B$ such that

$$b_3h_6 = b_9 + x_6 + \alpha_3,$$
$$b_3\bar{h}_6 = b_9 + y_6 + \beta_3.$$

It is easy to see that $\alpha_3 \neq \beta_3$. Since $z_{12} = h_6 + \bar{h}_6$, this means by Lemma 2.3 that

$$b_5\bar{b}_9 = x_6 + y_6 + 3b_9 + \alpha_3 + \beta_3.$$

Thus $(b_5\bar{b}_9, b_5\bar{b}_9) = 13$.

By Lemma 2.3, we have that

$$b_9\bar{b}_9 = 1 + 3b_8 + c_4 + \bar{c}_4 + h_6 + \bar{h}_6 + \bar{b}_3f_{12}.$$

But $b_5\bar{b}_5 = 1 + 2b_8 + c_4 + \bar{c}_4$. Since $(b_9\bar{b}_9, b_8) = 3$, we have that

$$13 = (b_5\bar{b}_9, b_5\bar{b}_9) = 9 + (c_4, \bar{b}_3f_{12}) + (\bar{c}_4, \bar{b}_3f_{12}).$$

We conclude that $(c_4, \bar{b}_3f_{12}) = (\bar{c}_4, \bar{b}_3f_{12}) = 2$. Thus

$$(b_3c_4, f_{12}) = (b_3\bar{c}_4, f_{12}) = 2.$$

From (2.53) and (2.54), it follows that

$$(x_6 + z_6, f_{12}) = (y_6 + p_6, f_{12}) = 2.$$

Since $x_6 \neq y_6$ by $(b_3\bar{b}_3, b_5\bar{b}_5) = 3$ and $z_6 \neq x_6$, $y_6 \neq p_6$ by Lemma 2.2, it follows that $f_{12} = 2z_6$ and $z_6 = p_6$. Hence $b_3c_7 = b_9 + 2z_6$ by Lemma 2.3. Therefore $(b_3c_7, b_3c_7) = 5$ and $(\bar{b}_3z_6, c_7) = 2$. So $\bar{b}_3z_6 = 2c_7 + c_4$. But

$$b_8c_7 = b_8 + h_6 + \bar{h}_6 + 2\bar{b}_3z_6 = b_8 + h_6 + \bar{h}_6 + 4c_7 + 2c_4,$$

This is impossible for $b_8 c_7$ is real and c_4 is not real.

Substep 2 There exists no NITA such that $h_6 \neq \bar{i}_6$.

Suppose that $h_6 \neq \bar{i}_6$. By (2.56), it follows that $(b_3 \bar{p}_6, h_6) = (p_6, b_3 \bar{h}_6) = (\bar{i}_6, \bar{b}_3 z_6) = (b_3 \bar{i}_6, z_6) = 1$. But $b_9 \in Supp\{b_3 \bar{h}_6\}$ by (2.51) and $(c_4, b_6 \bar{p}_6) = (c_4, \bar{b}_3 z_6) = 1$ by (2.53) and (2.54). By (2.56), (2.51) and (2.54), there exist δ_3, z_3 and $v_8 \in N^* B$ such that

$$b_3 \bar{h}_6 = b_9 + p_6 + \delta_3, \tag{2.57}$$

$$\bar{b}_3 z_6 = \bar{i}_6 + c_4 + v_8, \tag{2.58}$$

$$b_3 \bar{p}_6 = h_6 + c_4 + v_8. \tag{2.59}$$

Hence $\bar{b}_3 \delta_3 = \bar{h}_6 + \gamma_3$, some $\gamma_3 \in B$. Therefore $(b_3 \gamma_3, b_3 \gamma_3) \geq 2$. It follows that

$$\delta_3 \bar{\delta}_3 = \gamma_3 \bar{\gamma}_3 = 1 + b_8$$

by Lemma 2.2.

Since

$$(b_3 \bar{b}_3) c_4 = c_4 + c_4 b_8,$$

$$b_3 (\bar{b}_3 c_4) = b_3 \bar{y}_6 + b_3 \bar{p}_6$$

$$= c_4 + \bar{i}_6 + b_8 + h_6 + c_4 + v_8,$$

we have that $c_4 b_8 = c_4 + h_6 + \bar{i}_6 + v_8 + b_8$. Then $(c_4 b_8, c_4 b_8) \geq 5$. Let $c_4 \bar{c}_4 = 1 + b_8 + r_7$, $r \in N^* B$ and r_7 is real. By (2.52) and c_4 is not real, we have that $r_7 = c_7$ and then $(c_4 b_8, c_4 b_8) = 5$. Consequently $v_8 \in B$. By $b_3(\delta_3 \bar{\delta}_3) = (b_3 \bar{\delta}_3)\delta_3$, we have that

$$\delta_3 \bar{\gamma}_3 = b_3 + x_6 \text{ or } b_3 + y_6,$$

$$\delta_3 \bar{h}_6 = b_3 + b_9 + y_6 \text{ or } b_3 + b_9 + x_6.$$

Without loss of generality, let $\delta_3 \bar{\gamma}_3 = b_3 + x_6$. Then

$$1 + 2b_8 + b_8^2 = (\delta_3 \bar{\delta}_3)(\gamma_3 \bar{\gamma}_3) = (\delta_3 \bar{\gamma}_3)(\bar{\delta}_3 \gamma_3) = b_3 \bar{b}_3 + b_3 \bar{x}_6 + \bar{b}_3 x_6 + x_6 \bar{x}_6$$

$$= 1 + 3b_8 + c_4 + \bar{c}_4 + h_6 + \bar{h}_6 + x_6 \bar{x}_6.$$

It holds by (2.52) that

$$x_6 \bar{x}_6 = 1 + 2b_8 + c_7 + i_6 + \bar{i}_6.$$

Furthermore $(b_5 \bar{b}_5, x_6 \bar{x}_6) = 5$. On the other hand, the following equations hold:

$$b_3 (b_5 \bar{b}_5) = b_3 + 2b_3 b_8 + b_3 c_4 + b_3 \bar{c}_4$$

$$= 3b_3 + 2b_9 + 2x_6 + 2y_6 + x_6 + z_6 + y_6 + p_6,$$

$$(b_3\bar{b}_5)b_5 = \bar{b}_3 b_5 + \bar{x}_6 b_5 + \bar{y}_6 b_5$$
$$= b_3 + x_6 + y_6 + \bar{x}_6 b_5 + \bar{y}_6 b_5.$$

Hence

$$\bar{x}_6 b_5 + \bar{y}_6 b_5 = 2b_3 + 2b_9 + 2x_6 + 2y_6 + z_6 + p_6. \tag{2.60}$$

It is proved that $(b_5\bar{b}_5, x_6\bar{x}_6) = 5$. By the above equation and $(b_5\bar{x}_5, b_3) = (b_5\bar{y}_5, b_3) = 1$, we can only have that

$$b_5\bar{x}_6 = b_3 + b_9 + x_6 + y_6 + z_6 \text{ or } b_3 + b_9 + x_6 + y_6 + p_6,$$
$$b_5\bar{y}_6 = b_3 + b_9 + x_6 + y_6 + p_6 \text{ or } b_3 + b_9 + x_6 + y_6 + z_6.$$

Hence

$$\bar{b}_3(b_5\bar{x}_6) = \bar{b}_3 b_3 + \bar{b}_3 b_9 + \bar{b}_3 x_6 + \bar{b}_3 y_6 + \bar{b}_3 z_6$$
$$\text{or } \bar{b}_3 b_3 + \bar{b}_3 b_9 + \bar{b}_3 x_6 + \bar{b}_3 y_6 + \bar{b}_3 p_6$$
$$= 1 + b_8 + c_7 + b_8 + \bar{h}_6 + \bar{i}_6 + c_4 + h_6 + b_8$$
$$+ \bar{c}_4 + i_6 + b_8 + \bar{i}_6 + c_4 + v_8$$
$$\text{or } 1 + b_8 + c_7 + b_8 + \bar{h}_6 + \bar{i}_6 + c_4 + h_6 + b_8$$
$$+ \bar{c}_4 + i_6 + b_8 + \bar{h}_6 + \bar{c}_4 + \bar{v}_8,$$
$$(\bar{b}_3 b_5)\bar{x}_6 = b_3\bar{x}_6 + x_6\bar{x}_6 + y_6\bar{x}_6$$
$$= \bar{c}_4 + \bar{h}_6 + b_8 + 1 + 2b_8 + c_7 + i_6 + \bar{i}_6 + y_6\bar{x}_6.$$

Therefore

$$y_6\bar{x}_6 = 2c_4 + h_6 + b_8 + \bar{i}_6 + v_8 \text{ or } c_4 + \bar{c}_4 + h_6 + b_8 + \bar{h}_6 + \bar{v}_8.$$

If $y_6\bar{x}_6 = 2c_4 + h_6 + b_8 + \bar{i}_6 + v_8$, then $(c_4 x_6, y_6) = 2$. Hence $(c_4 x_6, c_4 x_6) \geq 5$, a contradiction. Therefore

$$y_6\bar{x}_6 = c_4 + \bar{c}_4 + h_6 + b_8 + \bar{h}_6 + \bar{v}_8. \tag{2.61}$$

Thus $(\bar{c}_4 x_6, y_6) = 1$.

Consider the following equations:

$$b_3(c_4\bar{c}_4) = b_3 + b_3 b_8 + b_3 c_7$$
$$= 2b_3 + b_9 + x_6 + y_6 + b_9 + f_{12},$$
$$(b_3 c_4)\bar{c}_4 = \bar{c}_4 x_6 + \bar{c}_4 z_6,$$
$$(b_3\bar{c}_4)c_4 = c_4 y_6 + c_4 p_6.$$

By the expression $x_6 \bar{x}_6$, we have that $(x_6, \bar{c}_4 x_6) = 0$. Consequently, $(x_6, \bar{c}_4 z_6) = 1$ by the above equations. Since $b_3 c_7 = b_9 + f_{12}$ and $L_1(B) = \{1\}$ and $L_2(B) = \emptyset$, so the constituents of f_{12} are of degrees larger than 3. Since $(\bar{c}_4 x_6, b_3) = (\bar{c}_4 z_6, b_3) = 1$ by (2.53), $(\bar{c}_4 x_6, \bar{c}_4 x_6) = 4$ and it is proved that $(\bar{c}_4 x_6, y_6) = 1$, we have that

$$\bar{c}_4 x_6 = b_3 + y_6 + b_9 + g_6, \quad g_6 \in B.$$

Consequently $\bar{c}_4 z_6 = b_3 + x_6 + b_9 + f_6$, where $f_{12} = f_6 + g_6$ and $f_6 \in N^* B$.
Let $\bar{b}_3 g_6 = c_7 + x_{11}$ and $\bar{b}_3 f_6 = c_7 + y_{11}, x_{11}, y_{11} \in N^*(B)$. Then

$$\bar{b}_3(\bar{c}_4 x_6) = \bar{b}_3 b_3 + \bar{b}_3 y_6 + \bar{b}_3 b_9 + \bar{b}_3 g_6$$
$$= 1 + b_8 + \bar{c}_4 + i_6 + b_8 + c_7 + b_8 + \bar{h}_6 + \bar{i}_6 + c_7 + x_{11},$$
$$\bar{c}_4(\bar{b}_3 x_6) = \bar{c}_4 c_4 + \bar{c}_4 h_6 + \bar{c}_4 b_8$$
$$= 1 + b_8 + c_7 + \bar{c}_4 h_6 + \bar{c}_4 + \bar{h}_6 + i_6 + \bar{v}_8 + b_8,$$
$$\bar{b}_3(\bar{c}_4 z_6) = \bar{b}_3 b_3 + \bar{b}_3 x_6 + \bar{b}_3 b_9 + \bar{b}_3 f_6$$
$$= 1 + b_8 + c_4 + h_6 + b_8 + c_7 + b_8 + \bar{h}_6 + \bar{i}_6 + c_7 + y_{11},$$
$$\bar{c}_4(\bar{b}_3 z_6) = \bar{c}_4 c_4 + \bar{c}_4 \bar{i}_6 + \bar{c}_4 v_8$$
$$= 1 + b_8 + c_7 + \bar{c}_4 \bar{i}_6 + \bar{c}_4 v_8.$$

Hence

$$\bar{c}_4 h_6 + \bar{v}_8 = b_8 + \bar{i}_6 + c_7 + x_{11}, \tag{2.62}$$
$$\bar{c}_4 \bar{i}_6 + \bar{c}_4 v_8 = c_4 + h_6 + 2b_8 + \bar{i} + \bar{h}_6 + c_7 + y_{11}. \tag{2.63}$$

We assert that $v_8 \neq b_8$. Otherwise, if $v_8 = b_8$, then $(c_4 b_8, b_8) = 2$. So $(b_8^2, c_4) = 2$, a contradiction by (2.52). Hence $\bar{v}_8 \in Supp\{x_{11}\}$ by (2.62). Let $x_{11} = \bar{v}_8 + x_3$, then it follows by (2.62) that

$$\bar{c}_4 h_6 = b_8 + \bar{i}_6 + c_7 + x_3.$$

By (2.63), we have that $\bar{h}_6 \in Supp\{\bar{c}_4 \bar{i}_6 + \bar{c}_4 v_8\}$. So $(\bar{c}_4 h_6, i_6) \geq 1$ or $(\bar{c}_4 h_6, \bar{v}_8) \geq 1$. The latter case will lead to $v_8 = b_8$, a contradiction. Hence $(\bar{c}_4 h_6, i_6) \geq 1$. Thus $i_6 = \bar{i}_6$ and $(\bar{c}_4 h_6, i_6) = 1$. Furthermore,

$$x_6 \bar{x}_6 = 1 + 2b_8 + c_7 + 2i_6.$$

And it follows by (2.49) and (2.51) that

$$b_3 i_6 = b_9 + y_6 + z_3, \quad \text{some } z_3 \in B.$$

Thus

$$b_3(x_6 \bar{x}_6) = b_3 + 2b_3 b_8 + b_3 c_7 + 2b_3 i_6$$
$$= 3b_3 + 2b_9 + 2x_6 + 2y_6 + b_9 + g_6 + f_6 + 2b_9 + 2y_6 + 2z_3,$$

$$x_6(b_3\bar{x}_6) = x_6\bar{c}_4 + x_6\bar{h}_6 + x_6b_8$$

$$= b_3 + y_6 + b_9 + g_6 + x_6\bar{h}_6 + x_6b_8.$$

Then $x_6\bar{h}_6 + x_6b_8 = 2b_3 + 2b_9 + 2x_6 + 3y_6 + f_6 + 2b_9 + 2z_3$.

By the expression $y_6\bar{x}_6$ and $v_8 \neq b_8$, we have that $(b_8x_6, y_6) = 1$. Hence $(x_6\bar{h}_6, y_6) = 2$ and so $(\bar{x}_6y_6, \bar{h}_6) = 2$. Again by the expression $y_6\bar{x}_6$, we have that $h_6 = \bar{h}_6$. Now (2.48) implies that $x_6 \in Supp\{b_3h_6\}$. Then $p_6 = x_6$ by (2.57). Consequently $b_3\bar{c}_4 = x_6 + y_6$, from which we have $(bc_4, \bar{b}_3x_6) = 1$. Thus $c_4 = \bar{c}_4$ by (2.48). Step 4 follows. This completes the proof of Lemma 2.6. □

Therefore $(b_5\bar{b}_5, b_8) = 1$ and $(b_3\bar{b}_3, b_5\bar{b}_5) = 2$. Consequently by Lemma 2.3 we have $x_{12} \in B$ and $(b_3b_8, b_3b_8) = 3$. Thus Theorem 2.2 holds. Furthermore, in Lemma 2.3 we proved that $c_7 \in B$.

2.4 General Information on NITA Generated by b_3 and Satisfying $b_3^2 = \bar{b}_3 + b_6$ and $b_3^2 = c_3 + b_6$

Since the structure of NITA generated by b_3 and satisfying $b_3^2 = \bar{b}_3 + b_6$ and $b_3^2 = c_3 + b_6$ are related, we investigate them simultaneously. Some results are written in the same proposition. The purpose of this section is to prove (3) and (4) of the Main Theorem 1. The following lemmas will be useful during our investigation.

Lemma 2.7

(1) *If* $b_3^2 = \bar{b}_3 + b_6$, *then* $b_3b_8 = b_6 + \bar{b}_3b_6$;
(2) *If* $b_3^2 = c_3 + b_6$, *then there exists an element* $x_6 \in B$ *such that* $\bar{b}_3c_3 = b_3 + x_6$ *and* $c_3\bar{c}_3 = 1 + b_8$. *Furthermore,* $b_3b_8 = x_6 + \bar{b}_3b_6$ *and* $c_3b_8 = b_6 + b_3x_6$.

Proof (1) follows from $(b_3\bar{b}_3)b_3 = b_3^2\bar{b}_3$.

(2) Since $b_3 \in Supp\{\bar{b}_3c_3\}$, $(b_3\bar{b}_3, c_3\bar{c}_3) \geq 2$. But $(b_3\bar{b}_3, c_3\bar{c}_3) \leq 2$, so $(b_3\bar{b}_3, c_3\bar{c}_3) = 2$ and

$$c_3\bar{c}_3 = 1 + b_8, \qquad \bar{b}_3c_3 = b_3 + x_6, \quad x_6 \in B.$$

Checking $(b_3\bar{b}_3)b_3 = b_3^2\bar{b}_3$ and $c_3(c_3\bar{c}_3) = b_3(\bar{b}_3c_3)$, one has (2). □

Lemma 2.8 *If* $(b_3b_8, b_3b_8) \geq 4$, *then any constituent of* b_3b_8 *and* $\bar{b}_3b_6$ *is either* b_3 *or of degree* > 3 *and* $(b_3b_8, b_3b_8) = 4, 5$.

Proof Since $(b_3b_8, b_3) = 1$, the degree of the sum of the remaining constituents of b_3b_8 must be 21.

On the other hand, any constituent y_n of b_3b_8 must have degree ≥ 4. Otherwise, $n = 3$ and $b_3\bar{y}_3 = 1 + b_8$, which implies that $b_3 = y_3$, a contradiction. □

Theorem 2.3 *There exists no NITA generated by b_3 and satisfying $b_3^2 = \bar{b}_3 + b_6$ or $b_3^2 = c_3 + b_6$ and $(b_3b_8, b_3b_8) = 2$.*

Proof Since $b_3\bar{b}_3 = 1 + b_8$ and $(b_3b_8, b_3b_8) = 2$, we may assume that $b_3b_8 = b_3 + b_{21}$, where $b_{21} \in B$. By Lemma 2.7, one has that

$$b_3 + b_{21} = b_6 + \bar{b}_3b_6 \text{ or } x_6 + \bar{b}_3b_6,$$

which is impossible. □

Theorem 2.4 *There is no NITA generated by b_3 and satisfying $b_3^2 = \bar{b}_3 + b_6$ or $b_3^2 = c_3 + b_6$ and $(b_3b_8, b_3b_8) = 5$.*

Proof (1) There is no NITA satisfying $b_3^2 = \bar{b}_3 + b_6$ and $(b_3b_8, b_3b_8) = 5$.

By $b_3^2\bar{b}_3 = (b_3\bar{b}_3)b_3$, we have that $b_3b_8 = b_6 + \bar{b}_3b_6$, which implies that $b_8 \in Supp\{\bar{b}_3b_6\}$. But $b_3 \in Supp\{\bar{b}_3b_6\}$. We need a sum with degree 7 of two constituents to make up $\bar{b}_3b_6$, which is impossible by Lemma 2.7.

(2) There is no NITA satisfying $b_3^2 = c_3 + b_6$ and $(b_3b_8, b_3b_8) = 5$.

By Lemma 2.7 and $(b_3b_8, b_3b_8) = 5$, we may assume that

(A) $b_3b_8 = b_3 + x_6 + x_4 + y_4 + z_7$,

(B) $b_3b_8 = b_3 + x_6 + x_4 + y_5 + z_6$,

(C) $b_3b_8 = b_3 + x_6 + x_5 + y_5 + z_5$.

Since

$$\bar{b}_3(\bar{b}_3c_3) = b_3\bar{b}_3 + x_6\bar{b}_3,$$
$$\bar{b}_3^2c_3 = c_3\bar{c}_3 + c_3\bar{b}_6$$
$$= 1 + b_8 + c_3\bar{b}_6,$$

we have that $\bar{b}_3x_6 = c_3\bar{b}_6$.

Since $(c_3b_8, c_3b_8) = (b_3b_8, b_3b_8) = 5$ by Lemma 2.7 and $c_3b_8 = b_6 + b_3x_6$ by Lemma 2.7, we have $(b_3x_6, b_3x_6) = 4$. Furthermore, $(\bar{b}_3x_6, \bar{b}_3x_6) = 4$. But $(\bar{b}_3x_6, b_8) = 1$. We may assume

$$c_3\bar{b}_6 = \bar{b}_3x_6 = b_8 + m_3 + n_3 + v_4,$$

where $m_3, n_3, v_4 \in B$ and m_3, n_3, v_4 are distinct. We may assume

$$c_3\bar{m}_3 = b_6 + m_3^*, \quad m_3^* \in B,$$
$$c_3\bar{n}_3 = b_6 + n_3^*.$$

Hence $(c_3\bar{m}_3, c_3\bar{m}_3) = 2$. So $m_3\bar{m}_3 = 1 + b_8$. Moreover,

$$(c_3\bar{c}_3)\bar{m}_3 = \bar{m}_3 + \bar{m}_3b_8,$$

$$(c_3 \bar{m}_3) \bar{c}_3 = \bar{c}_3 b_6 + \bar{c}_3 m_3^*$$
$$= b_8 + \bar{m}_3 + \bar{n}_3 + \bar{v}_4 + \bar{c}_3 m_3^*.$$

Then

$$\bar{m}_3 b_8 = b_8 + \bar{n}_3 + \bar{v}_4 + \bar{c}_3 m_3^*,$$

which implies that $m_3 \bar{n}_3 = 1 + b_8$. So $m_3 = n_3$, a contradiction. □

Theorem 2.5 *There exists no NITA generated by b_3 and satisfying $b_3^2 = \bar{b}_3 + b_6$, $(b_3 b_8, b_3 b_8) = 4$.*

Proof Since $b_3 b_8 = b_6 + \bar{b}_3 b_6$ by Lemma 2.7 and by assumption that $(b_3 b_8, b_3 b_8) = 4$, then $b_8 \in Supp\{\bar{b}_3 b_6\}$ and $(\bar{b}_3 b_6, \bar{b}_3 b_6) = 3$. But $b_3 \in Supp\{\bar{b}_3 b_6\}$. Thus there exists $p_7 \in B$ such that

$$\bar{b}_3 b_6 = b_3 + b_8 + p_7, \qquad b_3 b_8 = b_3 + b_6 + b_8 + p_7.$$

Hence

$$(b_3 \bar{b}_3) b_8 = b_8 + b_8^2,$$
$$\bar{b}_3 (b_3 b_8) = \bar{b}_3 b_3 + \bar{b}_3 b_6 + \bar{b}_3 b_8 + \bar{b}_3 p_7$$
$$= 1 + b_8 + b_3 + b_8 + p_7 + \bar{b}_3 + \bar{b}_6 + b_8 + \bar{p}_7 + \bar{b}_3 p_7.$$

Consequently, $b_8^2 = 1 + b_3 + \bar{b}_3 + p_7 + \bar{p}_7 + 2b_8 + \bar{b}_6 + \bar{b}_3 p_7$. On the other hand,

$$(b_3 \bar{b}_3)^2 = 1 + 2b_8 + b_8^2,$$
$$b_3^2 \bar{b}_3^2 = b_3 \bar{b}_3 + \bar{b}_3 b_6 + b_3 \bar{b}_6 + b_6 \bar{b}_6$$
$$= 1 + b_8 + b_3 + b_8 + p_7 + \bar{b}_3 + b_8 + \bar{p}_3 + b_6 \bar{b}_6.$$

We have that $b_8^2 = b_3 + \bar{b}_3 + p_7 + \bar{p}_7 + b_8 + b_6 \bar{b}_6$. Hence $b_6 \bar{b}_6 = 1 + \bar{b}_6 + b_8 + \bar{b}_3 p_7$.
 Since

$$\bar{b}_3^2 b_6 = b_3 b_6 + \bar{b}_6 b_6$$
$$= b_3 b_6 + 1 + b_8 + \bar{b}_6 + \bar{b}_3 p_7,$$
$$\bar{b}_3 (\bar{b}_3 b_6) = \bar{b}_3 b_3 + \bar{b}_3 b_8 + \bar{b}_3 p_7$$
$$= 1 + b_8 + \bar{b}_3 + \bar{b}_6 + b_8 + \bar{p}_7 + \bar{b}_3 p_7,$$

it holds that $b_3 b_6 = \bar{b}_3 + \bar{p}_7 + b_8$. Now the following equations follow:

$$b_3 (\bar{b}_3 b_6) = b_3^2 + b_3 b_8 + b_3 p_7$$
$$= \bar{b}_3 + b_6 + b_3 + b_6 + b_8 + p_7 + b_3 p_7,$$
$$(b_3 b_6) \bar{b}_3 = \bar{b}_3^2 + \bar{b}_3 \bar{p}_7 + b_8 \bar{b}_3$$
$$= b_3 + \bar{b}_6 + \bar{b}_3 + \bar{b}_6 + b_8 + \bar{p}_7 + \bar{b}_3 \bar{p}_7.$$

Then $2b_6 + p_7 + b_3 p_7 = 2\bar{b}_6 + \bar{p}_7 + \bar{b}_3 \bar{p}_7$. Since b_6 is nonreal, we have that $(b_3 p_7, \bar{b}_6) = 2$, a contradiction to the expression $\bar{b}_3 b_6$. The theorem follows. □

Theorem 2.6   *There exists no NITA satisfying $b_3^2 = c_3 + b_6$, $(b_3 b_8, b_3 b_8) = 4$ and $c_3^2 = r_4 + s_5$.*

Proof Let us investigate Sub-NITA generated by c_2. By assumptions, we have that $(b_3 \bar{b}_3, b_3 \bar{b}_3) = (c_3 \bar{c}_3, c_3 \bar{c}_3)$, hence $b_3 \bar{b}_3 = c_3 \bar{c}_3$. Therefore $(c_3 \bar{c}_3, b_8^2) = (b_3 \bar{b}_3, b_8^2) = 4$, thus $(c_3 b_8, c_3 b_8) = 4$, a contradiction to Theorem 2.2. □

2.5 NITA Generated by b_3 Satisfying $b_3^2 = \bar{b}_3 + b_6$ and b_6 Nonreal and $b_{10} \in B$ is Real

Now we discuss NITA satisfying the following hypothesis.

Hypothesis 2.1   *Let (A, B) be a NITA generated by a nonreal element $b_3 \in B$ satisfies $b_3 \bar{b}_3 = 1 + c_8$ and $b_3^2 = \bar{b}_3 + b_6$, where $b_6 \in B$ is nonreal. Here we still assume that $L(B) = 1$ and $L_2(B) = \emptyset$. In this section, we use the symbols b_i, c_i, d_i to denote elements of B with degree i, where $i \geq 2$.*

By the associative law and Hypothesis 2.1, we have $b_3(b_3 \bar{b}_3) = (b_3^2)\bar{b}_3$, $b_3 + c_8 b_3 = \bar{b}_3^2 + b_6 \bar{b}_3$, so

$$c_8 b_3 = \bar{b}_6 + b_6 \bar{b}_3.$$

Hence $(b_3 b_6, c_8) = (b_3 c_8, \bar{b}_6) = 1$, so

$$b_3 b_6 = c_8 + b_{10},$$

therefore $b_{10} \in B$ since $(b_3 c_8, b_3 c_8) = 3$.

Now we shall deal with the case $b_{10} = \bar{b}_{10} \in B$.

Lemma 2.9   *Let (A, B) satisfy Hypothesis 2.1 and $\bar{b}_{10} = b_{10} \in B$, then*

1) $b_3 b_6 = c_8 + b_{10}$;
2) $\bar{b}_3 b_6 = b_3 + b_{15}$, *where $b_{15} \in B$*;
3) $b_3 c_8 = b_3 + \bar{b}_6 + b_{15}$;
4) $b_3 b_{10} = b_{15} + \bar{b}_6 + \bar{c}_9$, *$c_9 \in B$*;
5) $b_3 b_{15} = 2\bar{b}_{15} + b_6 + c_9$;
6) $b_6^2 = 2\bar{b}_6 + \bar{c}_9 + b_{15}$;
7) $b_6 c_8 = \bar{b}_3 + 2\bar{b}_{15} + b_6 + c_9$;
8) $\bar{b}_6 b_6 = 1 + c_8 + x + y + z + w$, *where $x, y, z, w \in B$, $|x + y + z + w| = 27$, and the degrees of x, y, z and $w \geq 5$. Moreover, $\overline{x + y + z + w} = x + y + z + w$*;
9) $c_8^2 = 1 + 2c_8 + 2b_{10} + x + y + z + w$;

10) $b_3\bar{b}_{15} = b_{10} + c_8 + x + y + z + w$;

11) $(\bar{c}_9 b_3, c_9) = 1$ or $(b_6 c_9, \bar{c}_9) \geq 1$.

Proof We have seen that

$$c_8 b_3 = \bar{b}_6 + b_6 \bar{b}_3 \tag{1}$$

and

$$b_3 b_6 = c_8 + b_{10}, \tag{2}$$

where b_{10} is real. So we obtain $(\bar{b}_3 b_6, \bar{b}_3 b_6) = 2$ and consequently we obtain

$$\bar{b}_3 b_6 = b_3 + b_{15}, \tag{3}$$

where $b_{15} \in B$. Hence by (1)

$$c_8 b_3 = b_3 + \bar{b}_6 + b_{15}. \tag{4}$$

By the associative law and (2), (3), we have $b_3(\bar{b}_3 b_6) = (b_3 b_6)\bar{b}_3$ and $b_3^2 + b_{15} b_3 = c_8 \bar{b}_3 + b_{10}\bar{b}_3$, so by (4) and Hypothesis 2.1 we have

$$b_{15} b_3 = \bar{b}_{15} + \bar{b}_3 b_{10}. \tag{5}$$

If $x \in B$ and $(b_{15} b_3, x) = (b_{15}, \bar{b}_3 x)$, so $|x| \geq 5$, $\forall x \in B$. So every constituent of $\bar{b}_3 b_{10}$ has degrees ≥ 5. Since b_{10} is real, the constituents of $b_3 b_{10}$ also have degrees ≥ 5, so by the associative law and Hypothesis 2.1 and (4), (2):

$$(\bar{b}_3 b_3)c_8 = (b_3 c_8)\bar{b}_3,$$

$$c_8 + c_8^2 = b_3\bar{b}_3 + \bar{b}_6\bar{b}_3 + b_{15}\bar{b}_3,$$

$$c_8^2 = 1 + c_8 + b_{10} + \bar{b}_3 b_{15}. \tag{6}$$

By the associative law and Hypothesis 2.1 and (2):

$$(b_3\bar{b}_3)(b_3\bar{b}_3) = b_3^2\bar{b}_3^2,$$

$$1 + 2c_8 + c_8^2 = \bar{b}_3 b_3 + \bar{b}_3\bar{b}_6 + b_6 b_3 + b_6\bar{b}_6,$$

$$c_8^2 = 2b_{10} + c_8 + b_6\bar{b}_6; \tag{7}$$

so we get

$$1 = (\bar{b}_3 b_{15}, b_{10}) = (b_{15}, b_3 b_{10}).$$

By (2),

$$(b_3 b_{10}, \bar{b}_6) = (b_3 b_6, b_{10}) = 1.$$

$b_3 b_{10}$ have constituents of degree ≥ 5 as we have proved. Therefore

$$b_3 b_{10} = b_{15} + \bar{b}_6 + \bar{c}_9 \tag{8}$$

where $c_9 \in B$. So by (5),

$$b_3 b_{15} = 2\bar{b}_{15} + b_6 + c_9. \tag{9}$$

By the associative law and Hypothesis 2.1 and (2), (3), (4):

$$(b_3^2)b_6 = (b_3 b_6)b_3,$$
$$\bar{b}_3 b_6 + b_6^2 = c_8 b_3 + b_{10} b_3,$$

consequently

$$b_6^2 = 2\bar{b}_6 + b_{15} + \bar{c}_9.$$

Now

$$b_3(\bar{b}_3 b_6) = (b_3 \bar{b}_3)b_6,$$
$$b_3^2 + b_{15} b_3 = b_6 + b_6 b_8.$$

Thus

$$c_8 b_6 = \bar{b}_3 + 2\bar{b}_{15} + b_6 + c_9.$$

So

$$(\bar{b}_6 b_6, c_8) = (b_6, b_6 c_8) = 1.$$

Since

$$(b_6^2, b_6^2) = 6 = (\bar{b}_6 b_6, \bar{b}_6 b_6),$$

hence

$$b_6 \bar{b}_6 = 1 + c_8 + x + y + z + w$$

where $x, y, z, w \in B$. Therefore by (7), (6)

$$c_8^2 = 1 + 2c_8 + 2b_{10} + x + y + z + w,$$

and

$$b_3 \bar{b}_5 = b_{10} + c_8 + x + y + z + w.$$

So $(b_3 \bar{b}_{15}, m) = (b_3 \bar{m}, b_{15})$, so $|m| \geq 5$, hence

$$|x|, |y|, |z|, |w| \geq 5.$$

By (8),

$$(b_3 c_9, b_{10}) = (b_3 b_{10}, \bar{c}_9) = 1,$$

hence

$$b_3 c_9 = b_{10} + \alpha,$$

where $\alpha \in N^*B$. By the associative law and Hypothesis 2.1:

$$(b_3^2)c_9 = (b_3c_9)b_3,$$

$$c_9\bar{b}_3 + c_9b_6 = b_{10}b_3 + \alpha b_3 = b_{15} + \bar{b}_6 + \bar{c}_9 + \alpha b_3;$$

hence $(c_9\bar{b}_3, \bar{c}_9) \geq 1$, or $(b_6c_9, \bar{c}_9) \geq 1$. By (9) we conclude that either $(c_9\bar{b}_3, \bar{c}_9) = 1$ or $(b_6c_9, \bar{c}_9) \geq 1$. $\square$

We start to investigate the case $(\bar{c}_9b_3, c_9) = 1$ of Lemma 2.9.

Lemma 2.10 *Let (A, B) satisfy Hypothesis 2.1. Assume that $\bar{b}_{10} = b_{10} \in B$ and $(\bar{c}_9b_3, c_9) = 1$. Then we have the following:*

1) $b_3\bar{c}_9 = \bar{b}_{15} + c_9 + c_3, c_3, c_9 \in B$;
2) $b_6b_{10} = 3\bar{b}_{15} + \bar{b}_3 + c_3 + c_9, b_{15} \in B$;
3) $c_8b_{15} = 5b_{15} + b_3 + \bar{c}_3 + 2\bar{b}_6 + 3\bar{c}_9$;
4) $b_3\bar{c}_3 = c_9$;
5) $b_3c_3 = b_9, \bar{b}_9 = b_9 \in B$;
6) $b_3c_9 = b_9 + b_{10} + b_8, b_9 = \bar{b}_9, \bar{b}_8 = b_8 \in B$;
7) $\bar{c}_3b_6 = b_{10} + b_8$;
8) $b_6\bar{b}_{15} = 4b_{15} + \bar{b}_6 + b_3 + \bar{c}_3 + 2\bar{c}_9$;
9) $c_3\bar{c}_3 = 1 + b_8$;
10) $\bar{b}_6b_6 = 1 + c_8 + b_9 + b_8 + c_5 + b_5, \overline{c_5 + b_5} = c_5 + b_5$;
11) $c_8^2 = 1 + 2c_8 + 2b_{10} + b_9 + b_8 + b_5 + c_5$;
12) $b_3\bar{b}_{15} = b_{10} + c_8 + b_9 + b_8 + b_5 + c_5$;
13) $b_6b_{15} = 2c_8 + 3b_{10} + 2b_9 + 2b_8 + b_5 + c_5$;
14) $c_8b_{10} = 2c_8 + 2b_{10} + 2b_9 + 2b_8 + b_5 + c_5$;
15) $b_6\bar{c}_9 = b_{10} + c_8 + 2b_9 + b_8 + b_5 + c_5$;
16) $b_{10}^2 = 1 + 2c_8 + 2b_{10} + 3b_9 + 2b_8 + 2b_5 + 2c_5$.

Proof By 5) in Lemma 2.9, $(\bar{b}_3c_9, b_{15}) = (c_9, b_3b_{15}) = 1$. So by the assumption in the lemma, $(b_3\bar{c}_9, c_9) = 1$, hence

$$b_3\bar{c}_9 = \bar{b}_{15} + c_9 + c_3, \tag{1}$$

where $c_3 \in B$. By the associative law and Hypothesis 2.1 and 4) in Lemma 2.9

$$(b_3^2)b_{10} = (b_3b_{10})b_3,$$

$$\bar{b}_3b_{10} + b_6b_{10} = b_{15}b_3 + \bar{b}_6b_3 + \bar{c}_9b_3;$$

by 2), 4), 5) in Lemma 2.9 and (1),

$$b_6b_{10} = 3\bar{b}_{15} + \bar{b}_3 + c_3 + c_9. \tag{2}$$

By the associative law and Hypothesis 2.1 and 5) in Lemma 2.9:

$$(\bar{b}_3 b_3) b_{15} = (b_3 b_{15}) \bar{b}_3,$$
$$c_8 b_{15} + b_{15} = 2\bar{b}_{15} b_3 + \bar{b}_3 b_6 + c_9 \bar{b}_3;$$

by (1) and 2), 5) in Lemma 2.9,

$$c_8 b_{15} = 5b_{15} + b_3 + \bar{c}_3 + 2\bar{b}_6 + 3\bar{c}_9. \tag{3}$$

By (1),

$$(\bar{c}_3 b_3, c_9) = (b_3 \bar{c}_9, c_3) = 1,$$

hence

$$\bar{c}_3 b_3 = c_9. \tag{3'}$$

So

$$(c_3 b_3, c_3 b_3) = (b_3 \bar{c}_9, c_3) = 1,$$

hence

$$c_3 b_3 = b_9. \tag{4}$$

By the associative law and Hypothesis 2.1 and (3)

$$\bar{c}_3 (b_3^2) = (\bar{c}_3 b_3) b_3,$$
$$\bar{c}_3 \bar{b}_3 + \bar{c}_3 b_6 = c_9 b_3,$$

hence

$$c_9 b_3 = \bar{b}_9 + \bar{c}_3 b_6. \tag{5}$$

By (1) and (4) in Lemma 2.9, $(c_9 b_3, b_{10}) = (b_3 b_{10}, \bar{c}_9) = 1, (b_3 c_9, b_3 c_9) = (b_3 \bar{c}_9, b_3 \bar{c}_9) = 3$, hence

$$c_9 b_3 = \bar{b}_9 + b_{10} + b_8, \tag{6}$$

where $b_8 \in B$. So

$$\bar{c}_3 b_6 = b_{10} + b_8. \tag{7}$$

By the associative law and Hypothesis 2.1 and 4) in Lemma 2.9:

$$(\bar{b}_3 b_3) b_{10} = (b_3 b_{10}) \bar{b}_3,$$
$$b_{10} + c_8 b_{10} = b_{15} \bar{b}_3 + \bar{b}_6 \bar{b}_3 + \bar{c}_9 \bar{b}_3,$$

by 1), 10) in Lemma 2.9 and (6)

$$c_8 b_{10} = c_8 + \bar{x} + \bar{y} + \bar{z} + \bar{w} + \bar{b}_8. \tag{8}$$

b_{10}, c_8 are reals, so if $\bar{b}_8 \neq b_8$, then $(c_8 b_{10}, b_8) \neq 0$, and without loss of generality $\bar{x} = b_8$, so $x = \bar{b}_8$ and by 8) in Lemma 2.9 and without loss of generality $y = \bar{b}_8$, so by (8) $(c_8b_{10}, \bar{b}_8) = 2$, hence $(c_8b_{10}, b_8) = 2$, so without loss of generality $\bar{z} = b_8$, hence $z = \bar{b}_8$ and by 8) in Lemma 2.9 without loss of generality $w = b_8$ then $(c_8b_{10}, b_8) = 3$, a contradiction. Hence $b_8 = \bar{b}_8$. In the same way $b_9 = \bar{b}_9$.

By the associative law and 2), 6) in Lemma 2.9:

$$(b_6^2)\bar{b}_3 = (b_6\bar{b}_3)b_6,$$

$$2\bar{b}_6\bar{b}_3 + \bar{c}_9\bar{b}_3 + b_{15}\bar{b}_3 = b_3b_6 + b_{15}b_6,$$

and by Lemma 2.9 and (6)

$$b_6b_{15} = 2c_8 + 3b_{10} + b_8 + b_9 + x + y + w + z. \tag{9}$$

By the associative law and the Hypothesis 2.1 and 4) in Lemma 2.9:

$$b_{10}(b_3\bar{b}_3) = (\bar{b}_3b_{10})b_3,$$

$$b_{10} + c_8b_{10} = b_3\bar{b}_{15} + b_3b_6 + b_3c_9;$$

by (6) and 1), 10) in Lemma 2.9

$$c_8b_{10} = 2c_8 + 2b_{10} + x + y + z + w + b_8 + b_9. \tag{10}$$

By the associative law and 2), 3), 6) in Lemma 2.9 and (3):

$$c_8(\bar{b}_3b_6) = (\bar{b}_3c_8)b_6,$$

$$b_3c_8 + b_{15}c_8 = \bar{b}_3b_6 + b_6^2 + \bar{b}_{15}b_6,$$

$$b_6\bar{b}_{15} = 4b_{15} + \bar{b}_6 + b_3 + \bar{c}_3 + 2\bar{c}_9.$$

By the associative law and (1), (4), (6) and Hypothesis 2.1:

$$(b_3^2)\bar{c}_9 = (b_3\bar{c}_9)b_3,$$

$$\bar{b}_3\bar{c}_9 + b_6\bar{c}_9 = \bar{b}_{15}b_3 + c_9b_3 + c_3b_3,$$

$$b_6\bar{c}_9 = b_{10} + c_8 + x + y + z + w + b_9. \tag{11}$$

By the associative law and 1) in Lemma 2.9 and (2):

$$(b_3b_6)b_{10} = (b_6b_{10})b_3,$$

$$c_8b_{10} + b_{10}^2 = 3\bar{b}_{15}b_3 + \bar{b}_3b_3 + c_3b_3 + c_9b_3,$$

by (4), (6), (10) and Hypothesis 2.1

$$b_{10}^2 = 1 + 2c_8 + b_9 + 2b_{10} + 2x + 2y + 2z + 2w. \tag{12}$$

By (2), $(b_6b_{10}, b_6b_{10}) = 12$. But by 8) in Lemma 2.9 and (12) $(b_6\bar{b}_6, b_{10}^2) = 10$, so we obtain, without loss of generality,

$$x = b_9. \tag{13}$$

By (4) and that $b_9 = \bar{b}_9$, $(c_3b_3, \bar{c}_3\bar{b}_3) = (b_3^2, \bar{c}_3^2) = 1$, but by (3') $(c_3^2, b_3) = (c_3\bar{b}_3, \bar{c}_3) = 0$, so by Hypothesis 2.1

$$(\bar{c}_3^2, b_6) = 1. \tag{14}$$

Hence

$$(c_3\bar{c}_3, c_3\bar{c}_3) = 2,$$

so

$$c_3\bar{c}_3 = 1 + t_8. \tag{15}$$

By (4)

$$(c_3b_3, c_3b_3) = (c_3\bar{c}_3, b_3\bar{b}_3) = 1,$$

therefore by hypothesis $t_8 \neq c_8$. By (12), (13) and 9) in Lemma 2.9

$$(b_{10}^2, c_8^2) = 18 = (c_8b_{10}, c_8b_{10}).$$

So by (8) we conclude that, without loss of generality,

$$y = b_8. \tag{16}$$

So by (13), (16) and 8) in Lemma 2.9, we obtain $|z| = |w| = 5$, set $z = b_5$ and $w = c_5$. By (7),

$$(\bar{c}_3b_6, \bar{c}_3b_6) = (\bar{c}_3c_3, \bar{b}_6b_6) = 2,$$

so by (15) and 8) in Lemma 2.9 and by (16) $t_8 = b_8$. So $c_3\bar{c}_3 = 1 + b_8$. By 8), 9), 10) in Lemma 2.9 and (10), (11), (12), (13), (16), (9)

$$\bar{b}_6b_6 = 1 + c_8 + b_9 + b_8 + c_5 + b_5,$$
$$c_8^2 = 1 + 2c_8 + 2b_{10} + b_9 + b_8 + b_5 + c_5,$$
$$b_3\bar{b}_{15} = b_{10} + c_8 + b_9 + b_8 + b_5 + c_5,$$
$$b_6b_{15} = 2c_8 + 3b_{10} + 2b_9 + 2b_8 + b_5 + c_5,$$
$$c_8b_{10} = 2c_8 + 2b_{10} + 2b_9 + 2b_8 + b_5 + c_5,$$
$$b_6\bar{c}_9 = b_{10} + c_8 + 2b_9 + b_8 + b_5 + c_5,$$
$$b_{10}^2 = 1 + 2c_8 + 2b_{10} + 3b_9 + 2b_8 + 2b_5 + 2c_5.$$

□

Lemma 2.11 *Let (A, B) satisfy Hypothesis 2.1. Assume that $\bar{b}_{10} = b_{10} \in B$ and $(\bar{c}_9b_3, c_9) = 1$. Then we obtain the following equations:*

1) $b_3b_9 = b_{15} + \bar{c}_9 + \bar{c}_3$;
2) $c_3c_8 = c_9 + \bar{b}_{15}$;

3) $b_3b_8 = b_{15} + \bar{c}_9$;

4) $b_3c_5 = b_{15}$;

5) $b_3b_5 = b_{15}$;

6) $b_6b_5 = \bar{b}_{15} + b_6 + c_9$;

7) $b_6c_9 = 2b_{15} + 2\bar{c}_9 + \bar{b}_6$;

8) $c_3b_6 = \bar{c}_3 + b_{15}$;

9) $b_6b_9 = 2\bar{b}_{15} + b_6 + 2c_9$;

10) $b_6c_5 = \bar{b}_{15} + b_6 + c_9$;

11) $b_6b_8 = 2\bar{b}_{15} + b_6 + c_3 + c_9$;

12) $c_3b_{10} = \bar{b}_{15} + b_6 + c_9$;

13) $c_3c_9 = b_{15} + \bar{c}_9 + b_3$;

14) $c_3b_8 = c_3 + b_6 + \bar{b}_{15}$;

15) $c_3b_9 = \bar{b}_{15} + \bar{b}_3 + c_9$;

16) $c_3\bar{b}_{15} = 2b_{15} + \bar{b}_6 + \bar{c}_9$;

17) $b_8\bar{b}_{15} = 5\bar{b}_{15} + \bar{b}_3 + c_3 + 2b_6 + 3c_9$;

18) $b_8c_9 = 3\bar{b}_{15} + 2c_9 + b_6 + \bar{b}_3$.

Proof By 5), 6), 12), in Lemma 2.10

$$(b_3b_9, \bar{c}_3) = (b_3c_3, b_9) = 1,$$
$$(b_3b_9, \bar{c}_9) = (b_3c_9, b_9) = 1,$$
$$(b_3b_9, b_{15}) = (b_3\bar{b}_{15}, b_9) = 1,$$

then

$$b_3b_9 = b_{15} + \bar{c}_9 + \bar{c}_3. \tag{1}$$

By the associative law and 1), 4) in Lemma 2.10 and Hypothesis 2.1,

$$(\bar{b}_3c_3)b_3 = (\bar{b}_3b_3)c_3,$$
$$\bar{c}_9b_3 = c_3 + c_3c_8,$$

so

$$c_3c_8 = c_9 + \bar{b}_{15}. \tag{2}$$

By 6), 12) in Lemma 2.10,

$$(b_3b_8, b_{15}) = (b_3\bar{b}_{15}, b_8) = 1,$$
$$(b_3b_8, \bar{c}_9) = (b_3c_9, b_8) = 1,$$

then

$$b_3b_8 = b_{15} + \bar{c}_9. \tag{3}$$

By the associative law and 2) in Lemma 2.9 and 10) in Lemma 2.10:

$$(b_6\bar{b}_6)b_3 = (b_3\bar{b}_6)b_6,$$
$$b_3 + c_8b_3 + b_9b_3 + b_8b_3 + c_5b_3 + b_5b_3 = \bar{b}_3b_6 + \bar{b}_{15}b_6,$$

by 2), 3) in Lemma 2.9 and (3), (1) and 8) in Lemma 2.10:

$$c_5 b_3 + b_5 b_3 = 2 b_{15},$$

so

$$c_5 b_3 = b_{15}, \tag{4}$$
$$b_5 b_3 = b_{15}. \tag{5}$$

By the associative law and Hypothesis 2.1 and (5) and 5) in Lemma 2.9

$$(b_3^2) b_5 = (b_3 b_5) b_3 = b_3 b_{15},$$
$$\bar{b}_3 b_5 + b_6 b_5 = 2 \bar{b}_{15} + b_6 + c_9,$$

by (5)

$$(\bar{b}_3 b_5, \bar{b}_3 b_5) = (b_5 b_3, b_5 b_3) = 1.$$

So $\bar{b}_3 b_5 = \bar{b}_{15}$, therefore

$$b_6 b_5 = \bar{b}_{15} + b_6 + c_9. \tag{6}$$

By the associative law and 4), 7) in Lemma 2.10:

$$(b_3 \bar{c}_3) b_6 = (\bar{c}_3 b_6) b_3,$$
$$c_9 b_6 = b_{10} b_3 + b_8 b_3,$$

by 4) in Lemma 2.9 and by (3)

$$b_6 c_9 = 2 b_{15} + 2 \bar{c}_9 + \bar{b}_6. \tag{7}$$

By the associative law and the Hypothesis 2.1 and 5) in Lemma 2.10:

$$(b_3^2) c_3 = (b_3 c_3) b_3, \qquad \bar{b}_3 c_3 + b_6 c_3 = b_9 b_3,$$

so by (1) and 4) in Lemma 2.10

$$b_6 c_3 = \bar{c}_3 + b_{15}. \tag{8}$$

10), 13), 15) in Lemma 2.10,

$$(b_6 b_9, \bar{b}_{15}) = (b_6 b_{15}, b_9) = 2,$$
$$(b_6 b_9, b_6) = (b_9, \bar{b}_6 b_6) = 1,$$
$$(b_6 b_9, c_9) = (b_6 \bar{c}_9, b_9) = 2,$$

so

$$b_6 b_9 = 2 \bar{b}_{15} + b_6 + 2 c_9. \tag{9}$$

By the associative law and Hypothesis 2.1 and (4):

$$(b_3^2)c_5 = (b_3c_5)b_3, \qquad \bar{b}_3c_5 + b_6c_5 = b_{15}b_3,$$

by 5) in Lemma 2.9

$$\bar{b}_3c_5 + b_6c_5 = 2\bar{b}_{15} + b_6 + c_9,$$
$$(c_5\bar{b}_3, c_5\bar{b}_3) = (c_5b_3, c_5b_3) = 1,$$

so $\bar{b}_3c_5 = \bar{b}_{15}$, hence by 7), 10), 13), and 15) in Lemma 2.10

$$b_6c_5 = \bar{b}_{15} + b_6 + c_9, \tag{10}$$

$$(b_6b_8, c_3) = (b_6\bar{c}_3, b_8) = 1, \qquad (b_6b_8, b_6) = (b_8, \bar{b}_6b_6) = 1,$$
$$(b_6b_8, c_9) = (b_8, \bar{b}_6c_9) = 1, \qquad (b_6b_8, \bar{b}_{15}) = (b_6b_{15}, b_8) = 2,$$

so

$$b_6b_8 = 2\bar{b}_{15} + b_6 + c_3 + c_9. \tag{11}$$

By the associative law and 1) in Lemma 2.9 and (8)

$$c_3(b_3b_6) = (c_3b_6)b_3, \qquad c_3c_8 + c_3b_{10} = \bar{c}_3b_3 + b_{15}b_3,$$

by (2), 4) in Lemma 2.10 and 5) in Lemma 2.9

$$c_3b_{10} = \bar{b}_{15} + b_6 + c_9. \tag{12}$$

By the associative law and 4), 9) in Lemma 2.10:

$$(c_3\bar{c}_3)b_3 = (\bar{c}_3b_3)c_3, \qquad b_3 + b_8b_3 = c_9c_3,$$

by (3)

$$c_3c_9 = b_{15} + \bar{c}_9 + b_3. \tag{13}$$

By the associative law and 6) in Lemma 2.10 and (13)

$$(b_3c_9)c_3 = (c_3c_9)b_3,$$
$$b_9c_3 + b_{10}c_3 + b_8c_3 = b_{15}b_3 + \bar{c}_9b_3 + b_3^2;$$

by (12) and 5) in Lemma 2.9 and 1) in Lemma 2.10 and Hypothesis 2.1

$$b_9c_3 + b_8c_3 + b_{15} + b_6 + c_9 = 2\bar{b}_{15}b_6 + c_9 + \bar{b}_{15} + c_9 + c_3 + \bar{b}_3 + b_6,$$
$$b_9c_3 + b_8c_3 = 2\bar{b}_{15} + c_9 + c_3 + \bar{b}_3 + b_6;$$

by 5), 7), 9) in Lemma 2.10

$$(b_8c_3, \bar{b}_3) = (c_3b_3, b_8) = 0,$$
$$(b_8c_3, c_3) = (b_8, \bar{c}_3c_3) = 1,$$
$$(b_8c_3, b_6) = (b_8, \bar{c}_3b_6) = 1.$$

So

$$b_8c_3 = c_3 + b_6 + \bar{b}_{15}, \tag{14}$$

hence

$$b_9c_3 = \bar{b}_{15} + \bar{b}_3 + c_9. \tag{15}$$

By the associative law and 2) in Lemma 2.9 and 7) in Lemma 2.10:

$$c_3(b_3\bar{b}_6) = (c_3\bar{b}_6)b_3, \qquad c_3\bar{b}_3 + c_3\bar{b}_{15} = b_{10}b_3 + b_8b_3,$$

by 4) in Lemma 2.10 and 4) in Lemma 2.9 and (3):

$$c_3\bar{b}_{15} = 2b_{15} + \bar{b}_6 + \bar{c}_9. \tag{16}$$

By the associative law and 9) in Lemma 2.10:

$$(c_3\bar{b}_{15})\bar{c}_3 = (c_3\bar{c}_3)\bar{b}_{15}, \qquad 2b_{15}\bar{c}_3 + \bar{b}_6\bar{c}_3 + \bar{c}_9\bar{c}_3 = \bar{b}_{15} + b_8\bar{b}_{15},$$

by (16) and (8), (13)

$$b_8\bar{b}_{15} = 5\bar{b}_{15} + \bar{b}_3 + c_3 + 2b_6 + 3c_9. \tag{17}$$

By the associative law and (13) and 9) in Lemma 2.10:

$$(c_3\bar{c}_3)c_9 = (c_3c_9)\bar{c}_3, \qquad c_9 + b_8c_9 = b_{15}\bar{c}_3 + \bar{c}_9\bar{c}_3 + b_3\bar{c}_3,$$

by (13), (16) and 4) in Lemma 2.10:

$$c_9 + b_8c_9 = 2\bar{b}_{15} + b_6 + c_9 + \bar{b}_{15} + c_9 + \bar{b}_3 + c_9,$$

then

$$b_8c_9 = 3\bar{b}_{15} + 2c_9 + b_6 + \bar{b}_3. \tag{18}$$

$\square$

Lemma 2.12 *Let (A, B) satisfy Hypothesis 2.1. Assume that $\bar{b}_{10} = b_{10} \in B$ and $(\bar{c}_9b_3, c_9) = 1$. Then we obtain the following equations:*

1) $c_3\bar{c}_9 = b_{10} + b_9 + c_8$;
2) $c_3b_{15} = b_5 + c_5 + c_8 + b_8 + b_9 + b_{10}$;
3) $c_3^2 = \bar{c}_3 + \bar{b}_6$;
4) $\bar{b}_{15}b_{15} = 1 + 3b_5 + 3c_5 + 5c_8 + 5b_8 + 6b_6 + 6b_{10}$;
5) $b_{15}^2 = 9\bar{b}_{15} + 4b_6 + 2\bar{b}_3 + 2c_3 + 6c_9$;

6) $b_{10}b_{15} = 6b_{15} + 3\bar{b}_6 + b_3 + \bar{c}_3 + 4\bar{c}_9;$
7) $c_9b_{15} = 2b_5 + 2c_5 + 3c_8 + 3b_8 + 3b_9 + 4b_{10};$
8) $c_9\bar{b}_{15} = 6b_{15} + 2\bar{b}_6 + b_3 + \bar{c}_3 + 3\bar{c}_9;$
9) $c_9^2 = 3b_{15} + b_3 + \bar{c}_3 + 2\bar{b}_6 + 2\bar{c}_9;$
10) $c_9\bar{c}_9 = 1 + 2c_8 + 2b_9 + 2b_{10} + c_5 + b_5 + 2b_8;$
11) $c_8c_9 = 3\bar{b}_{15} + b_6 + c_3 + 2c_9;$
12) $b_9c_9 = \bar{b}_3 + 2c_9 + 3\bar{b}_{15} + c_3 + 2b_6;$
13) $c_9b_{10} = \bar{b}_3 + 2c_9 + 3\bar{b}_{15} + c_3 + 2b_6;$
14) $b_5c_9 = 2\bar{b}_{15} + b_6 + c_9;$
15) $c_5c_9 = 2\bar{b}_{15} + b_6 + c_9;$
16) $b_5 = \bar{b}_5, c_5 = \bar{c}_5;$
17) $c_3b_5 = \bar{b}_{15};$
18) $c_3c_5 = \bar{b}_{15}.$

Proof By 2), 12), 15) in Lemma 2.11, $(c_3\bar{c}_9, b_{10}) = (c_3b_{10}, c_9) = 1, (c_3\bar{c}_9, b_9) = (c_3b_9, c_9) = 1, (c_3\bar{c}_9, c_8) = (c_3c_8, c_9) = 1$, then

$$c_3\bar{c}_9 = b_{10} + b_9 + c_8. \tag{1}$$

By the associative law and 3) and 14) in Lemma 2.11:

$$(b_3b_8)c_3 = (c_3b_8)b_3, \qquad b_{15}c_3 + \bar{c}_9c_3 = c_3b_3 + b_6b_3 + \bar{b}_{15}b_3,$$

by 5), 12) in Lemma 2.10 and (1) and 1) in Lemma 2.9

$$c_3b_{15} = b_5 + c_5 + c_8 + b_8 + b_9 + b_{10}. \tag{2}$$

By the associative law and (1) and 1) Lemma 2.10:

$$(b_3\bar{c}_9)c_3 = (c_3\bar{c}_9)b_3, \qquad \bar{b}_{15}c_3 + c_9c_3 + c_3^2 = b_{10}b_3 + b_9b_3 + c_8b_3;$$

by 1), 13), 16) in Lemma 2.11 and 3), 4) in Lemma 2.9

$$c_3^2 = \bar{c}_3 + \bar{b}_6. \tag{3}$$

By the associative law and 2) in Lemma 2.9 and 8) in Lemma 2.10:

$$(b_3\bar{b}_6)b_{15} = (\bar{b}_6b_{15})b_3,$$

$$\bar{b}_3b_{15} + \bar{b}_{15}b_{15} = 4\bar{b}_{15}b_3 + b_6b_3 + \bar{b}_3b_3 + c_3b_3 + 2c_9b_3.$$

By 6), 5), 12) in Lemma 2.10 and 1) in Lemma 2.9 and Hypothesis 2.1, we obtain that

$$\bar{b}_{15}b_{15} = 1 + 3b_5 + 3c_5 + 5c_8 + 5b_8 + 6b_6 + 6b_{10}. \tag{4}$$

By the associative law and 2) in Lemma 2.9 and 13) in Lemma 2.10:

$$(b_3\bar{b}_6)\bar{b}_{15} = b_3(\bar{b}_6\bar{b}_{15}),$$

$$\bar{b}_3\bar{b}_{15} + \bar{b}_{15}^2 = 2c_8b_3 + 3b_{10}b_3 + 2b_9b_3 + 2b_8b_3 + b_5b_3 + c_5b_3;$$

by 3), 4) in Lemma 2.9 and 1), 3), 4), 5) in Lemma 2.11

$$\bar{b}_{15}^2 = 9\bar{b}_{15} + 4b_6 + 2\bar{b}_3 + 3c_3 + 6c_9. \tag{5}$$

By 4), 12), 13) in Lemma 2.10 and 12) in Lemma 2.11 and 5) in Lemma 2.9:

$$(b_{10}b_{15}, b_3) = (b_{10}, \bar{b}_{15}b_{15}) = 6, \qquad (b_{10}b_{15}, \bar{b}_6) = (b_{15}b_6, b_{10}) = 3,$$

$$(b_{10}b_{15}, b_3) = (\bar{b}_{15}b_3, b_{10}) = 1, \qquad (b_{10}b_{15}, \bar{c}_3) = (b_{10}c_3, \bar{b}_{15}) = 1,$$

$$(b_{10}b_{15}, \bar{b}_3c_3) = (b_{15}b_3, b_{10}c_3) = 4,$$

so $(b_{10}b_{15}, \bar{c}_9) = 4$. Hence

$$b_{10}b_{15} = 6\bar{b}_{15} + 3\bar{b}_6 + b_3 + \bar{c}_3 + 4\bar{c}_9. \tag{6}$$

By the associative law and 4) in Lemma 2.10 and 16) in Lemma 2.11:

$$(b_3\bar{c}_3)b_{15} = b_3(\bar{c}_3b_{15}), \qquad c_9b_{15} = 2\bar{b}_{15}b_3 + b_3b_6 + b_3c_9;$$

by 6), 12) in Lemma 2.10 and 1) in Lemma 2.9:

$$c_9b_{15} = 2b_5 + 2c_5 + 3c_8 + 3b_8 + 3b_9 + 4b_{10}. \tag{7}$$

By 4) and Lemma 2.10 and (2) we get:

$$(b_3\bar{c}_3)\bar{b}_{15} = (\bar{c}_3\bar{b}_{15})b_3,$$

$$c_9\bar{b}_{15} = b_5b_3 + c_5b_3 + c_8b_3 + b_8b_3 + b_9b_3 + b_{10}b_3;$$

by 1), 3), 4), 5) in Lemma 2.11 and 3), 4) in Lemma 2.9,

$$c_9\bar{b}_{15} = 6b_{15} + 2\bar{b}_6 + b_3 + \bar{c}_3 + 3\bar{c}_9. \tag{8}$$

By the associative law and 4) in Lemma 2.10 and (1):

$$(b_3\bar{c}_3)c_9 = b_3(\bar{c}_3c_9), \qquad c_9^2 = b_{10}b_3 + b_9b_3 + c_8b_3;$$

by 3), 4) in Lemma 2.9 and 1) in Lemma 2.11

$$c_9^2 = 3b_{15} + b_3 + \bar{c}_3 + 2\bar{b}_6 + 2\bar{c}_9. \tag{9}$$

By the associative law and 4) in Lemma 2.10 and 13) in Lemma 2.11:

$$(b_3\bar{c}_3)\bar{c}_9 = b_3(\bar{c}_3\bar{c}_9), \qquad c_9\bar{c}_9 = b_3\bar{b}_{15} + b_3c_9 + b_3\bar{b}_3;$$

by 6), 12) in Lemma 2.10 and Hypothesis 2.1:

$$c_9\bar{c}_9 = 1 + 2c_8 + 2b_9 + 2b_{10} + c_5 + b_5 + 2b_8. \tag{10}$$

By the associative law and 4) in Lemma 2.10 and 2) in Lemma 2.11:

$$(b_3\bar{c}_3)c_8 = (\bar{c}_3c_8)b_3, \qquad c_9c_8 = \bar{c}_9b_3 + b_{15}b_3;$$

by 1) in Lemma 2.10 and 5) in Lemma 2.9:

$$c_9c_8 = 3\bar{b}_{15} + b_6 + c_3 + 2c_9. \tag{11}$$

By the associative law and 5) in Lemma 2.10 and 13) in Lemma 2.11:

$$(b_3c_3)c_9 = (c_3c_9)b_3, \qquad b_9c_9 = b_{15}b_3 + \bar{c}_9b_3 + b_3^2;$$

by 5) in Lemma 2.9 and Hypothesis 2.1 and 1) in Lemma 2.10

$$b_9c_9 = \bar{b}_3 + 2c_9 + 3\bar{b}_{15} + c_3 + 2b_6. \tag{12}$$

By the associative law and 4) in Lemma 2.10 and 12) in Lemma 2.11:

$$(b_3\bar{c}_3)b_{10} = (\bar{c}_3b_{10})b_3, \qquad c_9b_{10} = b_{15}b_3 + \bar{b}_6b_3 + \bar{c}_9b_3;$$

by 2) and 5) in Lemma 2.9 and 1) in Lemma 2.10

$$c_9b_{10} = \bar{b}_3 + 2c_9 + 4\bar{b}_{15} + c_3 + b_6. \tag{13}$$

By (2),

$$(\bar{c}_3b_5, b_{15}) = (b_5, c_3b_{15}) = 1, \qquad (\bar{c}_3c_5, b_{15}) = (c_5, c_3b_{15}) = 1.$$

So

$$\bar{c}_3b_5 = b_{15}, \tag{14}$$
$$\bar{c}_3c_5 = b_{15}. \tag{15}$$

Hence by 4), 5) in Lemma 2.10 and the associative law:

$$(\bar{c}_3b_5)b_3 = (b_3\bar{c}_5)b_5, \qquad (\bar{c}_3c_5)b_3 = (b_3\bar{c}_3)c_5,$$
$$b_{15}b_3 = c_9b_5, \qquad b_{15}b_3 = c_9c_5.$$

By 5) in Lemma 2.9 we get:

$$b_5c_9 = 2\bar{b}_{15} + b_6 + c_9, \tag{16}$$
$$c_5c_9 = 2\bar{b}_{15} + b_6 + c_9. \tag{17}$$

By the associative law and 12) in Lemma 2.10 and 5) in Lemma 2.11:

$$(b_3\bar{b}_{15})b_5 = (b_3b_5)\bar{b}_{15} = b_{15}\bar{b}_{15},$$

$$b_{10}b_5 + c_8b_5 + b_9b_5 + b_8b_5 + b_5^2 + c_5b_5 = b_{15}\bar{b}_{15},$$

so b_5 is real and so c_5 is real. So by (14), (15) $c_3b_5 = \bar{b}_{15}, c_3b_5 = b_{15}$.   □

Lemma 2.13   *Let (A, B) satisfy Hypothesis 2.1. Assume that $\bar{b}_{10} = b_{10} \in B$ and $(\bar{c}_9b_3, c_9) = 1$. Then we obtain the following equations:*

1) $b_5c_8 = c_5 + c_8 + b_8 + b_9 + b_{10}$;
2) $b_5b_8 = c_5 + c_8 + b_8 + b_9 + b_{10}$;
3) $b_5b_9 = b_{10} + c_8 + b_9 + b_8 + b_5 + c_5$;
4) $b_5b_{10} = c_8 + b_8 + b_9 + b_5 + 2b_{10}$;
5) $b_5^2 = 1 + b_{10} + b_9 + b_5$;
6) $c_5b_5 = b_8 + c_8 + b_9$;
7) $c_8b_8 = b_5 + c_5 + c_8 + b_8 + 2b_9 + 2b_{10}$;
8) $c_8b_9 = 2b_8 + 2b_9 + 2b_{10} + b_5 + c_5 + c_8$;
9) $b_9^2 = 1 + b_5 + c_5 + 2c_8 + 2b_8 + 2b_9 + 2b_{10}$;
10) $b_8^2 = 1 + b_5 + c_5 + c_8 + 2b_8 + b_9 + 2b_{10}$;
11) $b_8b_9 = b_5 + c_5 + 2c_8 + b_8 + 2b_9 + 2b_{10}$;
12) $b_8b_{10} = b_5 + c_5 + 2c_8 + 2b_8 + 2b_9 + 2b_{10}$;
13) $c_5c_8 = b_5 + c_8 + b_8 + b_9 + b_{10}$;
14) $c_5b_8 = b_5 + c_8 + b_8 + b_9 + b_{10}$;
15) $c_5b_9 = b_{10} + c_8 + b_9 + b_8 + b_5 + c_5$;
16) $c_5b_{10} = b_8 + c_8 + b_9 + c_5 + 2b_{10}$;
17) $c_5^2 = 1 + b_9 + b_{10} + c_5$;
18) $b_5b_{15} = 3b_{15} + \bar{b}_6 + b_3 + \bar{c}_3 + 2\bar{c}_9$;
19) $c_5b_{15} = 3b_{15} + \bar{b}_6 + b_3 + \bar{c}_3 + 2\bar{c}_9$;
20) $b_9b_{15} = 6b_{15} + 3\bar{c}_9 + b_3 + 2\bar{b}_6 + \bar{c}_3$;
21) $b_9b_{10} = b_5 + c_5 + 2c_8 + 2b_8 + 2b_9 + 3b_{10}$.

Proof   By the associative law and Hypothesis 2.1 and 12) in Lemma 2.10:

$$b_5(b_3\bar{b}_3) = (\bar{b}_3b_5)b_3 = \bar{b}_{15}b_3,$$

$$b_5 + b_5c_8 = \bar{b}_{15}b_3 = b_{10} + c_8 + b_9 + b_8 + b_5 + c_5.$$

Thus

$$c_8b_5 = c_5 + c_8 + b_8 + b_9 + b_{10}. \tag{1}$$

By the associative law and 9) in Lemma 2.10 and 18) in Lemma 2.12:

$$(c_3\bar{c}_3)b_5 = (\bar{c}_3b_5)c_3, \qquad b_5 + b_5b_8 = b_{15}c_3;$$

by 2) in Lemma 2.12

$$b_5b_8 = c_5 + c_8 + b_8 + b_9 + b_{10}. \tag{2}$$

By the associative law and 5), 12) in Lemma 2.10 and 18) in Lemma 2.12:

$$(b_3 c_3) b_5 = (c_3 b_5) b_3,$$

$$b_9 b_5 = \bar{b}_{15} b_3 = b_{10} + c_8 + b_9 + b_8 + b_5 + c_5. \tag{3}$$

By the associative law and 1) in Lemma 2.9 and 5) in Lemma 2.11:

$$(b_3 b_6) b_5 = (b_3 b_5) b_6, \qquad c_8 b_5 + b_{10} b_5 = b_{15} b_6;$$

by (1) and 13) in Lemma 2.10:

$$b_5 b_{10} = c_8 + b_8 + b_9 + b_5 + 2 b_{10}. \tag{4}$$

By the associative law and 12) in Lemma 2.10 and 5) in Lemma 2.11:

$$(b_3 \bar{b}_{15}) b_5 = (b_3 b_5) \bar{b}_{15},$$

$$b_{10} b_5 + c_8 b_5 + b_9 b_5 + b_8 b_5 + b_5^2 + c_5 b_5 = b_{15} \bar{b}_{15};$$

so by 4) in Lemma 2.12 and (1), (2), (3), (4),

$$b_5^2 + c_5 b_5 = 1 + b_5 + 2 b_9 + b_{10} + b_8 + c_8,$$

and by (2), (4), (3), (1),

$$(b_5^2, b_{10}) = (b_5, b_5 b_{10}) = 1, \qquad (b_5^2, b_9) = (b_5, b_5 b_9) = 1.$$

Therefore

$$b_5^2 = 1 + b_{10} + b_9 + b_5, \tag{5}$$

$$c_5 b_5 = b_8 + c_8 + b_9. \tag{6}$$

By the associative law and Hypothesis 2.1 and 3) in Lemma 2.11:

$$(b_3 \bar{b}_3) b_8 = (\bar{b}_3 b_8) b_3, \qquad b_8 + c_8 b_8 = \bar{b}_{15} b_3 + c_9 b_3;$$

by 6), 12) in Lemma 2.10, we obtain

$$c_8 b_8 = b_5 + c_5 + c_8 + b_8 + 2 b_9 + 2 b_{10}. \tag{7}$$

By the associative law and Hypothesis 2.1 and 1) in Lemma 2.11:

$$(b_3 \bar{b}_3) b_9 = (\bar{b}_3 b_9) b_3, \qquad b_9 + c_8 b_9 = \bar{b}_{15} b_3 + c_9 b_3 + c_3 b_3;$$

by 5), 6), 12) in Lemma 2.10 we obtain

$$b_9 + c_8 b_9 = b_{10} + c_8 + b_9 + b_8 + b_5 + c_5 + b_9 + b_{10} + b_8 + b_9.$$

By 5) in Lemma 2.10 and 15) in Lemma 2.11 we obtain

$$c_8 b_9 = 2b_8 + 2b_9 + 2b_{10} + b_5 + c_5 + c_8, \tag{8}$$

$$(b_3 c_3) b_9 = (c_3 b_9) b_3, \qquad b_9^2 = \bar{b}_{15} b_3 + \bar{b}_3 b_3 + c_9 b_3;$$

by 12) in Lemma 2.10 and Hypothesis 2.1 and 6) in Lemma 2.10:

$$b_9^2 = 1 + b_5 + c_5 + 2c_8 + 2b_8 + 2b_9 + 2b_{10}. \tag{9}$$

By the associative law and 9) in Lemma 2.10 and 14) in Lemma 2.11:

$$(c_3 \bar{c}_3) b_8 = c_3 (\bar{c}_3 b_8), \qquad b_8 + b_8^2 = c_3 \bar{c}_3 + c_3 \bar{b}_6 + c_3 b_{15};$$

by 7), 9) in Lemma 2.10 and 2) in Lemma 2.12:

$$b_8^2 = 1 + b_5 + c_5 + c_8 + 2b_8 + b_9 + 2b_{10}. \tag{10}$$

By the associative law and 9) in Lemma 2.10 and 15) in Lemma 2.11:

$$(c_3 \bar{c}_3) b_9 = (\bar{c}_3 b_9) c_3, \qquad b_9 + b_8 b_9 = b_{15} c_3 + b_3 c_3 + \bar{c}_9 c_3;$$

by 1), 2) in Lemma 2.12 and 5) in Lemma 2.10:

$$b_9 + b_8 b_9 = b_5 + c_5 + c_8 + b_8 + b_9 + b_{10} + b_9 + b_{10} + b_9 + c_8,$$

$$b_8 b_9 = b_5 + c_5 + 2c_8 + b_8 + 2b_9 + 2b_{10}. \tag{11}$$

By the associative law and 9) in Lemma 2.10, 12) in Lemma 2.11:

$$(c_3 \bar{c}_3) b_{10} = (\bar{c}_3 b_{10}) c_3, \qquad b_{10} + b_8 b_{10} = b_{15} c_3 + \bar{b}_6 c_3 + \bar{c}_9 c_3,$$

so by 1), 2) in Lemma 2.12 and 7) in Lemma 2.10:

$$b_8 b_{10} = b_5 + c_5 + 2c_8 + 2b_8 + 2b_9 + 2b_{10}. \tag{12}$$

By the associative law and Hypothesis 2.1 and 4) in Lemma 2.11, 12) in Lemma 2.10:

$$(b_3 \bar{b}_3) c_5 = (\bar{b}_3 c_5) b_3,$$

$$c_5 + c_5 c_8 = \bar{b}_{15} b_3 = b_{10} + c_8 + b_9 + b_8 + b_5 + c_5,$$

$$c_5 c_8 = b_{10} + c_8 + b_9 + b_8 + b_5. \tag{13}$$

By the associative law and 9) in Lemma 2.10 and 2), 18) in Lemma 2.12:

$$(c_3 \bar{c}_3) c_5 = (\bar{c}_3 c_5) c_3,$$

$$c_5 + c_5 b_8 = b_{15} c_3 = b_5 + c_5 + c_8 + b_8 + b_9 + b_{10},$$

$$c_5 b_8 = b_5 + c_8 + b_8 + b_9 + b_{10}. \tag{14}$$

By the associative law and 5), 12) in Lemma 2.10, 18) in Lemma 2.12:

$$(b_3c_3)c_5 = (c_3c_5)b_3,$$

$$b_9c_5 = \bar{b}_{15}b_3 = b_{10} + c_8 + b_9 + b_8 + b_5 + c_5. \tag{15}$$

By the associative law and 1) in Lemma 2.9 and 10) in Lemma 2.11:

$$(b_3b_6)c_5 = (b_6c_5)b_3,$$

$$c_8c_5 + b_{10}c_5 = \bar{b}_{15}b_3 + b_3b_6 + b_3c_9.$$

By 6), 12) in Lemma 2.10 and (13) and 1) in Lemma 2.9 we get:

$$b_{10} + c_8 + b_9 + b_8 + b_5 + b_{10}c_5 = b_{10} + c_8 + b_9 + b_8 + b_5 + c_5 + c_8 + b_{10}$$
$$+ b_9 + b_{10} + b,$$

$$b_{10}c_5 = c_5 + c_8 + 2b_{10} + b_9 + b_8. \tag{16}$$

By the associative law and 10) in Lemma 2.10, 10) in Lemma 2.11:

$$(b_6\bar{b}_6)c_5 = \bar{b}_6(b_6c_5),$$

$$c_5 + c_5c_8 + b_9c_5 + b_8c_5 + c_5^2 + b_5c_5 = \bar{b}_6\bar{b}_{15} + \bar{b}_6b_6;$$

by (13), (15), (14), (6) and 10), 13) in Lemma 2.10:

$$c_5^2 = 1 + b_9 + b_{10} + c_5. \tag{17}$$

By the associative law and 18) in Lemma 2.12 and (5):

$$c_3b_5^2 = (c_3b_5)b_5,$$

$$c_3 + b_{10}c_3 + b_9c_3 + b_5c_3 = \bar{b}_{15}b_5;$$

so by 12), 15) in Lemma 2.11 and 17) in Lemma 2.12:

$$b_5\bar{b}_{15} = 3\bar{b}_{15} + b_6 + \bar{b}_3 + c_3 + 2c_9. \tag{18}$$

By the associative law and 18) in Lemma 2.12 and (17):

$$(c_5^2)c_3 = (c_5c_3)c_5, \qquad c_3 + b_9c_3 + b_{10}c_3 + c_5c_3 = \bar{b}_{15}c_5;$$

so by 12), 15) in Lemma 2.11 and 18) in Lemma 2.12:

$$\bar{b}_{15}c_5 = 3\bar{b}_{15} + b_6 + \bar{b}_3 + c_3 + 2c_9. \tag{19}$$

By the associative law and 4) in Lemma 2.11 and (15):

$$(b_3c_5)b_9 = b_3(c_5b_9),$$

$$b_{15}b_9 = b_3b_8 + c_8b_3 + b_5b_3 + c_5b_3 + b_9b_3 + b_{10}b_3;$$

by 3), 4) in Lemma 2.9 and 1), 3), 4), 5) in Lemma 2.11:

$$b_{15}b_9 = b_{15} + \bar{c}_9 + b_3 + \bar{b}_6 + b_{15} + 2b_{15} + 2b_{15} + 2\bar{c}_9 + \bar{c}_3 + \bar{b}_6,$$

$$b_{15}b_9 = 6b_{15} + 3\bar{c}_9 + b_3 + 2\bar{b}_6 + \bar{c}_3.$$

By the associative law and 5) in Lemma 2.10 and 4) in Lemma 2.9:

$$b_{10}(b_3c_3) = (b_{10}b_3)c_3,$$

$$b_{10}b_9 = b_{15}c_3 + \bar{b}_6c_3 + \bar{c}_9c_3,$$

$$b_{10}b_9 = b_5 + c_5 + c_8 + b_8 + b_9 + b_{10} + b_8 + c_8 + b_9 + b_{10},$$

$$b_{10}b_9 = b_5 + c_5 + 2c_8 + 2b_8 + 2b_9 + 3b_{10}.$$

$\square$

This completes the investigation of 11) in Lemma 2.9, subcase $(\bar{c}_9b_3, c_9) = 1$. Thus we have the remaining case that $(b_3\bar{c}_9, c_9) = 0$. In the remainder of this section, we deal with the investigation of 11) in Lemma 2.9, subcase $(b_6c_9, \bar{c}_9) \geq 1$.

Lemma 2.14 *Let (A, B) satisfy Hypothesis 2.1. Assume that $\bar{b}_{10} = b_{10} \in B$ and $(b_6c_9, \bar{c}_9) \geq 1$. Then we obtain:*

1) $b_3c_9 = b_{10} + \bar{x} + s, s \in B, |\bar{x} + s| = 17, |\bar{s}| = |s|$;
2) $(c_8b_{15}, \bar{c}_9) = (c_8c_9, \bar{b}_{15}) = 2$;
3) $\bar{b}_6\bar{c}_9 = 3c_9 + \bar{b}_{15} + b_6 + t_6, t_6 \in N^*B$;
4) $c_8b_{10} = 2c_8 + 2b_{10} + 2x + y + z + w + s$;
5) $b_6b_{15} = 3b_{10} + 2c_8 + x + y + z + w + \bar{x} + s$;
6) $\bar{b}_3c_9 = b_{15} + \varepsilon + \theta, \varepsilon, \theta \in B, |\varepsilon + \theta| = 12$;
7) $b_6b_{10} = b_3 + 3\bar{b}_{15} + \bar{\varepsilon} + \bar{\theta}$;
8) $c_8b_{15} = 5b_{15} + 2\bar{b}_6 + 2\bar{c}_9 + b_3 + \varepsilon + \theta$.

Proof By the associative law and Hypothesis 2.1 and 10) in Lemma 2.9:

$$(b_3^2)\bar{b}_{15} = (b_3\bar{b}_{15})b_3,$$

$$\bar{b}_3\bar{b}_{15} + b_6\bar{b}_{15} = b_{10}b_3 + c_8b_3 + b_3(x + y + z + w).$$

So by 5), 4), 3) in Lemma 2.9,

$$b_6\bar{b}_{15} = b_3 + \bar{b}_6 + b_3(x + y + z + w). \tag{1}$$

By the associative law and Hypothesis 2.1 and 4) in Lemma 2.9:

$$(b_3^2)b_{10} = (b_3b_{10})b_3,$$

$$\bar{b}_3b_{10} + b_6b_{10} = b_{15}b_3 + \bar{b}_6b_3 + \bar{c}_9b_3;$$

by 4), 5), 2) in Lemma 2.9:

$$b_6b_{10} = \bar{b}_3 + 2\bar{b}_{15} + b_3\bar{c}_9. \tag{2}$$

By the associative law and (2) and 4) in Lemma 2.9:

$$(b_3 b_{10})\bar{b}_6 = (\bar{b}_6 b_{10})b_3,$$

$$\bar{b}_6^2 + b_{15}\bar{b}_6 + \bar{b}_6\bar{c}_9 = b_3^2 + 2b_{15}b_3 + (\bar{b}_3 b_3)c_9.$$

So by 5), 6) in Lemma 2.9 and Hypothesis 2.1:

$$2b_6 + c_9 + \bar{b}_{15} + \bar{b}_3 + b_6 + \bar{b}_3(x + y + z + w) + \bar{b}_6\bar{c}_9$$

$$= \bar{b}_3 + b_6 + 4\bar{b}_{15} + 2b_6 + 2c_9 + c_9 + c_8 c_9,$$

$$\bar{b}_6\bar{c}_9 + \bar{b}_3(x + y + z + w) = 3\bar{b}_{15} + 2c_9 + c_8 c_9. \tag{3}$$

By the associative law and 6), 7) in Lemma 2.9:

$$(b_6^2)c_8 = (b_6 c_8)b_6,$$

$$2\bar{b}_6 c_8 + \bar{c}_9 c_8 + b_{15}c_8 = \bar{b}_3 b_6 + 2\bar{b}_{15}b_6 + b_6^2 + c_9 b_6;$$

by 7), 2), 6) in Lemma 2.9 and (3), (1)

$$c_8 b_{15} = b_3 + 2\bar{b}_6 + \bar{c}_9 + b_{15} + b_3(x + y + z + w). \tag{4}$$

By the associative law and Hypothesis 2.1:

$$(b_3\bar{b}_3)c_9 = (\bar{b}_3 c_9)b_3, \qquad c_9 + c_9 c_8 = (\bar{b}_3 c_9)b_3,$$

but by 5) in Lemma 2.9,

$$(\bar{b}_3 c_9, b_{15}) = (c_9, b_3 b_{15}) = 1.$$

So

$$c_9 + c_9 c_8 = b_{15}b_3 + \alpha_1 b_3;$$

hence by 5) in Lemma 2.9 we obtain that:

$$c_9 c_8 = 2\bar{b}_{15} + b_6 + \alpha_1 b_3. \tag{5}$$

Hence

$$(c_8 c_9, \bar{b}_{15}) \geq 2. \tag{6}$$

By 10) in Lemma 2.9,

$$(\bar{b}_3 x, \bar{b}_{15}) = 1, \qquad (\bar{b}_3 y, \bar{b}_{15}) = 1, \qquad (\bar{b}_3 z, \bar{b}_{15}) = 1, \qquad (\bar{b}_3 w, \bar{b}_{15}) = 1.$$

So by (3),

$$(\bar{b}_6\bar{c}_9, \bar{b}_{15}) \geq 1. \tag{7}$$

By (6), $(c_8 b_{15}, \bar{c}_9) \geq 2$, so by (4),

$$(b_3(x + y + z + w), \bar{c}_9) \neq 0.$$

Hence one of $b_3 x$, $b_3 y$, $b_3 z$, and $b_3 w$ contains $\bar{c}_9$; thus, without loss of generality, we may assume that

$$(b_3 x, \bar{c}_9) \neq 0. \tag{8}$$

By the associative law and Hypothesis 2.1 and 4) in Lemma 2.9:

$$(b_3 \bar{b}_3) b_{10} = (b_3 b_{10}) \bar{b}_3,$$

$$b_{10} + c_8 b_{10} = b_{15} \bar{b}_3 + \bar{b}_6 \bar{b}_3 + \bar{c}_9 \bar{b}_3;$$

by 10), 1) in Lemma 2.9

$$c_8 b_{10} = 2 c_8 + b_{10} + x + y + z + w + \bar{b}_3 \bar{c}_9. \tag{9}$$

By (8) and 4) in Lemma 2.9, $(b_3 c_9, b_{10}) = (b_3 b_{10}, \bar{c}_9) = 1$, $(b_3 c_9, \bar{x}) \neq 0$; so

$$b_3 c_9 = b_{10} + \bar{x} + \alpha, \quad \alpha \in N^* B. \tag{10}$$

Then $(b_3 x, \bar{c}_9) = (b_3 c_9, \bar{x}) \neq 0$. By (9) since c_8, b_{10}, $x + y + z + w$ are reals then $b_3 c_9$ is real. By 10) in Lemma 2.9, $(b_3 x, b_{15}) = 1$; therefore

$$b_3 x = \bar{c}_9 + b_{15} + \beta, \tag{11}$$

where $\beta \in N^* B$. So $|x| \geq 8$. Hence by (10)

$$3 \leq (b_3 c_9, b_3 c_9) \leq 5.$$

Thus

$$3 \leq (b_3 \bar{b}_3, c_9 \bar{c}_9) \leq 5.$$

So by Hypothesis $2 \leq (c_9 \bar{c}_9, c_8) \leq 4$. Thus

$$2 \leq (c_9, c_9 c_8) \leq 4. \tag{12}$$

If $(c_8 b_{15}, \bar{c}_9) \geq 4$ then by (4), we have that $(b_3(x + y + z + w), \bar{c}_9) \geq 3$. Since $1 + (c_8, b_{15} \bar{b}_{15}) = (1 + c_8, b_{15} \bar{b}_{15}) = (b_3 \bar{b}_3, b_{15} \bar{b}_{15}) = (b_3 b_{15}, b_3 b_{15}) = 6$ by 10) in Lemma 2.9, we have that $(c_8 b_{15}, b_{15}) = 5$. Hence we obtain, without loss of generality, by (11) that

$$b_3 y = \bar{c}_9 + b_{15} + \beta_1, \qquad b_3 z = \bar{c}_9 + b_{15} + \beta_2.$$

So the degrees of x, y and z are all larger or equal to 8, but $|x + y + z + w| = 27$; hence $|x| = |y| = |z| = 8$ by (11) and $(b_3 c_9, x) = (b_3 c_9, y) = (b_3 c_9, z) = 1$, a contradiction to (10). If $(c_8 b_{15}, \bar{c}_9) = 3$. Then by (5) and (12)

$$c_8 c_9 = 3 \bar{b}_{15} + b_6 + 2 c_9 + c_3, \tag{13}$$

where $c_3 \in B$. By (4), $(b_3(x+y+z+w), \bar{c}_9) = 2$, so by (3) $(\bar{b}_6\bar{c}_9, c_9) = 2$ and by 10), 6) in Lemma 2.9, $(\bar{b}_6\bar{c}_9, b_6) = 1$, $(\bar{b}_3(x+y+z+w), \bar{b}_{15}) = 4$, so by (3), (13) $(\bar{b}_6\bar{c}_9, \bar{b}_{15}) = 2$. Hence

$$b_6c_9 = \bar{b}_6 + 2\bar{c}_9 + 2b_{15}. \tag{14}$$

So by (3),

$$\bar{b}_3(x+y+z+w) = 4\bar{b}_{15} + 2c_9 + c_3.$$

Then we obtain, without loss of generality, that the only possibility is

$$x = b_8, \qquad y = b_9, \qquad z = b_5, \qquad w = c_5,$$

when b_8, b_9, b_5, c_5 are reals. So

$$\bar{b}_3b_8 = \bar{b}_{15} + c_9, \tag{15}$$

$$\bar{b}_3b_9 = \bar{b}_{15} + c_9 + c_3, \tag{16}$$

$$\bar{b}_3b_5 = \bar{b}_{15}, \tag{17}$$

$$\bar{b}_3c_5 = \bar{b}_{15}. \tag{18}$$

So $(b_3c_9, b_8) = 1$, and by 4) in Lemma 2.9, $(b_3c_9, b_{10}) = 1$. Thus

$$b_3c_9 = b_{10} + b_8 + b_9. \tag{19}$$

So by the associative law and Hypothesis 2.1

$$(b_3^2)c_9 = (b_3c_9)b_3,$$

$$c_9\bar{b}_3 + c_9b_6 = b_{10}b_3 + b_8b_3 + b_9b_3.$$

Therefore by (15), (16), (14) $c_9\bar{b}_3 = b_{15} + \bar{c}_9 + \bar{c}_3$ which contradicts the assumption in the lemma, we conclude that $(c_8b_{15}, \bar{c}_9) \le 2$ and by (8) and (4),

$$(c_8b_{15}, \bar{c}_9) = 2. \tag{20}$$

By (4)

$$(b_3(x+y+z+w), c_9) = 1, \tag{21}$$

and by (3), (12) we obtain $3 \le (\bar{b}_6\bar{c}_9, c_9) \le 5$, by (20) $(c_8c_9, \bar{b}_{15}) = 2$. So by (3) and that $(\bar{b}_3(x+y+z+w), \bar{b}_{15}) = 4$ we get $(\bar{b}_6\bar{c}_9, \bar{b}_{15}) = 1$. And by 6) in Lemma 2.9, $(\bar{b}_6\bar{c}_9, b_6) = 1$. So $(\bar{b}_6\bar{c}_9, c_9) = 3$. Then by (3), (21), $(c_8c_9, c_9) = 2$. So $(c_8, \bar{c}_9c_9) = 2$, and by Hypothesis 2.1 $(\bar{c}_9c_9, \bar{b}_3b_3) = 3$. Thus $(b_3c_9, b_3c_9) = 3$. So by 4) in Lemma 2.9 and (10), we obtain that

$$b_3c_9 = b_{10} + \bar{x} + s, \tag{22}$$

$s \in B$. Also

$$\bar{b}_6\bar{c}_9 = 3c_9 + \bar{b}_{15} + b_6 + t_6, \tag{23}$$

where $t_6 \in N^*B$, $t_6 \neq b_6, \bar{b}_6$. By (9) and (22),

$$c_8 b_{10} = 2c_8 + 2b_{10} + 2x + y + z + w + \bar{S}. \tag{24}$$

By the associative law and 5) in Lemma 2.9:

$$(b_3^2)b_{15} = (b_3 b_{15})b_3,$$
$$\bar{b}_3 b_{15} + b_6 b_{15} = 2\bar{b}_{15}b_3 + b_6 b_3 + c_9 b_3;$$

by (22) and 1), 10) in Lemma 2.9:

$$b_6 b_{15} = 3b_{10} + 2c_8 + x + y + z + w + \bar{x} + S. \tag{25}$$

By the associative law and Hypothesis 2.1 and 5) in Lemma 2.9:
Since $b_{15}(b_3\bar{b}_3) = \bar{b}_3(b_{15}b_3)$, then

$$c_8 b_{15} = 4b_{15} + 2\bar{b}_6 + 2\bar{c}_9 + b_3 + \bar{b}_3 c_9. \tag{26}$$

By (22) and 5) in Lemma 2.9:

$$\bar{b}_3 c_9 = b_{15} + \varepsilon + \theta, \tag{27}$$

where $\varepsilon, \theta \in B$, $|\varepsilon + \theta| = 12$, so by (26)

$$c_8 b_{15} = 5b_{15} + 2\bar{b}_6 + 2\bar{c}_9 + b_3 + \varepsilon + \theta.$$

By the associative law and Hypothesis 2.1 and 4) in Lemma 2.9:

$$(b_3^2)b_{10} = b_3(b_3 b_{10}),$$
$$b_{10}\bar{b}_3 + b_{10}b_6 = b_{15}b_3 + \bar{b}_6 b_3 + \bar{c}_9 b_3;$$

so by (27) and 5), 2), 4) in Lemma 2.9:

$$b_6 b_{10} = \bar{b}_3 + 3\bar{b}_{15} + \bar{\varepsilon} + \bar{\theta}.$$

□

Lemma 2.15 *Let (A, B) satisfy Hypothesis 2.1. Assume that $\bar{b}_{10} = b_{10} \in B$ and $(b_3\bar{c}_9, c_9) = 0$. Then $(b_6 c_9, \bar{c}_9) \geq 1$ is impossible.*

Proof We assume that our assumption is possible. By Hypothesis 2.1 and 10) in Lemma 2.9 and the associative law:

$$(b_3\bar{b}_3)b_{15} = b_3(\bar{b}_3 b_{15}),$$
$$b_{15} + c_8 b_{15} = b_3 b_{10} + b_3 c_8 + b_3\bar{x} + b_3\bar{y} + b_3\bar{z} + b_3\bar{w}.$$

By 3), 4) in Lemma 2.9, 8) in Lemma 2.14:

$$b_{15} + 5b_{15} + 2\bar{b}_6 + 2\bar{c}_9 + b_3 + \varepsilon + \theta$$

$$= b_{15} + \bar{b}_6 + \bar{c}_9 + b_3 + \bar{b}_6 + b_{15} + b_3\bar{x} + b_3\bar{y} + b_3\bar{z} + b_3\bar{w},$$

$$4b_{15} + \bar{c}_9 + \varepsilon + \theta = b_3\bar{x} + b_3\bar{y} + b_3\bar{z} + b_3\bar{w}. \tag{1}$$

By 10) in Lemma 2.9, and 1) in Lemma 2.14:

$$(b_3\bar{x}, b_{15}) = (\bar{b}_{15}b_3, x) = 1, \qquad (b_3\bar{y}, b_{15}) = (b_3\bar{b}_{15}, y) = 1,$$

$$(b_3\bar{z}, b_{15}) = (\bar{b}_{15}b_3, z) = 1, \qquad (b_3\bar{w}, b_{15}) = (b_3\bar{b}_{15}, w) = 1,$$

$(b_3\bar{x}, \bar{c}_9) = (b_3c_9, x) = 1$, so $x = \bar{x}$. If otherwise two of x, y, z, w have degrees ≥ 8, this is impossible. Then

$$b_3x = \bar{c}_9 + b_{15}, \tag{2}$$

and then $x = x_8$. So $|\bar{y} + \bar{z} + \bar{w}| = 19$, and also we get $5 \leq |\bar{y}|, |\bar{z}|, |\bar{w}| \leq 9$ and also, without loss of generality, $|\bar{y}| = 5$, hence $\bar{y} = \bar{y}_5$, so

$$b_3\bar{y}_5 = b_{15}. \tag{3}$$

So

$$|\bar{z} + \bar{w}| = 14. \tag{4}$$

By 1) in Lemma 2.14 and the fact that $x = x_8$, we obtain that

$$b_3c_9 = b_{10} + x_8 + s_9. \tag{5}$$

So by the associative law and Hypothesis 2.1 and (5):

$$(b_3^2)c_9 = b_3(b_3c_9),$$

$$c_9\bar{b}_3 + b_6c_9 = b_3b_{10} + b_3x_8 + s_9b_3.$$

Then by (2) and (6), 3) in Lemma 2.14, 4) in Lemma 2.9:

$$b_{15} + \varepsilon + \theta + 3\bar{c}_9 + b_{15} + \bar{b}_6 + \bar{t}_6 = b_{15} + \bar{b}_6 + \bar{c}_9 + b_{15} + \bar{c}_9 + s_9b_3,$$

$$s_9b_3 = \varepsilon + \theta + \bar{c}_9 + \bar{t}_6. \tag{6}$$

If $|z| = 5$, then by (4) $|w| = 9$, hence $z = \bar{z}_5$, $w = \bar{w}_9$. So

$$b_3z_5 = b_{15},$$

$$b_3w_9 = b_{15} + \varepsilon + \theta. \tag{7}$$

So by (6)

$$(\bar{b}_3 \varepsilon, s_9) = (\varepsilon, b_3 s_9) = 1, \qquad (\bar{b}_3 \theta, s_9) = (\theta, b_3 s_9) = 1,$$
$$(\bar{b}_3 \varepsilon, w_9) = (\varepsilon, b_3 w_9) = 1, \qquad (\bar{b}_3 \theta, w_9) = (\theta, b_3 w_9) = 1.$$

Hence

$$\bar{b}_3 \varepsilon = s_9 + w_9, \tag{8}$$
$$\bar{b}_3 \theta = s_9 + w_9, \tag{9}$$

and hence $\varepsilon = \varepsilon_6$, $\theta = \theta_6$, by 4) in the Lemma 2.14 $s_9 = \bar{s}_9$. So by the associative law and Hypothesis 2.1:

$$\bar{\varepsilon}_6 (b_3^2) = b_3 (\bar{\varepsilon}_6 b_3),$$
$$\bar{\varepsilon}_6 \bar{b}_3 + \bar{\varepsilon}_6 b_6 = s_9 b_3 + w_9 b_3;$$

so by (6), (7)

$$\bar{\varepsilon}_6 \bar{b}_3 + \bar{\varepsilon}_6 b_6 = \varepsilon_6 + \theta_6 + \bar{c}_9 + \bar{t}_6 + b_{15} + \varepsilon_6 + \theta_6,$$
$$\bar{\varepsilon}_6 \bar{b}_3 + \bar{\varepsilon}_6 b_6 = 2\varepsilon_6 + 2\theta_6 + \bar{c}_9 + \bar{t}_6 + b_{15}.$$

By 6) in Lemma 2.14 $(\bar{\varepsilon}_6 \bar{b}_3, \bar{c}_9) = (\varepsilon_6, \bar{b}_3 c_9) = 1$ and $(\bar{\varepsilon}_6 \bar{b}_3, \delta) = 1$, where $\delta = \varepsilon_6$ or θ_6, so $\bar{\varepsilon}_6 \bar{b}_3 = \delta + \bar{c}_9 + m_3$, so $\bar{t}_6 = m_3 + k_3$, and $(b_3 m_3, \bar{\varepsilon}_6) = (\bar{b}_3 \bar{\varepsilon}_6, m_3) = 1$. Therefore

$$(b_3 m_3, b_3 m_3) = (b_3 \bar{b}_3, m_3 \bar{m}_3) > 1.$$

In (6) we get

$$(\bar{b}_3 m_3, s_9) = (m_3, b_3 s_9) = 1,$$

then

$$(\bar{b}_3 m_3, \bar{b}_3 m_3) = (\bar{b}_3 b_3, \bar{m}_3 m_3) = 1,$$

a contradiction.

If $|\bar{z}| = 6$ then by (4) $|\bar{w}| = 8$, hence $z = \bar{z}$, $w = \bar{w}$. So

$$b_3 z_6 = b_{15} + \varepsilon_3, \, b_3 w_8 = b_{15} + \theta_9.$$

So $(\varepsilon_3 \bar{b}_3, z_6) = (\varepsilon_3, b_3 z_6) = 1$, hence $(\varepsilon_3 \bar{b}_3, \varepsilon_3 \bar{b}_3) > 1$. By (6), $(s_9 b_3, \varepsilon_3) = (s_9, \bar{b}_3 z_6) = 1$, hence $\bar{b}_3 \varepsilon_3 = s_9$, so $(\varepsilon_3 \bar{b}_3, \varepsilon_3 \bar{b}_3) = 1$, a contradiction. If $|\bar{z}| = 7$ then by (4) $|\bar{w}| = 7$, so

$$z = \bar{z}, \qquad w = \bar{w} \quad \text{or} \quad \bar{z} = w,$$
$$b_3 \bar{z}_7 = b_{15} + \varepsilon_6, \tag{10}$$
$$b_3 \bar{w}_7 = b_{15} + \theta_6. \tag{11}$$

So $(\bar{b}_3 \varepsilon_6, \bar{z}_7) = (\varepsilon_6, b_3 \bar{z}_7) = 1$. By (6) $(\bar{b}_3 \varepsilon_6, s_9) = (\varepsilon_6, b_3 s_9) = 1$, hence

$$\bar{b}_3 \varepsilon_6 = \bar{z}_7 + s_9 + m_2,$$

a contradiction. □

Consequently, 11) in Lemma 2.9 subcase $(b_6c_9, \bar{c}_9) \geq 1$ is impossible and we have only the case $(b_3\bar{c}_9, c_9) = 1$. We can now state Theorem 2.7.

Theorem 2.7 *Let* (A, B) *be a NITA generated by a nonreal element* $b_3 \in B$ *satisfying* $b_3\bar{b}_3 = 1 + c_8$ *and* $b_3^2 = \bar{b}_3 + b_6$, *where* $b_6 \in B$ *is nonreal and satisfying Hypothesis 2.1. Then* $b_3b_6 = c_8 + b_{10}$ *and if* $b_{10} = \bar{b}_{10}$ *then* $(A, B) \cong (CH(3A_6), Irr(3A_6))$. (A, B) *is a Table Algebra of dimension 17:* $B = \{1, b_3, \bar{b}_3, c_3, \bar{c}_3, b_6, \bar{b}_6, c_9, \bar{c}_9, b_{15},$ $\bar{b}_{15}, b_5, c_5, b_8, c_8, b_9, b_{10}\}$ *and* B *has an increasing series of table subsets* $\{b_1\} \subseteq$ $\{b_1, b_5, c_5, b_8, c_8, b_9, b_{10}\} \subseteq B$ *defined by*

1) $b_3\bar{b}_3 = 1 + c_8$;
2) $b_3^2 = \bar{b}_3 + b_6$;
3) $b_3c_3 = b_9$;
4) $b_3\bar{c}_3 = c_9$;
5) $b_3b_6 = c_8 + b_{10}$;
6) $b_3\bar{b}_6 = \bar{b}_3 + \bar{b}_{15}$;
7) $b_3c_9 = b_9 + b_{10} + b_8$;
8) $b_3\bar{c}_9 = \bar{b}_{15} + c_9 + c_3$;
9) $b_3b_{15} = 2\bar{b}_{15} + b_6 + c_9$;
10) $b_3\bar{b}_{15} = b_{10} + c_8 + b_9 + b_8 + b_5 + c_5$;
11) $b_3b_5 = b_{15}$;
12) $b_3c_5 = b_{15}$;
13) $b_3b_8 = b_{15} + \bar{c}_9$;
14) $b_3c_8 = b_3 + \bar{b}_6 + b_{15}$;
15) $b_3b_9 = b_{15} + \bar{c}_9 + \bar{c}_3$;
16) $b_3b_{10} = b_{15} + \bar{b}_6 + \bar{c}_9$;
18) $c_3^2 = \bar{c}_3 + \bar{b}_6$;
19) $c_3b_6 = \bar{c}_3 + b_{15}$;
20) $c_3\bar{b}_6 = b_{10} + b_8$;
21) $c_3c_9 = b_3 + b_{15} + \bar{c}_9$;
22) $c_3\bar{c}_9 = c_8 + b_9 + b_{10}$;
23) $c_3b_{15} = b_5 + c_5 + c_8 + b_8 + b_9 + b_{10}$;
24) $c_3\bar{b}_{15} = 2b_{15} + \bar{b}_6 + \bar{c}_9$;
25) $c_3b_5 = \bar{b}_{15}$;
26) $c_3c_5 = \bar{b}_{15}$;
27) $c_3b_8 = c_3 + b_6 + \bar{b}_{15}$;
28) $c_3c_8 = c_9 + \bar{b}_{15}$;
29) $c_3b_9 = b_{15} + \bar{b}_3 + c_9$;
30) $c_3b_{10} = \bar{b}_{15} + b_6 + c_9$;
31) $b_6\bar{b}_6 = 1 + c_8 + b_9 + b_8 + c_5 + b_5$;
32) $b_6c_9 = 2b_{15} + 2\bar{c}_9 + \bar{b}_6$;
33) $b_6\bar{c}_9 = b_{10} + c_8 + 2b_9 + b_8 + b_5 + c_5$;
34) $b_6b_{15} = 2c_8 + 3b_{10} + 2b_9 + 2b_8 + b_5 + c_5$;
35) $b_6\bar{b}_{15} = 4b_{15} + \bar{b}_6 + b_3 + \bar{c}_3 + 2\bar{c}_9$;
36) $b_6b_5 = \bar{b}_{15} + b_6 + c_9$;
37) $b_6c_5 = \bar{b}_{15} + b_6 + c_9$;

38) $b_6 b_8 = 2\bar{b}_{15} + b_6 + c_3 + c_9;$

39) $b_6 c_8 = \bar{b}_3 + 2\bar{b}_{15} + b_6 + c_9;$

40) $b_6 b_9 = b_6 + 2c_9 + 2\bar{b}_{15};$

41) $b_6 b_{10} = 3\bar{b}_{15} + \bar{b}_3 + c_3 + c_9;$

42) $b_6^2 = 2\bar{b}_6 + \bar{c}_9 + b_{15};$

43) $c_9 \bar{c}_9 = 1 + 2c_8 + 2b_9 + 2b_{10} + c_5 + b_5 + 2b_8;$

44) $c_9 b_{15} = 2b_5 + 2c_5 + 3c_8 + 3b_8 + 3b_9 + 4b_{10};$

45) $c_9 \bar{b}_{15} = 6b_{15} + 2\bar{b}_6 + b_3 + \bar{c}_3 + 3\bar{c}_9;$

46) $c_9 b_5 = 2\bar{b}_{15} + b_6 + c_9;$

47) $c_9 c_5 = 2\bar{b}_{15} + b_6 + c_9;$

48) $c_9 b_8 = 3\bar{b}_{15} + \bar{b}_3 + b_6 + 2c_9;$

49) $c_9 c_8 = 3\bar{b}_{15} + c_3 + b_6 + 2c_9;$

50) $c_9 b_9 = \bar{b}_3 + 2c_9 + 3\bar{b}_{15} + c_3 + 2b_6;$

51) $c_9 b_{10} = \bar{b}_3 + 2c_9 + 4\bar{b}_{15} + c_3 + b_6;$

52) $c_9^2 = 3b_{15} + b_3 + \bar{c}_3 + 2\bar{b}_6 + 2\bar{c}_9;$

53) $b_{15} \bar{b}_{15} = 1 + 3b_5 + 3c_5 + 5c_8 + 5b_8 + 6b_9 + 6b_{10};$

54) $b_{15}^2 = 9\bar{b}_{15} + 4b_6 + 2\bar{b}_3 + 2c_3 + 6c_9;$

55) $b_{15} b_5 = 3b_{15} + \bar{b}_6 + b_3 + \bar{c}_3 + 2\bar{c}_9;$

56) $b_{15} c_5 = 3b_{15} + \bar{b}_6 + b_3 + \bar{c}_3 + 2\bar{c}_9;$

57) $b_{15} b_8 = 5b_{15} + b_3 + \bar{c}_3 + 2\bar{b}_6 + 3\bar{c}_9;$

58) $b_{15} c_8 = 5b_{15} + b_3 + \bar{c}_3 + 2\bar{b}_6 + 3\bar{c}_9;$

59) $b_{15} b_9 = 6b_{15} + 3\bar{c}_9 + b_3 + 2\bar{b}_6 + \bar{c}_3;$

60) $b_{15} b_{10} = 6b_{15} + 3\bar{b}_6 + + b_3 + \bar{c}_3 + 4\bar{c}_9;$

61) $b_5^2 = 1 + b_{10} + b_9 + b_5;$

62) $b_5 c_5 = b_8 + c_8 + b_9;$

63) $b_5 b_8 = c_5 + c_8 + b_8 + b_9 + b_{10};$

64) $b_5 c_8 = c_5 + b_8 + b_9 + b_{10} + c_8;$

65) $b_5 b_9 = c_8 + c_5 + b_8 + b_5 + b_9 + b_{10};$

66) $b_5 b_{10} = c_8 + b_8 + b_9 + b_5 + 2b_{10};$

67) $c_5^2 = 1 + b_9 + b_{10} + c_5;$

68) $c_5 b_8 = b_5 + c_8 + b_8 + b_9 + b_{10};$

69) $c_5 c_8 = b_5 + c_8 + b_8 + b_9 + b_{10};$

70) $c_5 b_9 = b_8 + c_8 + b_5 + c_5 + b_9 + b_{10};$

71) $c_5 b_{10} = b_8 + c_8 + b_9 + c_5 + 2b_{10};$

72) $b_8^2 = 1 + b_5 + c_5 + c_8 + 2b_8 + b_9 + 2b_{10};$

73) $b_8 c_8 = b_5 + c_5 + c_8 + b_8 + 2b_9 + 2b_{10};$

74) $b_8 b_9 = b_5 + c_5 + 2c_8 + b_8 + 2b_9 + 2b_{10};$

75) $b_8 b_{10} = b_5 + c_5 + 2c_8 + 2b_8 + 2b_9 + 2b_{10};$

76) $c_8^2 = 1 + 2c_8 + 2b_{10} + b_9 + b_8 + b_5 + c_5;$

77) $c_8 b_9 = b_5 + c_5 + c_8 + 2b_8 + 2b_9 + 2b_{10};$

78) $c_8 b_{10} = 2c_8 + 2b_{10} + 2b_9 + 2b_8 + b_5 + c_5;$

79) $b_9 b_{10} = b_5 + c_5 + 2c_8 + 2b_8 + 2b_9 + 3b_{10};$

80) $b_9^2 = 1 + b_5 + c_5 + 2c_8 + 2b_8 + 2b_9 + 2b_{10};$

81) $b_{10}^2 = 1 + 2c_8 + 2b_{10} + 3b_9 + 2b_8 + 2b_5 + 2c_5.$

Proof The theorem follows by Lemmas 2.9–2.15.   □

In this section, we do not have information about what NITA satisfying b_{10} nonreal. But we conjecture:

Conjecture 2.1 *There exists no NITA satisfying Hypothesis 2.1 and $b_3 b_6 = c_8 + b_{10}$, where b_{10} is nonreal.*

2.6 NITA Generated by b_3 Satisfying $b_3^2 = c_3 + b_6$, $c_3 \neq b_3$, $\bar{b}_3$, b_6 Non-real, $(b_3 b_8, b_3 b_8) = 4$ and $c_3^2 = r_3 + s_6$

Now we discuss NITA satisfying the following hypothesis.

Hypothesis 2.2 *Let (A, B) be a NITA generated by a nonreal element $b_3 \in B$ satisfying $b_3^2 = c_3 + b_6$, $c_3 \neq b_3$, $\bar{b}_3$, b_6 nonreal, $(b_3 b_8, b_3 b_8) = 4$ and $c_3^2 = r_3 + s_6$. Here we still assume that $L(B) = 1$ and $L_2(B) = \emptyset$.*

In this section, we are not able to classify NITAs satisfying Hypothesis 2.2, but we have the following theorem.

Theorem 2.8 *There exists no NITA satisfying Hypothesis 2.2 and c_3 generates a sub-NITA isomorphic to one of the four known NITAs in the Main Theorem 1.*

Proof By assumptions we have that (1) $b_3 \bar{b}_3 = 1 + b_8 = c_3 \bar{c}_3$; (2) $b_3^2 = c_3 + b_6$; (3) $(b_3 b_8, b_3 b_8) = 4$ and (4) $c_3^2 = r_3 + s_6$.

Step 1 There exists no NITA satisfying hypotheses (1) to (4) and containing a NITA strictly isomorphic to $(Ch(PSL(2, 7)), Irr(PSL(2, 7)))$.

By [CA], $(Ch(PSL(2, 7)), Irr(PSL(2, 7)))$ the sub-NITA generated by c_3 which is of dimension 6 and has base elements: $1, c_3, \bar{c}_3, s_6, b_7, b_8$. Moreover, we have the following equations:

(a1) $c_3 b_8 = c_3 + s_6 + b_7 + b_8$,
(a2) $s_6^2 = 1 + 2s_6 + b_7 + b_8$,
(a3) $c_3 s_6 = \bar{c}_3 + b_7 + b_8$.

It follows by (1) that $(b_3 \bar{c}_3, b_3 \bar{c}_3) = (b_3 \bar{b}_3, c_3 \bar{c}_3) = 2$, hence we obtain equation (a4): $b_3 \bar{c}_3 = \bar{b}_3 + t_6$, where $t_6 \in B$. By the associative law and equations (a1) and (a4), we have

$$(b_3 \bar{b}_3) \bar{c}_3 = \bar{c}_3 + \bar{c}_3 b_8 = \bar{c}_3 + \bar{c}_3 + s_6 + b_7 + b_8,$$

$$(b_3 \bar{c}_3) \bar{b}_3 = \bar{b}_3^2 + \bar{b}_3 t_6 = \bar{c}_3 + \bar{b}_6 + \bar{b}_3 t_6.$$

Then we get $\bar{b}_6 = s_6$, so b_6 is real. Moreover we have equation (a5): $\bar{b}_3 t_6 = \bar{c}_3 + s_6 + b_7 + b_8$. By $(b_3 \bar{b}_3) b_3 = b_3^2 \bar{b}_3$, we arrive at $b_3 b_8 = \bar{t}_6 + \bar{b}_3 s_6$. Now by

(3) we obtain equation (a6): $\bar{b}_3 s_6 = b_3 + x + y$, where distinct b_3, x and y such that $x + y$ is of degree 15. Hence we have equation (a7): $b_3 b_8 = b_3 + \bar{t}_6 + x + y$. Therefore by (a2), (a3), (a5) and (a6)

$$b_3^2 s_6 = (c_3 + s_6)s_6 = c_3 s_6 + s_6 s_6 = \bar{c}_3 + b_7 + b_8 + 1 + 2s_6 + b_7 + b_8,$$

$$b_3(b_3 s_6) = b_3(\bar{b}_3 + \bar{x} + \bar{y}) = 1 + b_8 + b_3 \bar{x} + b_3 \bar{y}.$$

So $b_3 \bar{x} + b_3 \bar{y} = \bar{c}_3 + 2s_6 + 2b_7 + b_8$, but by (a7), the left hand side of this equation contains two b_8, and the right side only one, a contradiction.

Step 2 There exists no NITA satisfying hypotheses (1) to (4) and containing a sub-NITA generated by c_3 which is a NITA of dimension 17, 32 or 22.

By [CA], we see that for the NITAs of dimensions 17, 32 and 22, one can find new base elements: b_5, c_5, s_6, x_8, b_9 and b_{10} such that the following equations hold:

(a1) $c_3 b_8 = c_3 + \bar{s}_6 + b_{15}$,
(a2) $s_6 \bar{s}_6 = 1 + b_5 + c_5 + b_8 + x_8 + b_9$,
(a3) $c_3 s_6 = b_8 + b_{10}$.

Suppose $b_3 \bar{c}_3 = \bar{b}_3 + t_6$, where $t_6 \in B$. Then by hypothesis and equation (a1) it follows that $(b_3 \bar{b}_3)\bar{c}_3 = \bar{c}_3 + \bar{c}_3 b_8 = \bar{c}_3 + \bar{c}_3 + s_6 + \bar{b}_{15}$, $(b_3 \bar{c}_3)\bar{b}_3 = \bar{b}_3^2 + \bar{b}_3 t_6 = \bar{c}_3 + \bar{b}_6 + \bar{b}_3 t_6$. We get $s_6 = \bar{b}_6$ and equation (a4): $\bar{b}_3 t_6 = \bar{c}_3 + \bar{b}_{15}$. Hence $b_3 + b_3 b_8 = (b_3 \bar{b}_3)b_3 = b_3^2 \bar{b}_3 = \bar{b}_3 c_3 + b_3 \bar{s}_6$, and we obtain the equation (a5): $b_3 b_8 = \bar{t}_6 + b_3 \bar{s}_6$. Now by (3) we obtain equation (a6): $b_3 \bar{s}_6 = b_3 + x + y$, where distinct b_3, x and y such that $x + y$ is of degree 15. Hence we have equation (a7): $b_3 b_8 = b_3 + \bar{t}_6 + x + y$. Therefore

$$b_3^2 s_6 = (c_3 + \bar{s}_6)s_6 = c_3 s_6 + s_6 \bar{s}_6 = b_8 + b_{10} + 1 + b_5 + c_5 + b_8 + x_8 + b_9,$$

$$b_3(b_3 s_6) = b_3(\bar{b}_3 + \bar{x} + \bar{y}) = 1 + b_8 + b_3 \bar{x} + b_3 \bar{y}.$$

So $b_3 \bar{x} + b_3 \bar{y} = 1 + b_5 + c_5 + b_8 + x_8 + b_9$, which implies $x_8 = b_8$ by (a7), a contradiction. This concludes the proof. □

2.7 Structure of NITA Generated by b_3 and Satisfying $b_3^2 = c_3 + b_6$, $c_3 \neq b_3$, $\bar{b}_3$, $(b_3 b_8, b_3 b_8) = 3$ and c_3 Non-real

In this section we classify NITA generated by b_3 and satisfying $b_3^2 = c_3 + b_6$, $c_3 \neq b_3$, $\bar{b}_3$ and c_3 nonreal and $(b_3 b_8, b_3 b_8) = 3$. First, we state Hypothesis 2.3.

Hypothesis 2.3 *Let (A, B) be a NITA generated by a nonreal element $b_3 \in B$ such that $b_3 \bar{b}_3 = 1 + b_8$ and $b_3^2 = c_3 + b_6$, c_3, $b_6 \in B$, $c_3 \neq b_3$, $c_3 \neq \bar{c}_3$ and $(b_3 b_8, b_3 b_8) = 3$.*

Lemma 2.16 *Let (A, B) be a NITA satisfying Hypothesis 2.3. Then $c_3\bar{c}_3 = b_3\bar{b}_3 = 1 + b_8$ and $b_3b_8 = b_3 + x_6 + y_{15}$ and $c_3b_8 = c_3 + y_6 + z_{15}$, where $b_8, x_6, y_{15} \in B$, and there exist $r_3, s_6, u_3, v_6, w_3, z_6 \in B$ such that*

$$b_3^2 = r_3 + s_6, \qquad \bar{b}_3 r_3 = b_3 + x_6, \qquad \bar{b}_3 s_6 = b_3 + y_{15},$$

$$b_3 c_3 = u_3 + v_6, \qquad \bar{b}_3 u_3 = c_3 + y_6, \qquad \bar{b}_3 v_6 = c_3 + z_{15},$$

$$\bar{b}_3 c_3 = w_3 + z_6, \qquad b_3 w_3 = c_3 + y_6, \qquad b_3 z_6 = c_3 + z_{15}.$$

Proof It is obvious that $(b_3^2, b_3^2) = 2$. So $b_3^2 = r_3 + s_6$ or $r_4 + s_5$. Hence

$$b_3^2 \bar{b}_3 = \bar{b}_3 r_3 + \bar{b}_3 s_6 \text{ or } \bar{b}_3 r_4 + \bar{b}_3 s_6,$$

$$(b_3 \bar{b}_3) b_3 = b_3 + b_3 b_8$$

$$= 2b_3 + x_6 + y_{15}.$$

Since $b_3 \in Supp\{\bar{b}_3 r_3\}$ and $Supp\{\bar{b}_3 s_6\}$ or $b_3 \in Supp\{\bar{b}_3 r_4\}$ and $Supp\{\bar{b}_3 s_5\}$, the first part of the lemma follows.

Since $b_3 \bar{b}_3 = c_3 \bar{c}_3 = 1 + b_8$, it follows that $b_3 c_3 = u_3 + v_6$ or $u_4 + v_5$ and $\bar{b}_3 c_3 = w_3 + z_6$ or $w_4 + z_5$. The second and third parts of the lemma follow from $(b_3 \bar{b}_3) c_3 = (b_3 c_3) \bar{b}_3 = (\bar{b}_3 c_3) b_3$. □

Lemma 2.17 *Let (A, B) satisfy Hypothesis 2.3. Then there are nonreal base elements x_6, y_{15}, x_{15} and real basis elements $r_3, s_6, d_9, x_{10}, t_{15}$ such that the following equations hold:*

$$
\begin{array}{ll}
b_3\bar{b}_3 = 1 + b_8, & b_3^2 = c_3 + b_6, \\
b_3\bar{c}_3 = \bar{b}_3 + \bar{x}_6, & b_3 x_6 = c_3 + y_{15}, \\
b_3 b_8 = b_3 + x_6 + x_{15}, & b_3\bar{x}_6 = b_8 + x_{10}, \\
b_3\bar{b}_6 = \bar{b}_3 + \bar{x}_{15}, & b_3 c_6 = \bar{c}_3 + \bar{y}_{15}, \\
b_3 c_3 = r_3 + s_6, & b_3 r_3 = \bar{c}_3 + \bar{b}_6, \\
b_3 s_6 = \bar{c}_3 + \bar{y}_{15}, & b_3 x_{10} = x_6 + x_{15} + b_9, \\
b_3\bar{y}_{15} = \bar{x}_6 + 2\bar{x}_{15} + \bar{b}_9, & b_3 y_{15} = s_6 + 2t_{15} + d_9, \\
b_3 b_6 = r_3 + t_{15}, & b_3 x_{15} = b_6 + 2y_{15} + c_9, \\
b_3 t_{15} = \bar{b}_6 + \bar{c}_9 + 2\bar{y}_{15}, &
\end{array}
$$

$$
\begin{array}{ll}
c_3\bar{c}_3 = 1 + b_8, & c_3 r_3 = \bar{b}_3 + \bar{x}_6, \\
c_3 s_6 = \bar{b}_3 + \bar{x}_{15}, & c_3^2 = \bar{c}_3 + \bar{b}_6, \\
c_3 b_8 = c_3 + b_6 + y_{15}, & c_3\bar{x}_6 = b_3 + x_{15}, \\
c_3 x_6 = r_3 + t_{15}, & c_3 x_{10} = b_6 + y_{15} + c_9, \\
c_3 y_{15} = \bar{b}_6 + \bar{c}_9 + 2\bar{y}_{15}, & c_3 b_6 = \bar{c}_3 + \bar{y}_{15}, \\
c_3 t_{15} = \bar{x}_6 + \bar{b}_9 + 2\bar{x}_{15}, & c_3 x_{15} = 2t_{15} + s_6 + d_9, \\
c_3\bar{x}_{15} = 2x_{15} + x_6 + b_9, & c_3 b_6 = b_8 + x_{10},
\end{array}
$$

$$r_3\bar{c}_3 = c_3 + x_6,$$
$$r_3\bar{x}_6 = c_3 + y_{15},$$
$$r_3b_6 = \bar{x}_{15} + \bar{b}_3,$$
$$r_3\bar{x}_{15} = b_6 + c_9 + 2y_{15},$$
$$r_3x_{10} = s_6 + t_{15} + d_9,$$

$$r_3\bar{b}_3 = c_3 + b_6,$$
$$r_3y_{15} = \bar{x}_6 + 2\bar{x}_{15} + \bar{b}_9,$$
$$r_3b_8 = s_6 + r_3 + t_{15},$$

$$x_6^2 = 2b_6 + y_{15} + c_9,$$

$$x_6b_8 = b_3 + x_6 + 2x_{15} + b_9,$$

$$b_6^2 = 2\bar{b}_6 + \bar{y}_{15} + \bar{c}_9,$$
$$b_6x_6 = 2s_6 + t_{15} + d_9,$$

$$b_6\bar{x}_6 = 2x_6 + b_9 + x_{15},$$
$$b_6b_8 = c_3 + b_6 + c_9 + 2y_{15},$$

$$s_6\bar{b}_6 = 2x_6 + x_{15} + b_9,$$
$$s_6\bar{c}_3 = b_3 + x_{15},$$

$$s_6b_8 = r_3 + s_6 + 2t_{15} + d_9,$$

$$b_8y_{15} = c_3 + 2b_6 + 2c_9 + 4y_{15} + \bar{b}_3d_9.$$

Proof By Hypothesis 2.3 $(b_3b_8, b_3b_8) = 3$. Then the following equations hold by Lemma 2.7:

$$b_3\bar{b}_3 = 1 + b_8, \qquad c_3\bar{c}_3 = 1 + b_8,$$
$$b_3^2 = c_3 + b_6, \qquad \bar{b}_3c_3 = b_3 + x_6.$$

Let $b_3x_6 = c_3 + y_{15}$, then

$$(b_3\bar{b}_3)c_3 = c_3 + c_3b_8,$$
$$(\bar{b}_3c_3)b_3 = b_3^2 + b_3x_6$$
$$= c_3 + b_6 + c_3 + y_{15}.$$

Hence

$$c_3b_8 = c_3 + b_6 + y_{15}.$$

Therefore $y_{15} \in B$ by Lemma 2.7.

Let $c_3\bar{x}_6 = b_3 + x_{15}$, $x_{15} \in N^*B$. Since $(c_3\bar{c}_3)b_3 = (b_3\bar{c}_3)c_3$, it follows that $x_{15} \in B$ and

$$b_3b_8 = b_3 + x_6 + x_{15}$$

by $b_3^2\bar{b}_3 = (b_3\bar{b}_3)b_3$. Since $y_{15} \in B$, so $(\bar{b}_3x_6, \bar{b}_3x_6) = (b_3x_6, b_3x_6) = 2$. Thus

$$\bar{b}_3x_6 = b_8 + x_{10}, \qquad x_{10} \in B.$$

Since $b_3^2\bar{c}_3 = (b_3\bar{c}_3)b_3$ and $b_3^2\bar{b}_3 = (b_3\bar{b}_3)b_3$, one has that

$$c_3\bar{b}_6 = b_8 + x_{10}, \qquad \bar{b}_3b_6 = b_3 + x_{15}.$$

By Lemma 2.16, there exist $w_3, z_6, r_3, s_6 \in B$ such that

$$c_3^2 = w_3 + z_6, \qquad w_3 \bar{c}_3 = c_3 + b_6, \qquad z_6 \bar{c}_3 = c_3 + y_{15},$$

$$b_3 c_3 = r_3 + s_6, \qquad b_3 r_3 = w_3 + z_6, \qquad b_3 s_6 = w_3 + z_{15}.$$

By Lemma 2.16 again, it follows that

$$r_3 \bar{c}_3 = b_3 + x_6, \qquad \bar{c}_3 s_6 = b_3 + x_{15},$$

$$r_3 \bar{b}_3 = c_3 + b_6, \qquad \bar{b}_3 s_6 = c_3 + y_{15}.$$

Thus $(\bar{c}_3 r_3, \bar{c}_3 r_3) = 2$, which implies that $r_3 \bar{r}_3 = 1 + b_8$.

Since

$$(b_3 r_3)\bar{c}_3 = w_3 \bar{c}_3 + z_6 \bar{c}_3$$

$$= 2c_3 + b_6 + y_{15},$$

$$(b_3 \bar{c}_3) r_3 = r_3 \bar{b}_3 + r_3 \bar{x}_6$$

$$= c_3 + b_6 + r_3 \bar{x}_6,$$

which implies that $r_3 \bar{x}_6 = c_3 + y_{15}$.

We assert that x_{10} is real. In fact,

$$\bar{b}_3 (r_3 \bar{c}_3) = b_3 \bar{b}_3 + x_6 \bar{b}_3$$

$$= 1 + b_8 + b_8 + x_{10},$$

$$(\bar{b}_3 r_3)\bar{c}_3 = c_3 \bar{c}_3 + b_6 \bar{c}_3$$

$$= 1 + b_8 + b_8 + \bar{x}_{10},$$

so it follows that x_{10} is real.

Since

$$\bar{b}_3 (w_3 \bar{c}_3) = c_3 \bar{b}_3 + b_6 \bar{b}_3$$

$$= b_3 + x_6 + b_3 + x_{15},$$

$$(\bar{b}_3 \bar{c}_3) w_3 = \bar{r}_3 w_3 + \bar{s}_6 w_3,$$

we have that

$$\bar{r}_3 w_3 = b_3 + x_6, \qquad w_3 \bar{s}_6 = b_3 + x_{15}.$$

Consequently $\bar{b}_3 w_3 = r_3 + s_6$.

Now we have that $(b_3 \bar{b}_3) w_3 = (\bar{b}_3 w_3) b_3$ and

$$w_3 b_8 = w_3 + z_6 + z_{15}.$$

Then $w_3 \bar{w}_3 = 1 + b_8$.

Since

$$(b_3\bar{b}_3)^2 = 1 + 2b_8 + b_8^2,$$
$$b_3^2\bar{b}_3^2 = (c_3 + b_6)(\bar{c}_3 + \bar{b}_6)$$
$$= 1 + b_8 + c_3\bar{b}_6 + \bar{c}_3b_6 + b_6\bar{b}_6$$
$$= 1 + 3b_8 + 2x_{10} + b_6\bar{b}_6,$$

so

$$b_8^2 = 2x_{10} + b_8 + b_6\bar{b}_6.$$

By $b_3^2\bar{b}_6 = (b_3\bar{b}_6)b_3$, we have that $x_{10} + b_6\bar{b}_6 = 1 + b_3\bar{x}_{15}$. Then $(b_3x_{10}, x_{15}) = 1$. But $(b_3x_{10}, x_6) = 1$. There exists $b_9 \in B$ such that

$$b_3x_{10} = x_6 + x_{15} + b_9, \quad b_9 \in N^*B.$$

It is easy to see that b_3x_{10} cannot have constituents of degree 3 and 4 by $L_1(B) = \{1\}$ and $L_2(B) = \emptyset$. Thus $b_9 \in B$.

Since

$$(\bar{b}_3x_6)b_3 = b_3b_8 + b_3x_{10}$$
$$= b_3 + x_6 + x_{15} + x_6 + x_{15} + b_9,$$
$$(b_3\bar{b}_3)x_6 = x_6 + x_6b_8,$$

we have

$$x_6b_8 = b_3 + x_6 + 2x_{15} + b_9.$$

Multiplying both sides of the equation $x_{10} + b_6\bar{b}_6 = 1 + b_3\bar{x}_{15}$ by b_3, we have that

$$b_3x_{10} + (b_3\bar{b}_6)b_6 = b_3 + b_3^2\bar{x}_{15},$$

so

$$x_6 + x_{15} + b_9 + b_6\bar{b}_3 + b_6\bar{x}_{15} = c_3\bar{x}_{15} + b_6\bar{x}_{15}.$$

Hence

$$c_3\bar{x}_{15} = 2x_{15} + x_6 + b_9.$$

Now checking the associative law of $(b_3\bar{c}_3)b_6 = (\bar{c}_3b_6)b_3$, $(r_3\bar{r}_3)x_6 = (\bar{r}_3x_6)r_3$ and $(r_3\bar{b}_3)\bar{x}_6 = (r_3\bar{x}_6)\bar{b}_3$, we obtain

$$b_6\bar{x}_6 = 2x_6 + b_9 + x_{15},$$
$$r_3\bar{y}_{15} = x_6 + 2x_{15} + b_9,$$
$$\bar{b}_3y_{15} = x_6 + 2x_{15} + b_9.$$

Hence

$$c_3\bar{y}_{15} + b_6\bar{y}_{15} = b_3^2\bar{y}_{15} = b_3(b_3\bar{y}_{15}) = b_3\bar{x}_6 + 2b_3\bar{x}_{15} + b_3\bar{b}_9.$$

Since $(b_3\bar{x}_6, b_8) = (b_3\bar{x}_6, x_{10}) = (b_3\bar{x}_{15}, b_8) = (b_3\bar{x}_{15}, x_{10}) = (b_3\bar{b}_9, x_{10}) = 1$ and $(b_3\bar{b}_9, b_8) = 0$, we have that

$$(c_3\bar{y}_{15} + b_6\bar{y}_{15}, b_8) = 3 \quad \text{and} \quad (c_3\bar{y}_{15} + b_6\bar{y}_{15}, x_{10}) = 4.$$

From the degrees, we know that $(c_3\bar{y}_{15}, x_{10}) \le 1$. If $(c_3\bar{y}_{15}, x_{10}) = 0$, then $(b_6\bar{y}_{15}, x_{10}) = 4$, so $b_6x_{10} = 4y_{15}$. But $(b_6x_{10}, c_3) = 1$, a contradiction. Hence $(c_3\bar{y}_{15}, x_{10}) = 1$ and $(b_6x_{10}, y_{15}) = (b_6\bar{y}_{15}, x_{10}) = 3$.

Since $(c_3x_{10}, c_3x_{10}) = (b_3x_{10}, b_3x_{10}) = 3$ and $(c_3x_{10}, b_6) = 1$, there exists $c_9 \in B$ such that

$$c_3x_{10} = b_6 + y_{15} + c_9.$$

Then

$$(\bar{b}_3c_3)x_6 = b_3x_6 + x_6^2$$
$$= c_3 + y_{15} + x_6^2,$$
$$(\bar{b}_3x_6)c_3 = c_3b_8 + c_3x_{10}$$
$$= c_3 + b_6 + y_{15} + b_6 + y_{15} + c_9,$$
$$(c_3\bar{c}_3)b_6 = b_6 + b_6b_8,$$
$$(\bar{c}_3b_6)c_3 = b_8c_3 + x_{10}c_3$$
$$= c_3 + b_6 + y_{15} + y_{15} + b_6 + c_9,$$
$$(c_3b_6)\bar{c}_3 = c_3 + b_6 + \bar{c}_3z_{15},$$

which implies that

$$x_6^2 = 2b_6 + y_{15} + c_9,$$
$$b_6b_8 = c_3 + b_6 + c_9 + 2y_{15},$$
$$c_3\bar{z}_{15} = \bar{b}_6 + \bar{c}_9 + 2\bar{y}_{15}.$$

Since

$$(b_3\bar{c}_3)^2 = (\bar{b}_3 + \bar{x}_6)^2$$
$$= \bar{b}_3^2 + 2\bar{b}_3\bar{x}_6 + \bar{x}_6^2$$
$$= 2\bar{c}_3 + 3\bar{b}_6 + 3\bar{y}_{15} + \bar{c}_9,$$
$$b_3^2\bar{c}_3^2 = c_3\bar{w}_3 + c_3\bar{z}_6 + \bar{w}_3b_6 + b_6\bar{z}_6$$
$$= \bar{c}_3 + \bar{b}_6 + \bar{c}_3 + \bar{y}_{15} + b_6\bar{w}_3 + b_6\bar{z}_6,$$

but $(\bar{w}_3 b_6, \bar{c}_3) = 1$, we have that

$$w_3\bar{b}_6 = c_3 + y_{15},$$
$$b_6\bar{z}_6 = 2\bar{b}_6 + \bar{y}_{15} + \bar{c}_9.$$

Since

$$(b_3\bar{r}_3)\bar{b}_6 = \bar{c}_3\bar{b}_6 + \bar{b}_6^2$$
$$= \bar{w}_3 + \bar{z}_{15} + \bar{b}_6^2,$$
$$(b_3\bar{b}_6)\bar{r}_3 = \bar{b}_3\bar{r}_3 + \bar{x}_{15}\bar{r}_3$$
$$= \bar{w}_3 + \bar{z}_6 + \bar{r}_3\bar{x}_{15},$$

and $(\bar{b}_6^2, \bar{z}_6) = 2$, we have that $(\bar{r}_3\bar{x}_{15}, \bar{z}_6) = 1$. Hence $(r_3\bar{z}_6, \bar{x}_{15}) = 1$. But $(r_3\bar{z}_6, \bar{b}_3) = 1$ by $b_3r_3 = w_3 + z_6$. Thus

$$\bar{r}_3 z_6 = b_3 + x_{15}.$$

From the following equations

$$\bar{b}_3(w_3\bar{b}_6) = \bar{b}_3(c_3 + y_{15})$$
$$= b_3 + x_6 + x_6 + 2x_{15} + b_9,$$
$$(\bar{b}_3 w_3)\bar{b}_6 = r_3\bar{b}_6 + s_6\bar{b}_6,$$

and $(r_3\bar{b}_6, r_3\bar{b}_6) = (b_3\bar{b}_6, b_3\bar{b}_6) = 2$, $(r_3\bar{b}_6, b_3) = 1$, we obtain

$$r_3\bar{b}_6 = b_3 + x_{15}, \qquad s_6\bar{b}_6 = 2x_6 + x_{15} + b_9.$$

By $(b_3 c_3)\bar{r}_3 = (c_3\bar{r}_3)b_3$ and $\bar{c}_3(w_3 b_8) = (\bar{c}_3 b_8)w_3$, we have that

$$\bar{r}_3 s_6 = b_8 + x_{10}, \qquad w_3\bar{y}_{15} = b_6 + c_9 + 2y_{15}.$$

Since $(\bar{r}_3 b_8, \bar{r}_3 b_8) = (r_3\bar{r}_3, b_8^2) = (b_3\bar{b}_3, b_8^2) = (b_3 b_8, b_3 b_8) = 3$ by Lemma 2.16 and $(\bar{r}_3 b_8, \bar{s}_6) = (\bar{r}_3 b_8, \bar{r}_3) = 1$, there exists $t_{15} \in B$ such that

$$\bar{r}_3 b_8 = \bar{s}_6 + \bar{r}_3 + t_{15}.$$

Since

$$(w_3 b_8)\bar{r}_3 = w_3\bar{r}_3 + z_6\bar{r}_3 + z_{15}\bar{r}_3$$
$$= b_3 + x_6 + b_3 + x_{15} + \bar{r}_3 z_{15},$$
$$(\bar{r}_3 w_3)b_8 = b_3 b_8 + x_6 b_8$$
$$= b_3 + x_6 + x_{15} + b_3 + x_6 + 2x_{15} + b_9,$$
$$(\bar{r}_3 b_8)w_3 = \bar{s}_6 w_3 + \bar{r}_3 w_3 + t_{15}w_3$$
$$= b_3 + x_{15} + b_3 + x_6 + w_3 t_{15},$$

we have that

$$r_3 \bar{z}_{15} = \bar{x}_6 + 2\bar{x}_{15} + \bar{b}_9,$$

$$w_3 t_{15} = x_6 + b_9 + 2x_{15}.$$

Since

$$(\bar{b}_3 r_3)b_8 = c_3 b_8 + b_8 b_6$$

$$= c_3 + b_6 + y_{15} + 2y_{15} + b_6 + c_3 + c_9,$$

$$(\bar{b}_3 b_8)r_3 = r_3 \bar{b}_3 + r_3 \bar{x}_6 + r_3 \bar{x}_{15}$$

$$= c_3 + b_6 + c_3 + y_{15} + r_3 \bar{x}_{15},$$

$$(r_3 b_8)\bar{b}_3 = r_3 \bar{b}_3 + s_6 \bar{b}_3 + \bar{t}_{15} \bar{b}_3$$

$$= c_3 + b_6 + c_3 + y_{15} + \bar{b}_3 \bar{t}_{15},$$

we have that $w_3 = \bar{c}_3$, $z_6 = \bar{b}_6$, $z_{15} = \bar{y}_{15}$ and

$$r_3 \bar{x}_{15} = b_6 + c_9 + 2y_{15},$$

$$b_3 t_{15} = \bar{b}_6 + \bar{c}_9 + 2\bar{y}_{15}.$$

Furthermore, $b_3 r_3 = \bar{c}_3 + \bar{b}_6$, and we have that $(b_3 c_3, \bar{r}_3) = 1$. But $(b_3 c_3, r_3) = 1$; hence $r_3 = \bar{r}_3$. Moreover $\bar{r}_3 + \bar{s}_6 = b_3 \bar{w}_3 = b_3 c_3 = r_3 + s_6$, which implies that $s_6 = \bar{s}_6$. Therefore t_{15} is real by the expression $r_3 b_8$, and

$$c_3 t_{15} = \bar{x}_6 + \bar{c}_9 + 2\bar{x}_{15}.$$

Now we have that $t_{15} \in Supp\{c_3 x_6\}$. Thus

$$c_3 x_6 = r_3 + t_{15}.$$

Since

$$(c_3 r_3)\bar{b}_6 = \bar{b}_3 \bar{b}_6 + \bar{x}_6 \bar{b}_6,$$

$$(r_3 \bar{b}_6)c_3 = b_3 c_3 + c_3 x_{15}$$

$$= r_3 + s_6 + c_3 x_{15},$$

$$(c_3 \bar{b}_6)r_3 = r_3 b_8 + r_3 x_{10}$$

$$= r_3 + s_6 + t_{15} + r_3 x_{10}$$

and $(r_3 x_{10}, s_6) = 1$, we have $s_6 \in Supp\{c_3 x_{15}\}$. But $(c_3 x_{15}, t_{15}) = (c_3 t_{15}, \bar{x}_{15}) = 2$. Let

$$c_3 x_{15} = 2t_{15} + s_6 + d_9, \quad d_9 \in N^* B.$$

Then $r_3 x_{10} = s_6 + t_{15} + d_9$ and by $(c_3 r_3)\bar{b}_6 = (c_3 \bar{b}_6)r_3$ we obtain that

$$\bar{b}_3 \bar{b}_6 + \bar{x}_6 \bar{b}_6 = r_3 + s_6 + t_{15} + s_6 + t_{15} + d_9.$$

Furthermore d_9 is real for r_3 and x_{10} are real, and

$$b_3b_6 = r_3 + t_{15},$$
$$x_6b_6 = 2s_6 + t_{15} + d_9.$$

Since

$$(b_3\bar{b}_3)b_6 = b_6 + b_6b_8$$
$$= b_6 + c_3 + b_6 + c_9 + 2y_{15},$$
$$b_3(\bar{b}_3b_6) = b_3^2 + b_3x_{15}$$
$$= c_3 + b_6 + b_3x_{15},$$
$$(b_3b_6)\bar{b}_3 = r_3\bar{b}_3 + t_{15}\bar{b}_3$$
$$= c_3 + b_6 + \bar{b}_3t_{15},$$

we have that

$$b_3x_{15} = b_6 + 2y_{15} + c_9,$$
$$b_3t_{15} = \bar{b}_6 + \bar{c}_9 + 2\bar{y}_{15}.$$

Since

$$b_3^2x_6 = c_3x_6 + b_6x_6$$
$$= r_3 + t_{15} + 2s_6 + t_{15} + d_9,$$
$$(b_3x_6)b_3 = b_3c_3 + b_3y_{15}$$
$$= r_3 + s_6 + b_3y_{15},$$

we have that

$$b_3y_{15} = 2t_{15} + s_6 + d_9.$$

Since $(b_3y_{15}, b_3y_{15}) = (b_3\bar{y}_{15}, b_3\bar{y}_{15}) = 6$, we have that $d_9 \in B$. Checking the associative law of $(b_3\bar{b}_3)s_6 = (b_3s_6)\bar{b}_3$, we have

$$s_6b_8 = r_3 + s_6 + 2t_{15} + d_9.$$

Since

$$(b_3\bar{b}_3)y_{15} = y_{15} + b_8y_{15},$$
$$(b_3y_{15})\bar{b}_3 = \bar{b}_3s_6 + 2\bar{b}_3t_{15} + \bar{b}_3d_9$$
$$= c_3 + y_{15} + 2b_6 + 2c_9 + 4y_{15} + \bar{b}_3d_9,$$

it follows that $b_8y_{15} = c_3 + 2b_6 + 2c_9 + 4y_{15} + \bar{b}_3d_9$.
This completes the proof of the lemma. $\square$

Remark It always follows that $(c_3c_9, \bar{y}_{15}) = 1$ by Lemma 2.17. So $(c_3c_9, c_3c_9) \geq 2$. In the following, we shall investigate the expression c_3c_9.

Lemma 2.18 *There exist no $x_4, y_8 \in B$ such that $c_3c_9 = x_4 + y_8 + \bar{y}_{15}$.*

Proof If there exist no $x_4, y_8 \in B$ such that

$$c_3c_9 = x_4 + y_8 + \bar{y}_{15},$$

then there exists $z_3 \in B$ such that $c_3\bar{x}_4 = \bar{c}_9 + z_3$. Furthermore, there is $y_5 \in B$ such that $c_3\bar{z}_3 = x_4 + y_5$. Thus $z_3\bar{z}_3 = 1 + b_8$. By Lemma 2.16 we have that

$$c_3(z_3\bar{z}_3) = c_3 + c_3b_8$$
$$= 2c_3 + b_6 + y_{15},$$
$$z_3(c_3\bar{z}_3) = z_3x_4 + z_3y_5.$$

It must follow that $z_3x_4 = 2c_3 + b_6$, which implies that $(z_3\bar{c}_3, x_4) = 2$, a contradiction. □

Lemma 2.19 *There exist no $x_5, y_7 \in B$ such that $c_3c_9 = x_5 + y_7 + \bar{y}_{15}$.*

Proof If some $x_5, y_7 \in B$ such that $c_3c_9 = x_5 + y_7 + \bar{y}_{15}$. Then $(c_3\bar{x}_5, \bar{c}_9) = 1$. There are three possibilities:

$$\text{(I)} \quad c_3\bar{x}_5 = \bar{c}_9 + 2x_3,$$
$$\text{(II)} \quad c_3\bar{x}_5 = \bar{c}_9 + x_3 + y_3,$$
$$\text{(III)} \quad c_3\bar{x}_5 = \bar{c}_9 + y_6.$$

It is easy to see that (I) will lead to a contradiction.

If (II) follows, then $c_3\bar{x}_5 = \bar{c}_9 + x_3 + y_3$. Set $c_3\bar{x}_3 = x_5 + z_4$, $c_3\bar{y}_3 = x_5 + t_4$, where $z_4, t_4 \in B$. Hence

$$\bar{c}_3(c_3\bar{x}_3) = \bar{c}_3x_5 + \bar{c}_3z_4$$
$$= c_9 + \bar{x}_3 + \bar{y}_3 + \bar{c}_3z_4,$$
$$(c_3\bar{c}_3)\bar{x}_3 = \bar{x}_3 + \bar{x}_3b_8.$$

Therefore $(x_3\bar{y}_3, b_8) = 1$, which implies that $\bar{y}_3x_3 = 1 + b_8$. So $y_3 = x_3$, a contradiction.

If (III) follows, then

$$\bar{c}_3(c_3\bar{x}_5) = \bar{c}_3\bar{c}_9 + \bar{c}_3y_6$$
$$= y_{15} + \bar{x}_5 + \bar{y}_7 + \bar{c}_3y_6,$$
$$(c_3\bar{c}_3)\bar{x}_5 = \bar{x}_5 + \bar{x}_5b_8.$$

Hence $\bar{x}_5 b_8 = y_{15} + \bar{y}_7 + \bar{c}_3 y_6$. Therefore $(b_8 y_{15}, \bar{x}_5) = 1$. By Lemma 2.17, we have that $(\bar{x}_5, \bar{b}_3 d_9) = 1$. Since $(\bar{b}_3 d_9, y_{15}) = 1$, we may set

$$\bar{b}_3 d_9 = \bar{x}_5 + y_{15} + e_7, \quad e_7 \in N^* B.$$

If $e_7 = m_3 + n_4$, then $b_3 n_4 = d_9 + p_3$, some $p_3 \in B$. Let $\bar{b}_3 p_3 = n_4 + z_5$, some $z_5 \in B$. Then $p_3 \bar{p}_3 = 1 + b_8$. Furthermore,

$$b_3 (p_3 \bar{p}_3) = b_3 + b_3 b_8$$
$$= 2b_3 + x_6 + x_{15},$$
$$(b_3 \bar{p}_3) p_3 = \bar{n}_4 p_3 + \bar{z}_5 p_3.$$

Note that $b_3 \in Supp\{\bar{n}_4 p_3\}$ and $b_3 \in Supp\{\bar{z}_5 p_3\}$, we come to a contradiction.

If $e_7 \in B$, then $\bar{b}_3 d_9 = \bar{x}_5 + y_{15} + e_7$. Let $b_3 \bar{x}_5 = d_9 + u_6$.

We assert that $u_6 \in B$. Otherwise, let $u_6 = r_3 + s_3$, where $r_3, s_3 \in B$, then

$$b_3 \bar{r}_3 = x_5 + u_4.$$

Hence $r_3 \bar{r}_3 = 1 + b_8$. Moreover,

$$b_3 (r_3 \bar{r}_3) = b_3 + b_3 b_8$$
$$= 2b_3 + x_6 + x_{15},$$
$$r_3 (b_3 \bar{r}_3) = r_3 x_5 + r_3 u_4,$$

which is impossible for $b_3 \in Supp\{r_3 x_5\}$ and $b_3 \in Supp\{r_3 u_4\}$.

Now we have that

$$\bar{b}_3 (c_3 \bar{x}_5) = \bar{b}_3 \bar{c}_9 + \bar{b}_3 y_6,$$
$$(\bar{b}_3 c_3) \bar{x}_5 = b_3 \bar{x}_5 + x_6 \bar{x}_5$$
$$= d_9 + u_6 + x_6 \bar{x}_5.$$

If $d_9, u_6 \in Supp\{\bar{b}_3 y_6\}$, then $\bar{b}_3 y_6 = d_9 + u_6 + t_3$. Let $b_3 t_3 = y_6 + p_3$. Hence $t_3 \bar{t}_3 = 1 + b_8$. Moreover,

$$b_3 (t_3 \bar{t}_3) = 2b_3 + x_6 + x_{15},$$
$$\bar{t}_3 (b_3 t_3) = \bar{t}_3 y_6 + \bar{t}_3 p_3.$$

So $\bar{t}_3 y_6 = b_3 + x_{15}$. But $3 = (\bar{b}_3 y_6, \bar{b}_3 y_6) = (\bar{t}_3 y_6, \bar{t}_3 y_6) = 2$ by Lemma 2.1, a contradiction.

If $u_6 \in Supp\{\bar{b}_3 y_6\}$, $d_9 \in Supp\{\bar{b}_3 \bar{c}_9\}$ or $u_6 \in Supp\{\bar{b}_3 \bar{c}_9\}$, $d_9 \in Supp\{\bar{b}_3 y_6\}$.

If the former one follows, then there exists a q_3 such that

$$b_3 c_9 = t_{15} + d_9 + q_3.$$

Hence $c_9 \in Supp\{\bar{b}_3 d_9\}$, a contradiction.

If $u_6 \in Supp\{\bar{b}_3\bar{c}_9\}$, $d_9 \in Supp\{\bar{b}_3 y_6\}$, then

$$b_3 d_9 = \bar{y}_{15} + y_6 + r_6,$$

where $r_6 \in N^*B$. But d_9 is real, $\bar{b}_3 d_9 = \bar{x}_5 + y_{15} + e_7$, a contradiction. This completes the proof.   □

Lemma 2.20 *There exist no m_6 and n_6 such that $c_3 c_9 = \bar{y}_{15} + m_6 + n_6$.*

Proof If there exist m_6 and n_6 such that $c_3 c_9 = \bar{y}_{15} + m_6 + n_6$. Then $m_6 \neq b_6, \bar{b}_6$, $n_6 \neq b_6, \bar{b}_6$ by Lemma 2.17 and $c_3 \bar{m}_6$ has three possibilities:

(I) $c_3 \bar{m}_6 = \bar{c}_9 + x_3 + y_3 + z_3$, x_3, y_3, z_3 distinct
(II) $c_3 \bar{m}_6 = \bar{c}_9 + x_3 + y_6$, x_3 and $y_6 \in B$,
(III) $c_3 \bar{m}_6 = \bar{c}_9 + x_4 + y_5$, x_4 and $y_5 \in B$.

Suppose (I) follows. We may assume that $c_3 \bar{x}_3 = m_6 + r_3$, where $r_3 \in B$. Then

$$\bar{x}_3 + \bar{x}_3 b_8 = \bar{x}_3(c_3 \bar{c}_3) = (\bar{x}_3 c_3)\bar{c}_3 = m_6 \bar{c}_3 + r_3 \bar{c}_3 = c_9 + \bar{x}_3 + \bar{y}_3 + \bar{z}_3 + r_3 \bar{c}_3.$$

Hence $\bar{y}_3 \in Supp\{\bar{x}_3 b_8\}$, which implies that $x_3 = y_3$, a contradiction.

If (II) follows, then $(c_3 \bar{x}_3, c_3 \bar{x}_3) = 2$, from which $x_3 \bar{x}_3 = 1 + b_8$ follows. Also we may set $c_3 \bar{m}_3 = x_3 + t_6$, where $t_6 \in B$. Hence

$$(c_3 \bar{c}_3)\bar{x}_3 = \bar{x}_3 + \bar{x}_3 b_8,$$

$$\bar{c}_3(c_3 \bar{x}_3) = \bar{c}_3 m_6 + \bar{c}_3 m_3$$

$$= \bar{x}_4 + \bar{y}_5 + c_9 + \bar{x}_3 + \bar{t}_6.$$

Therefore $x_3 b_8 = x_4 + y_5 + \bar{c}_9 + t_6$. We have that $(x_3 b_8, x_3 b_8) = 4$. But by Lemma 2.8, $(x_3 b_8, x_3 b_8) = (b_3 b_8, b_3 b_8) = 3$, a contradiction.

If (III) follows, set $c_3 \bar{x}_4 = m_6 + p_6$, where $p_6 \in B$. By $\bar{c}_3(c_3 \bar{m}_6) = (c_3 \bar{c}_3)\bar{m}_6$, one has that

$$\bar{m}_6 b_8 = y_{15} + \bar{m}_6 + \bar{n}_6 + \bar{p}_6 + \bar{c}_3 y_5.$$

Hence $(b_8 y_{15}, \bar{m}_6) = 1$. By the expression $b_8 y_{15}$ in Lemma 2.17, $(\bar{m}_6, \bar{b}_3 d_9) = 1$. But $(y_{15}, \bar{b}_3 d_9) = 1$. There exists $q_6 \in N^*B$ such that

$$\bar{b}_3 d_9 = y_{15} + \bar{m}_6 + q_6. \tag{2.64}$$

If $q_6 \notin B$, then $q = 2z_3$ or $q_6 = w_3 + u_3$. If the first one holds, then $(\bar{b}_3 d_9, z_3) = 2$ and $(b_3 z_3, d_9) = 2$, which is impossible. So the second one follows. Consequently $b_3 w_3 = d_9$. Hence

$$w_3 + w_3 b_8 = w_3(b_3 \bar{b}_3) = \bar{b}_3(b_3 w_3) = \bar{b}_3 d_9 = y_{15} + \bar{m}_6 + w_3 + u_3.$$

We can see that $w_3\bar{u}_3 = 1 + b_8$. Then $w_3 = u_3$, a contradiction. Hence $p_6 \in B$. It follows by Lemma 2.17 and (2.64) that

$$b_8 y_{15} = 5y_{15} + 2c_9 + 2b_6 + c_3 + \bar{m}_6 + q_6.$$

Since

$$c_3(\bar{b}_3 d_9) = c_3 y_{15} + c_3 \bar{m}_6 + c_3 q_6$$
$$= \bar{b}_6 + \bar{c}_9 + 2\bar{y}_{15} + x_4 + y_5 + \bar{c}_9 + c_3 q_6,$$
$$(c_3 \bar{b}_3) d_9 = b_3 d_9 + x_6 d_9$$
$$= \bar{y}_{15} + m_6 + \bar{q}_6 + x_6 d_9,$$

we have that $(c_3 q_6, m_6) \geq 1$. Hence $(c_3 \bar{m}_6, \bar{q}_6) \geq 1$, a contradiction to our assumption. This completes the proof.   □

Lemma 2.21   *There exists no NITA such that $(c_3 c_9, c_3 c_9) \geq 4$.*

Proof The proof is given in three steps.

Step 1 The following equations hold:

$$
\begin{aligned}
&\text{(i)} && c_3 c_9 = \bar{y}_{15} + y_3 + z_3 + w_3 + u_3,\\
&\text{(ii)} && c_3 c_9 = \bar{y}_{15} + y_3 + y_3 + y_6,\\
&\text{(iii)} && c_3 c_9 = \bar{y}_{15} + y_3 + y_4 + y_5,\\
&\text{(iv)} && c_3 c_9 = \bar{y}_{15} + y_4 + z_4 + w_4.
\end{aligned}
$$

For any $z \in Supp\{c_3 c_9\} \setminus \{\bar{y}_{15}\}$, it is sufficient to prove that $(c_3 c_9, z) = 1$. Otherwise $(c_3 c_9, z) \geq 2$. Since $c_3 c_9 - \bar{y}_{15}$ is of degree 12, one has that $(c_3 c_9, z) = 2$ and z is of degree 6. So $c_3 c_9 = \bar{y}_{15} + 2z_6$, $c_3 \bar{z}_6 = 2\bar{c}_9$. Hence

$$\bar{z}_6 + \bar{z}_6 b_8 = (c_3 \bar{c}_3) \bar{z}_6 = \bar{c}_3 (c_3 \bar{z}_6) = 2\bar{c}_3 \bar{c}_9 = 2y_{15} + 4\bar{z}_6.$$

Thus $\bar{z}_6 b_8 = 2y_{15} + 3\bar{z}_6$. Consequently $(b_8 y_{15}, \bar{z}_6) = 2$. By the expression $b_6 b_8$ in Lemma 2.17, $z_6 \neq b_6, \bar{b}_6$ and the expression $b_8 y_{15}$ in Lemma 2.17, we have that

$$\bar{b}_3 d_9 = y_{15} + 2\bar{z}_6,$$
$$b_8 y_{15} = 5y_{15} + 2c_9 + 2b_6 + c_3 + 2\bar{z}_6.$$

Therefore $b_3 \bar{z}_6 = 2d_9$.
   Since

$$(b_3 \bar{b}_3) d_9 = d_9 + b_8 d_9,$$
$$b_3(\bar{b}_3 d_9) = b_3 y_{15} + 2b_3 \bar{z}_6$$
$$= s_6 + 2t_{15} + 5d_9,$$

it follows that $b_8 d_9 = s_6 + 2t_{15} + 4d_9$. Consequently,

$$b_8(b_3\bar{z}_6) = 2b_8 b_9$$
$$= 2s_6 + 4t_{15} + 8d_9,$$
$$(b_3 b_8)\bar{z}_6 = b_3\bar{z}_6 + x_6\bar{z}_6 + x_{15}\bar{z}_6$$
$$= 2d_9 + x_6\bar{z}_6 + x_{15}\bar{z}_6.$$

Then $x_6\bar{z}_6 + x_{15}\bar{z}_6 = 2s_6 + 4t_{15} + 6d_9$. We have exactly three possibilities:

$$
\begin{array}{lll}
\text{(I)} & x_6\bar{z}_6 = 4d_9, & x_{15}\bar{z}_6 = 2s_6 + 4t_{15} + 2d_9, \\
\text{(II)} & x_6\bar{z}_6 = s_6 + 2t_{15}, & x_{15}\bar{z}_6 = s_6 + 2t_{15} + 6d_9, \\
\text{(III)} & x_6\bar{z}_6 = 2s_6 + t_{15} + d_9, & x_{15}\bar{z}_6 = 3t_{15} + 5d_9.
\end{array}
$$

The first case implies that

$$\bar{b}_3(x_6\bar{z}_6) = 4\bar{b}_3 d_9$$
$$= 4y_{15} + 8\bar{z}_6,$$
$$(\bar{b}_3 x_6)\bar{z}_6 = b_8\bar{z}_6 + x_{10}\bar{z}_6$$
$$= 2y_{15} + 3\bar{z}_6 + x_{10}\bar{z}_6.$$

So $x_{10}\bar{z}_6 = 2y_{15} + 5\bar{z}_6$. Thus $(z_6\bar{z}_6, x_{10}) = 5$, a contradiction.

The second and the third cases mean $(z_6 d_9, x_{15}) = 6$ and $(z_6 d_9, x_{15}) = 5$, respectively. This is impossible.

Step 2 No NITA satisfies either (i), (ii) or (iii).

It is sufficient to show that $y_3 \in Supp\{c_3 c_9\}$. If $y_3 \in Supp\{c_3 c_9\}$, then $\bar{c}_3 y_3 = c_9$. Hence $(\bar{b}_3 y_3, \bar{b}_3 y_3) = (\bar{c}_3 y_3, \bar{c}_3 y_3) = 1$ by Lemma 2.1. Let $\bar{b}_3 y_3 = g_9$, $g_9 \in B$. Then

$$b_3 c_9 = b_3(\bar{c}_3 y_3) = (b_3\bar{c}_3)y_3 = \bar{b}_3 y_3 + \bar{x}_6 y_3 = g_9 + \bar{x}_6 y_3.$$

Since $(t_{15}, b_3 c_9) = 1$, we have that $b_3 c_9 = g_9 + t_{15} + \alpha_3$, some $\alpha_3 \in B$. Therefore $(b_3 c_9, b_3 c_9) = 3$. But $(b_3 c_9, b_3 c_9) = (c_3 c_9, c_3 c_9)$ by Lemma 2.1, which is impossible for $(c_3 c_9, c_3 c_9) \geq 4$.

Step 3 No NITA satisfies (iv).

If (iv) holds, then there exists m_3 such that $\bar{c}_3 y_4 = c_9 + m_3$. Thus $c_3 m_3 = y_4 + k_5$, some $k_5 \in B$. So $(c_3 m_3, c_3 m_3) = 2$, which implies that $m_3\bar{m}_3 = 1 + b_8$. Hence

$$\bar{m}_3 y_4 + \bar{m}_3 k_5 = c_3(m_3\bar{m}_3) = c_3 + c_3 b_8 = (c_3\bar{c}_3)\bar{m}_3 = 2c_3 + b_6 + y_{15},$$

which is impossible for $c_3 \in Supp\{\bar{m}_3 y_4\}$ and $Supp\{\bar{m}_3 k_5\}$. The lemma follows. $\square$

Lemma 2.22 *There exists no NITA such that* $(c_3 c_9, c_3 c_9) = 2$.

Proof If $(c_3c_9, c_3c_9) = 2$, then $c_3c_9 = \bar{y}_{15} + b_{12}$, $b_{12} \in B$ and $c_3\bar{c}_9 = x_{10} + x_{17}$, $x_{17} \in B$.

Since $\bar{c}_3(c_3c_9) = \bar{c}_3\bar{y}_{15} + \bar{c}_3b_{12} = b_6 + c_9 + 2y_{15} + \bar{c}_3b_{12}$, we have that $b_8c_9 = b_6 + 2y_{15} + \bar{c}_3b_{12}$. On the other hand, since

$$c_9 + b_8c_9 = (c_3\bar{c}_3)c_9 = c_3(\bar{c}_3c_9) = c_3(\bar{x}_{10} + \bar{x}_{17}) = b_6 + y_{15} + c_9 + c_3\bar{x}_{17},$$

we have that $b_8c_9 = b_6 + y_{15} + c_3\bar{x}_{17}$. By the expression b_8y_{15} in Lemma 2.17, $(b_8c_9, y_{15}) = 2$. Hence $y_{15} \in Supp\{c_3\bar{x}_{17}\}$, which implies that $x_{17} \in Supp\{c_3\bar{y}_{15}\}$. By the expressions of c_3b_8 and c_3x_{10} in Lemma 2.17, $(c_3\bar{y}_{15}, b_8) = (c_3\bar{y}_{15}, x_{10}) = 1$. There exists $t_{10} \in N^*B$ such that $c_3\bar{y}_{15} = x_{17} + b_8 + x_{10} + t_{10}$, some $t_{10} \in N^*B$. The expression c_3y_{15} in Lemma 2.17 means that $(c_3y_{15}, c_3y_{15}) = 6$. Hence $(c_3\bar{y}_{15}, c_3\bar{y}_{15}) = 6$. So $t_{10} = x_3 + y_3 + z_4$. This leads to $y_{15} \in Supp\{c_3\bar{y}_3\}$, a contradiction. The lemma follows. □

Lemma 2.23 $c_3c_9 = d_3 + \bar{c}_9 + \bar{y}_{15}$.

Proof First, by the previous lemma in this section, we know that $(c_3c_9, c_3c_9) = 3$ and c_3c_9 is a sum of irreducible base elements of degree 3, 9 and 15. So we may assume that $c_3c_9 = d_3 + y_9 + \bar{y}_{15}$, where $d_3, y_9 \in B$. Then $c_3\bar{d}_3 = \bar{c}_9$. Checking the associative law for $\bar{c}_3(c_3\bar{d}_3) = (c_3\bar{c}_3)\bar{d}_3$, it follows that

$$\bar{d}_3b_8 = \bar{y}_9 + y_{15}.$$

Since

$$
\begin{aligned}
b_6(b_3\bar{x}_6) &= b_6b_8 + b_6x_{10} \\
&= c_3 + b_6 + c_9 + 2y_{15} + b_6x_{10}, \\
(b_6\bar{x}_6)b_3 &= 2b_3x_6 + b_3b_9 + b_3x_{15} \\
&= 2c_3 + 2y_{15} + b_3b_9 + b_6 + 2y_{15} + c_9,
\end{aligned}
$$

we have that

$$b_6x_{10} = b_3b_9 + c_3 + 2y_{15}.$$

Since $(c_3^2)\bar{d}_3 = \bar{c}_3\bar{d}_3 + \bar{b}_6\bar{d}_3$, $c_3(c_3\bar{d}_3) = c_3\bar{c}_9$ and $x_{10} \in Supp\{c_3\bar{c}_9\}$ by Lemma 2.17, we obtain $x_{10} \in Supp\{\bar{b}_6\bar{d}_3\}$. So $(b_6x_{10}, \bar{d}_3) = 1$. Furthermore $(b_3b_9, \bar{d}_3) = 1$. Thus

$$b_3\bar{d}_3 = \bar{b}_9.$$

Therefore

$$\bar{b}_3\bar{b}_9 = \bar{b}_3(b_3\bar{d}_3) = (\bar{b}_3b_3)\bar{d}_3 = d_3 + d_3b_8 = d_3 + y_9 + \bar{y}_{15},$$

which implies that

$$
\begin{aligned}
b_3b_9 &= \bar{d}_3 + \bar{y}_9 + y_{15}, \\
b_6x_{10} &= \bar{d}_3 + c_3 + \bar{y}_9 + 3y_{15}.
\end{aligned}
$$

Since $(r_3 b_3)d_3 = c_9 + \bar{b}_6 d_3$ and $(b_3 d_3)r_3 = r_3 \bar{b}_9$, we have that $c_9 \in Supp\{r_3 \bar{b}_9\}$. But $y_{15} \in Supp\{r_3 \bar{b}_9\}$ by Lemma 2.17. Thus there exists $m_3 \in B$ such that

$$r_3 \bar{b}_9 = m_3 + c_9 + y_{15}.$$

Hence $b_9 \in Supp\{r_3 \bar{c}_9\}$. But $(r_3 \bar{c}_9, x_{15}) = 1$ by Lemma 2.17. Therefore

$$r_3 \bar{c}_9 = n_3 + b_9 + x_{15}, \quad n_3 \in B.$$

By r_3 real, it follows that

$$r_3 n_3 = \bar{c}_9.$$

Furthermore,

$$r_3(r_3 \bar{c}_9) = r_3 x_{15} + r_3 b_9 + r_3 n_3$$
$$= \bar{b}_6 + \bar{c}_9 + 2\bar{y}_{15} + \bar{c}_9 + \bar{y}_{15} + \bar{m}_3 + \bar{c}_9,$$
$$r_3^2 \bar{c}_9 = \bar{c}_9 + b_8 \bar{c}_9,$$

so that

$$b_8 \bar{c}_9 = \bar{b}_6 + 2\bar{c}_9 + 3\bar{y}_{15} + \bar{m}_3.$$

Hence $(b_8 y_{15}, c_9) = 3$.

By the expression $b_8 y_{15}$ in Lemma 2.17, we have $(\bar{b}_3 d_9, c_9) = 1$. From the expression $b_3 t_{15}$ in Lemma 2.17, one has that $(t_{15}, b_3 c_9) = 1$. Hence there exists $u_3 \in B$ such that

$$b_3 c_9 = u_3 + d_9 + t_{15}.$$

Thus $\bar{b}_3 u_3 = c_9$. Moreover,

$$c_3 b_9 = (\bar{b}_3 \bar{d}_3)c_3 = \bar{b}_3(c_3 \bar{d}_3) = \bar{b}_3 \bar{c}_9 = \bar{u}_3 + d_9 + t_{15}.$$

Thus $u_3 c_3 = \bar{b}_9$.

Since

$$\bar{c}_3(b_3 \bar{u}_3) = \bar{c}_3 \bar{c}_9$$
$$= \bar{d}_3 + \bar{y}_9 + y_{15},$$
$$(b_3 \bar{c}_3)\bar{u}_3 = \bar{b}_3 \bar{u}_3 + \bar{x}_6 \bar{u}_3,$$

we have that

$$b_3 u_3 = y_9 \quad \text{and} \quad u_3 x_6 = \bar{y}_{15} + \bar{d}_3.$$

Hence

$$c_3 y_9 = c_3(b_3 u_3) = b_3(u_3 c_3) = b_3 \bar{b}_9.$$

Since $(x_{10}, b_3 \bar{b}_9) = 1$ by Lemma 2.17, one has that $(x_{10}, c_3 y_9) = 1$, so $(\bar{y}_9, c_3 x_{10}) = 1$. Therefore $\bar{y}_9 = c_9$ by the expression $c_3 x_{10}$ (see Lemma 2.17). The lemma follows. $\qquad \square$

Lemma 2.24 *A NITA generated by c_3 is isomorphic to the algebra of characters of $3 \cdot A_6$: $(Ch(3 \cdot A_6), Irr(3 \cdot A_6))$. Furthermore, we have all the equations of products of base elements in the table subset D listed in Sect. 2.2.*

Proof By Lemmas 2.17 and 2.23 the sub-algebra generated by c_3 satisfies the following equations:

$$c_3 \bar{c}_3 = 1 + b_8, \qquad c_3^2 = \bar{c}_3 + \bar{b}_6, \qquad c_3 b_6 = \bar{c}_3 + \bar{y}_{15},$$
$$c_3 \bar{b}_6 = b_8 + x_{10}, \ x_{10} \ \text{real}, \qquad c_3 x_{10} = b_6 + c_9 + y_{15}, \qquad c_3 y_{15} = \bar{b}_6 + \bar{c}_9 + 2\bar{y}_{15}.$$

Changing c_3 into b_3, b_6 into $\bar{b}_6$, y_{15} into b_{15}, x_{10} into b_{10}, we can see that the above equations are the same as those in Hypothesis 2.1 and Lemma 2.9. Hence the NITA generated by c_3 is exactly the Table Algebra of characters of $3 \cdot A_6$ by Theorem 2.7. Of course, we have equations of products of base elements in the table subset D listed in Sect. 2.2. □

Now we continue our calculations, and we shall construct a new NITA which is not derived from groups.

Theorem 2.9 *If c_3 is nonreal, $(b_3 b_8, b_3 b_8) = 3$. Then the NITA generated by b_3 with $b_3^2 = c_3 + b_6$ is isomorphic to the NITA in $(A(3 \cdot A_6 \cdot 2), B_{32})$, the NITA of dimension 32 in Sect. 2.2.*

Proof By Lemma 2.24, we may say that we have found a table subset: $D = \{1, b_8, x_{10}, b_5, c_5, c_8, x_9, c_3, \bar{c}_3, d_3, \bar{d}_3, c_9, \bar{c}_9, b_6, \bar{b}_6, y_5, \bar{y}_{15}\}$. To prove that the NITA is isomorphic to NITA of dimension 32 in Sect. 2.2, it is sufficient to find all the basis elements outside D and the remaining expressions of all products of basis elements. Until now we have found many products of basis elements in Lemmas 2.16, 2.17 and 2.23. It remains to find the rest of them. In order to make the proof read easily, we shall divide it into several steps:

Step 1 $b_3 b_9 = y_{15} + c_9 + \bar{d}_3$, $b_3 d_3 = \bar{b}_9$, $b_3 \bar{b}_9 = x_9 + c_8 + x_{10}$, $b_3 d_9 = \bar{c}_9 + d_3 + \bar{y}_{15}$, $b_3 \bar{d}_3 = d_9$, $b_3 \bar{c}_9 = \bar{b}_9 + \bar{x}_{15} + z_3$, $b_3 \bar{z}_3 = c_9$, $b_3 \bar{x}_{15} = b_5 + c_5 + b_8 + c_8 + x_9 + x_{10}$, $b_3 z_3 = x_9$, $b_3 y_3 = \bar{c}_9$.

Since by Lemma 2.17

$$b_3^2 x_{10} = c_3 x_{10} + b_6 x_{10}$$
$$= b_6 + c_9 + y_{15} + 3y_{15} + c_3 + \bar{d}_3 + c_9,$$
$$(b_3 x_{10}) b_3 = x_6 b_3 + x_{15} b_3 + b_9 b_3$$
$$= c_3 + y_{15} + b_6 + 2y_{15} + c_9 + b_3 b_9,$$

we have that $b_3 b_9 = y_{15} + c_9 + \bar{d}_3$. Furthermore, $b_3 d_3 = \bar{b}_9$. So

$$b_3 \bar{b}_9 = (b_3 d_3) b_3 = b_3^2 d_3 = c_3 d_3 + b_6 d_3 = x_9 + c_8 + x_{10}.$$

Set $b_3\bar{d}_3 = y_9$. Since $b_3 d_3 = \bar{b}_9$, then $y_9 \in B$. And

$$b_3 y_9 = (b_3 \bar{d}_3)b_3 = b_3^2 \bar{d}_3 = c_3 \bar{d}_3 + b_6 \bar{d}_3 = \bar{c}_9 + d_3 + \bar{y}_{15}.$$

We have that $(b_3 y_{15}, \bar{y}_9) = 1$. But $b_3 y_{15} = s_6 + 2t_{15} + d_9$, thus $d_9 = \bar{y}_9$ and $\bar{b}_3 d_3 = d_9$.

Since $(b_3 x_{15}, c_9) = (b_3 b_9, c_9) = 1$, there exists $z_3 \in B$ such that

$$b_3 \bar{c}_9 = \bar{b}_9 + \bar{x}_{15} + z_3.$$

Hence $b_3 \bar{z}_3 = c_9$.

Since

$$b_3^2 \bar{b}_6 = c_3 \bar{b}_6 + b_6 \bar{b}_6$$
$$= b_8 + x_{10} + 1 + b_8 + b_5 + c_5 + c_8 + x_9,$$
$$(b_3 \bar{b}_6)b_3 = b_3 \bar{b}_3 + b_3 \bar{x}_{15}$$
$$= 1 + b_8 + b_3 \bar{x}_{15},$$

we have that

$$b_3 \bar{x}_{15} = b_5 + c_5 + b_8 + c_8 + x_9 + x_{10}.$$

Now we have

$$b_3^2 \bar{c}_9 = c_3 \bar{c}_9 + b_6 \bar{c}_9$$
$$= c_8 + x_9 + x_{10} + 2x_9 + b_8 + x_{10} + c_8 + b_5 + c_5,$$
$$(b_3 \bar{c}_9)b_3 = b_3 \bar{b}_9 + b_3 \bar{x}_{15} + b_3 z_3$$
$$= c_8 + x_9 + x_{10} + b_5 + c_5 + b_8 + c_8 + x_9 + x_{10} + b_3 z_3,$$

which imply $b_3 z_3 = x_9$.

Step 2 $b_3 c_9 = t_{15} + d_9 + y_3$, y_3 real, $d_3 \bar{x}_6 = t_{15} + \bar{y}_3$, $c_3 b_9 = t_{15} + d_9 + y_3$, $c_3 y_3 = \bar{b}_9$, $b_3 y_3 = \bar{c}_9$ and $b_8 t_{15} = r_3 + \bar{y}_3 + 5t_{15} + 2s_6 + 3d_9$.

We have that

$$b_3 c_9 = b_3(d_3 \bar{c}_3) = (b_3 \bar{c}_3)d_3 = \bar{b}_3 d_3 + \bar{x}_3 d_3 = d_9 + \bar{x}_6 d_3.$$

Hence $(b_3 c_9, d_9) = 1$. But by Lemma 2.17 $(b_3 t_{15}, \bar{c}_9) = 1$. Then there exists $y_3 \in B$ such that

$$b_3 c_9 = t_{15} + d_9 + y_3, \qquad d_3 \bar{x}_6 = t_{15} + \bar{y}_3.$$

Consequently $b_3 \bar{y}_3 = \bar{c}_9$. Since $b_3 c_9 = (d_3 \bar{c}_3)b_3 = (b_3 d_3)\bar{c}_3 = \bar{c}_3 \bar{b}_9$, we have that

$$c_3 b_9 = t_{15} + d_9 + y_3.$$

Consequently $c_3\bar{y}_3 = \bar{b}_9$. Since

$$b_3^2 c_9 = c_3 c_9 + b_6 c_9$$
$$= d_3 + \bar{c}_9 + \bar{y}_{15} + \bar{b}_6 + 2\bar{c}_9 + 2\bar{y}_{15},$$
$$(b_3 c_9) b_3 = b_3 t_{15} + b_3 d_9 + b_3 y_3$$
$$= \bar{b}_6 + \bar{c}_9 + 2\bar{y}_{15} + d_3 + \bar{c}_9 + \bar{y}_{15} + b_3 y_3,$$

then $b_3 y_3 = \bar{c}_9$.

Since we have that

$$(b_3\bar{b}_3) t_{15} = t_{15} + b_8 t_{15}$$
$$(b_3 t_{15})\bar{b}_3 = \bar{b}_3\bar{b}_6 + \bar{b}_3\bar{c}_9 + 2\bar{b}_3\bar{y}_{15}$$
$$= r_3 + t_{15} + d_9 + \bar{y}_3 + t_{15} + 2s_6 + 4t_{15} + 2d_9,$$

so $b_8 t_{15} = r_3 + \bar{y}_3 + 5t_{15} + 2s_6 + 3d_9$, which implies that y_3 is real.

Step 3 $b_3 d_9 = \bar{c}_9 + \bar{y}_{15} + d_3$, $r_3 t_{15} = b_5 + c_5 + b_8 + c_8 + x_9 + x_{10}$, $r_3 d_9 = c_8 + x_9 + x_{10}$.

Since

$$b_3^2 y_{15} = c_3 y_{15} + b_6 y_{15}$$
$$= \bar{b}_6 + \bar{c}_9 + 2\bar{y}_{15} + \bar{c}_3 + d_3 + \bar{b}_6 + 2\bar{c}_9 + 4\bar{y}_{15},$$
$$(b_3 y_{15}) b_3 = b_3 s_6 + 2b_3 t_{15} + b_3 d_9$$
$$= \bar{c}_3 + \bar{y}_{15} + b_3 d_9 + 2\bar{b}_6 + 2\bar{c}_9 + 4\bar{y}_{15},$$

then $b_3 d_9 = \bar{c}_9 + \bar{y}_{15} + d_3$.

Since $r_3^2 b_8 = (r_3 b_8) r_3$ and $r_3^2 x_{10} = (r_3 x_{10}) r_3$, we have that

$$r_3 t_{15} = b_5 + c_5 + b_8 + c_8 + x_9 + x_{10},$$
$$r_3 d_9 = c_8 + x_9 + x_{10}.$$

Step 4 $r_3 d_3 = b_9$ and $r_3 b_9 = d_3 + \bar{c}_9 + \bar{y}_{15}$.

There are three possibilities: $r_3 d_3 = t_9, m_3 + n_6, m_4 + n_5$. Thus

$$r_3^2 d_3 = d_3 + d_3 b_8$$
$$= d_3 + \bar{c}_9 + \bar{y}_{15},$$
$$(r_3 d_3) r_3 = r_3 t_9$$
$$\text{or } m_3 r_3 + n_6 r_3$$
$$\text{or } m_4 r_3 + n_5 r_3,$$

If $r_3 d_3 = m_3 + n_6$, then $m_3 r_3 = \bar{c}_9$ and $r_3 n_6 = d_3 + \bar{y}_{15}$. But $r_3\bar{y}_{15} = x_6 + 2x_{15} + b_9$. Hence $n_6 = x_6$, so that $(r_3 x_6, d_3) = 1$, a contradiction.

If $r_3d_3 = m_4 + n_5$, then $m_4r_3 = d_3 + \bar{c}_9$ and $r_3n_5 = \bar{y}_{15}$. But $(r_3\bar{y}_{15}, n_5) = 0$, a contradiction. Hence $r_3d_3 = t_9$ and $r_3t_9 = d_3 + \bar{c}_9 + \bar{y}_{15}$, which implies that $(r_3\bar{y}_{15}, t_9) = 1$. Then $t_9 = b_9$.

Step 5 $x_6y_{15} = r_3 + 4t_{15} + s_6 + 2d_9 + y_3$, $b_8x_{15} = b_3 + \bar{z}_3 + 2x_6 + 5x_{15} + 3b_9$, $b_8b_9 = \bar{z}_3 + x_6 + 2b_9 + 3x_{15}$, $b_8d_9 = y_3 + s_6 + 2d_9 + 3t_{15}$, $y_3b_8 = d_9 + t_{15}$ and $b_6b_9 = s_6 + 2d_9 + 2t_{15}$ and $b_6x_{15} = r_3 + s_6 + y_3 + 2d_9 + 4t_{15}$.

The above equations follow from $b_3x_6^2 = (b_3x_6)x_6$, $(b_3\bar{b}_3)x_{15} = (b_3x_{15})\bar{b}_3$, $(b_3\bar{b}_3)b_9 = b_3b_9)\bar{b}_3$, $(b_3\bar{b}_3)d_9 = (b_3d_9)\bar{b}_3$, $(b_3\bar{b}_3)y_3 = (b_3y_3)\bar{b}_3$, $b_3^2b_9 = (b_3b_9)b_3$ and $b_3^2x_{15} = (b_3x_{15})b_3$.

Step 6 $c_3d_9 = \bar{b}_9 + z_3 + \bar{x}_{15}$, $b_6d_9 = \bar{x}_6 + 3\bar{b}_9 + 2\bar{x}_{15}$, $z_3\bar{c}_3 = d_9$.

Since

$$b_3^2d_9 = c_3d_9 + b_6d_9,$$

$$(b_3d_9)b_3 = b_3d_3 + b_3\bar{c}_9 + b_3\bar{y}_{15}$$

$$= 2\bar{b}_9 + 3\bar{x}_{15} + z_3 + \bar{x}_6,$$

and $(c_3x_{15}, d_9) = 1$, from which Step 6 follows.

Step 7 $b_8x_{15} = b_3 + \bar{z}_3 + 2x_6 + 5x_{15} + 3b_9$, $b_3x_9 = \bar{z}_3 + b_9 + x_{15}$, $b_3c_8 = x_{15} + b_9$, $b_3b_5 = x_{15}$ and $b_3c_5 = x_{15}$.

By Lemma 2.17 and Step 1, we have

$$(b_3x_{15})\bar{b}_3 = b_6\bar{b}_3 + 2y_{15}\bar{b}_3 + c_9\bar{b}_3 = b_3 + x_{15} + x_6 + 2x_{15} + b_9 + b_9 + x_{15} + \bar{z}_3,$$

$$(b_3\bar{b}_3)x_{15} = x_{15} + b_8x_{15},$$

hence

$$b_8x_{15} = b_3 + \bar{z}_3 + 2x_6 + 5x_{15} + 3b_9.$$

Consequently, we have the following equations:

$$b_8^2b_3 = b_3 + 2b_3b_8 + 2b_3x_{10} + b_3b_5 + b_3c_5 + b_3c_8 + b_3x_9$$

$$= 3b_3 + 2x_6 + 2x_{15} + 2x_6 + 2x_{15} + 2b_9 + b_3b_5 + b_3c_5 + b_3c_8 + b_3x_9,$$

$$(b_3b_8)b_8 = b_3b_8 + x_6b_8 + x_{15}b_8$$

$$= b_3 + x_6 + x_{15} + b_3 + x_6 + b_9 + 2x_{15} + b_3 + \bar{z}_3 + 2x_6 + 5x_{15} + 3b_9,$$

which imply that

$$b_3b_5 + b_3c_5 + b_3c_8 + b_3x_9 = 4x_{15} + 2b_9 + \bar{z}_3.$$

By the expression $b_3\bar{x}_{15}$ in Step 3, it can be easily shown that Step 7 follows.

Step 8 $r_3b_5 = t_{15}$, $r_3c_5 = t_{15}$, $r_3c_8 = t_{15} + d_9$, $r_3x_9 = t_{15} + d_9 + y_3$ and $r_3y_3 = x_9$.

Since

$$b_8^2 r_3 = r_3 + 2r_3 b_8 + 2r_3 x_{10} + r_3 b_5 + r_3 c_5 + r_3 c_8 + r_3 x_9$$

$$= r_3 + 2r_3 + 2s_6 + 2t_{15} + 2s_6 + 2t_{15} + 2d_9 + r_3 b_5 + r_3 c_5 + r_3 c_8 + r_3 x_9,$$

$$(r_3 b_8)b_8 = r_3 b_8 + b_8 s_6 + b_8 t_{15}$$

$$= r_3 + s_6 + t_{15} + r_3 + s_6 + 2t_{15} + d_9 + r_3 + y_3 + 5t_{15} + 2s_6 + 3d_9,$$

it holds that

$$r_3 b_5 + r_3 c_5 = r_3 c_8 + r_3 x_9 = 4t_{15} + 2d_9 + y_3.$$

But $(r_3 t_{15}, x_9) = 1$. The first four equations follow from the above equation. Consequently, $r_3 y_3 = x_9$.

Step 9 $r_3 c_9 = \bar{b}_9 + \bar{x}_{15} + z_3$, $r_3 z_3 = c_9$, $x_6 \bar{y}_{15} = \bar{b}_3 + z_3 + \bar{x}_6 + 4\bar{x}_{15} + 2\bar{b}_9$, $x_6 c_9 = 2t_{15} + 2d_9 + s_6$, $x_6 x_{10} = b_9 + b_3 + \bar{z}_3 + 3x_{15}$, $x_6 \bar{x}_6 = 1 + b_8 + b_5 + c_5 + c_8 + x_9$, $x_6 s_6 = 2\bar{b}_6 + \bar{c}_9 + \bar{y}_{15}$ and $x_6 c_9 = 2\bar{x}_{15} + \bar{x}_6 + 2\bar{b}_9$.

Since $r_3^2 b_9 = (r_3 b_9)r_3$, we have the first equation, from which the second equation follows. The rest of the equations follow from associativity of $(c_3 b_8)\bar{x}_6 = b_8(c_3 \bar{x}_6)$, $(d_3 b_8)\bar{x}_6 = b_8(d_3 \bar{x}_6)$, $(b_3 \bar{x}_6)x_6 = (b_3 x_6)\bar{x}_6$, $(b_3 \bar{c}_3)x_6 = (b_3 x_6)\bar{c}_3$, $(b_3 c_3)x_6 = b_3(c_3 x_6)$ and $(r_3 c_3)c_9 = r_3(c_3 c_9)$, respectively.

Step 10 $x_6 y_3 = d_3 + \bar{y}_{15}$.

By known equations, we have that $(x_6 y_3, \bar{y}_{15}) = (x_6 y_{15}, y_3) = 1$ and $(x_6 y_3, d_3) = (x_6 \bar{d}_3, y_3) = 1$, from which Step 10 follows.

Step 11 $x_6 b_5 = x_6 + b_9 + x_{15}$, $x_6 c_5 = x_6 + b_9 + x_{15}$, $x_6 c_8 = x_6 + b_9 + 2x_{15} + z_3$ and $z_3 x_6 = x_{10} + c_8$.

Since

$$b_8^2 x_6 = x_6 + 2x_6 b_8 + 2x_6 x_{10} + x_6 b_5 + x_6 c_5 + x_6 c_8 + x_6 x_9$$

$$= x_6 + 2b_3 + 2x_6 + 2b_9 + 4x_{15} + 2b_9 + 2b_3 + 2\bar{z}_3 + 6x_{15} +$$

$$+ x_6 b_5 + x_6 c_5 + x_6 c_8 + x_6 + 2b_9 + 2x_{15},$$

$$(b_8 x_6)b_8 = b_3 b_8 + x_6 b_8 + b_9 b_8 + 2b_8 x_{15}$$

$$= b_3 + x_6 + x_{15} + b_3 + x_6 + b_9 + 2x_{15} + \bar{z}_3 + x_6 + 2b_9 +$$

$$+ 3x_{15} + 2b_3 + 2\bar{z}_3 + 4x_6 + 10x_{15} + 6b_9,$$

we have that

$$x_6 b_5 + x_6 c_5 + x_6 c_8 = 3x_6 + 3b_9 + \bar{z}_3 + 4x_{15}.$$

Table 2.3 Further calculations

Values of inner products	New equations
$(z_3b_9, b_8) = (b_8b_9, \bar{z}_3) = 1$	
$(z_3x_{15}, b_8) = (b_8x_{15}, \bar{z}_3) = 1$	$z_3b_8 = \bar{b}_9 + \bar{x}_{15}$
$(x_6x_{10}, b_9) = (x_6c_5, b_9) = (x_6b_5, b_9) = 1$	
$(x_6b_8, b_9) = (x_6c_8, b_9) = 1, (x_6x_9, b_9) = 2$	$x_6\bar{b}_9 = b_5 + c_5 + x_{10} + b_8 + c_8 + 2x_9$
$(x_6b_8, x_{15}) = 2, (x_6x_{10}, x_{15}) = 3,$	
$(x_6b_5, x_{15}) = (x_6c_5, x_{15}) = 1$	$x_6\bar{x}_{15} = b_5 + c_5 + 2b_8 + 3x_{10} + 2c_8 + 2x_9$
$(x_6c_8, x_{15}) = (x_6x_9, x_{15}) = 2$	
$(x_6c_3, t_{15}) = (x_6\bar{d}_3, t_{15}) = (x_6b_6, t_{15}) = 1$	
$(x_6c_9, t_{15}) = 2, (x_6y_{15}, t_{15}) = 4$	$x_6t_{15} = \bar{c}_3 + d_3 + 2\bar{c}_9 + 4\bar{y}_{15} + \bar{b}_6$
$(x_6\bar{z}_3, \bar{d}_3) = (x_6\bar{y}_{15}, z_3) = 1$	$x_6\bar{z}_3 = \bar{d}_3 + y_{15}$
$(x_6d_3, \bar{x}_{15}) = (x_6\bar{b}_6, \bar{x}_{15}) = (x_6\bar{c}_3, \bar{x}_{15}) = 1$	
$(x_6\bar{c}_9, \bar{x}_{15}) = 2, (x_6\bar{y}_{15}, \bar{x}_{15}) = 4$	$x_6x_{15} = 4y_{15} + b_6 + 2c_9 + d_3 + c_3$

By the expression $x_6\bar{x}_6$ in Step 9, we have that $(x_6b_5, x_6) = (x_6c_5, x_6) = (x_6c_8, x_6) = 1$. The first three equations of Step 11 follow by comparing the degrees. Hence $(x_6z_3, c_8) = 1$. In addition to $(z_3x_6, x_{10}) = (x_6x_{10}, \bar{z}_3) = 1$, from which the last equation follows.

Step 12 $x_6x_9 = x_6 + 2b_9 + 2x_{15}$, $x_6d_9 = \bar{b}_6 + 2\bar{c}_9 + 2\bar{y}_{15}$, $x_6d_3 = z_3 + \bar{x}_{15}$ and $x_6b_9 = 2\bar{c}_9 + \bar{b}_6 + 2\bar{y}_{15}$.

It follows from $(b_3z_3)x_6 = b_3(z_3x_6)$, $(z_3\bar{c}_3)x_6 = \bar{c}_3(z_3x_6)$, $(r_3c_3)\bar{d}_3 = r_3(c_3\bar{d}_3)$ and $(r_3c_3)\bar{b}_9 = (r_3\bar{b}_9)c_3$, respectively.

Step 13 Based on the equations known by us, we can check the inner products of some products of basis elements and obtain Table 2.3.

Step 14 Repeatedly considering the associativity of basis elements, we can obtain new equations. Since the process is trivial and repeated, the concrete process is omitted and the correspondence between the associativity of basis elements and new equations are listed in the following table. Occasionally, we need to check the inner products based on the old and new equations to obtain a newer equation. Table 2.4 is an explanation. □

Table 2.4 Further calculations

Associativity checked or inner products	New equations
$(b_3c_3)\bar{d}_3 = b_3(c_3\bar{d}_3)$	$d_3s_6 = \bar{z}_3 + x_{15}$
$(b_3c_3)y_3 = b_3(c_3y_3)$	$y_3s_6 = c_8 + x_{10}$
$(b_3c_3)z_3 = (b_3z_3)c_3$	$z_3s_6 = \bar{d}_3 + y_{15}$
$(b_3c_3)\bar{z}_3 = b_3(c_3\bar{z}_3)$	$\bar{z}_3s_6 = d_3 + \bar{y}_{15}$
$(b_3c_3)s_6 = b_3(c_3s_6)$	$s_6^2 = 1 + b_5 + c_5 + b_8 + c_8 + x_9$
$(b_3c_3)b_9 = b_3(c_3b_9)$	$b_9s_6 = \bar{b}_6 + 2\bar{c}_9 + 2\bar{y}_{15}$
$(b_3c_3)x_{15} = b_3(c_3x_{15})$	$s_6x_{15} = \bar{c}_3 + d_3 + \bar{b}_6 + 2\bar{c}_9 + 4\bar{y}_{15}$
$(b_3c_3)t_{15} = b_3(c_3t_{15})$	$s_6t_{15} = b_5 + c_5 + 2b_8 + 2c_8 + 2x_9 + 3x_{10}$
$(b_3c_3)d_9 = b_3(c_3d_9)$	$s_6d_9 = b_5 + c_5 + 2x_9 + b_8 + c_8 + x_{10}$
$(b_3c_3)x_{10} = b_3(c_3x_{10})$	$s_6x_{10} = r_3 + y_3 + d_9 + 3t_{15}$
$(b_3c_3)b_5 = b_3(c_3b_5)$	$s_6b_5 = s_6 + d_9 + t_{15}$
$(b_3c_3)c_5 = b_3(c_3c_5)$	$s_6c_5 = s_6 + t_{15} + d_9$
$(b_3c_3)c_8 = b_3(c_3c_8)$	$s_6c_8 = y_3 + s_6 + d_9 + 2t_{15}$
$(b_3c_3)x_9 = b_3(c_3x_9)$	$s_6x_9 = s_6 + 2d_9 + 2t_{15}$
$(b_3c_3)c_9 = b_3(c_3c_9)$	$s_6c_9 = \bar{x}_6 + 2\bar{b}_9 + 2\bar{x}_{15}$
$(s_6b_3, \bar{y}_{15}) = 1$	
$(s_6b_9, \bar{y}_{15}) = 2, (s_6x_{15}, \bar{y}_{15}) = 4$	
$(s_6\bar{z}_3, \bar{y}_{15}) = (s_6x_6, \bar{y}_{15}) = 1$	$s_6y_{15} = 2\bar{b}_9 + 4\bar{x}_{15} + z_3 + \bar{x}_6 + \bar{b}_3$
$(b_3d_3)b_9 = d_3(b_3b_9)$	$b_9\bar{b}_9 = 1 + b_5 + c_5 + 2b_8 + 2c_8 + 2x_9 + 2x_{10}$
$(b_3d_3)\bar{b}_9 = d_3(b_3\bar{b}_9)$	$b_9^2 = \bar{d}_3 + c_3 + 2b_6 + 2c_9 + 3y_{15}$
$(b_3d_3)\bar{x}_{15} = d_3(b_3\bar{x}_{15})$	$b_9x_{15} = 6y_{15} + 3c_9 + \bar{d}_3 + 2b_6 + c_3$
$(b_3d_3)x_{15} = d_3(b_3x_{15})$	$b_9\bar{x}_{15} = 2b_5 + 2c_5 + 3c_8 + 3b_8 + 4x_{10} + 3x_9$
$(b_3d_3)t_{15} = d_3(b_3t_{15})$	$b_9t_{15} = d_3 + \bar{c}_3 + 6\bar{y}_{15} + 3\bar{c}_9 + 2\bar{b}_6$
$(b_3d_3)d_9 = d_3(b_3d_9)$	$b_9d_9 = \bar{c}_3 + d_3 + 2\bar{b}_6 + 2\bar{c}_9 + 3\bar{y}_{15}$
$(b_3d_3)y_3 = d_3(b_3y_3)$	$b_9y_3 = \bar{c}_3 + \bar{y}_{15} + \bar{c}_9$
$(b_3d_3)z_3 = d_3(b_3z_3)$	$b_9\bar{z}_3 = c_9 + c_3 + y_{15}$
$(b_3d_3)\bar{z}_3 = d_3(b_3\bar{z}_3)$	$b_9z_3 = b_8 + x_9 + x_{10}$
$(b_3d_3)x_{10} = b_3(d_3x_{10})$	$b_9x_{10} = \bar{z}_3 + b_3 + 2b_9 + x_6 + 4x_{15}$
$(b_3d_3)b_5 = b_3(d_3b_5)$	$\bar{b}_9c_9 = b_3 + \bar{z}_3 + 2x_6 + 3x_{15} + 2b_9$
$(b_3d_3)c_5 = b_3(d_3c_5)$	$b_9c_9 = r_3 + y_3 + 2s_6 + 2d_9 + 3t_{15}$
$(b_3d_3)b_6 = b_3(d_3b_6)$	$\bar{b}_9b_6 = 2b_9 + 2x_{15} + x_6$
$(b_3d_3)c_8 = b_3(d_3c_8)$	$\bar{b}_9y_{15} = 6x_{15} + b_3 + \bar{z}_3 + 2x_6 + 3b_9$
$(b_3d_3)x_9 = b_3(d_3x_9)$	$\bar{b}_9x_9 = 2\bar{b}_9 + 3\bar{x}_{15} + z_3 + \bar{b}_3 + 2\bar{x}_6$
$(b_3d_3)\bar{y}_{15} = b_3(d_3\bar{y}_{15})$	$\bar{b}_9\bar{y}_{15} = r_3 + y_3 + 2s_6 + 6t_{15} + 3d_9$
$(b_3d_3)d_3 = b_3(d_3^2)$	$\bar{b}_9d_3 = r_3 + t_{15} + d_9$
$(b_3d_3)c_3 = b_3(d_3c_3)$	$\bar{b}_9c_3 = \bar{z}_3 + b_9 + x_{15}$

(continued on the next page)

Table 2.4 (Continued)

Associativity checked or inner products	New equations
$(b_3d_3)\bar{d}_3 = b_3(d_3\bar{d}_3)$	$b_9d_3 = \bar{b}_3 + \bar{b}_9 + \bar{x}_{15}$
$(b_3d_3)c_3 = b_3(d_3c_3)$	$b_9\bar{c}_3 = z_3 + \bar{b}_9 + \bar{x}_{15}$
$(b_3b_5)x_{15} = (b_3x_{15})b_5$	$x_{15}^2 = 2c_3 + 2\bar{d}_3 + 4b_6 + 6c_9 + 9y_{15}$
$(b_3b_5)\bar{x}_{15} = (b_3\bar{x}_{15})b_5$	$x_{15}\bar{x}_{15} = 1 + 6x_9 + 3b_5 + 3c_5 + 5b_8 + +5c_8 + 6x_{10}$
$(b_3b_5)t_{15} = (b_3t_{15})b_5$	$x_{15}t_{15} = 4\bar{b}_6 + 6\bar{c}_9 + 9\bar{y}_{15} + +2\bar{c}_3 + 2d_3$
$(b_3b_5)d_9 = (b_3d_9)b_5$	$x_{15}d_9 = \bar{c}_3 + d_3 + 2\bar{b}_6 + 3\bar{c}_9 + 6\bar{y}_{15}$
$(b_3b_5)z_3 = (b_3z_3)b_5$	$x_{15}z_3 = b_5 + c_5 + b_8 + c_8 + x_9 + x_{10}$
$(b_3b_5)b_5 = b_3b_5^2$	$x_{15}b_5 = \bar{z}_3 + b_3 + x_6 + 2b_9 + 3x_{15}$
$(b_3b_5)x_{10} = (b_3x_{10})b_5$	$x_{15}x_{10} = b_3 + \bar{z}_3 + 3x_6 + 4b_9 + 6x_{15}$
$(b_3b_5)c_5 = (b_5c_5)b_3$	$x_{15}c_5 = b_3 + \bar{z}_3 + x_6 + 2b_9 + 3x_{15}$
$(b_3b_5)c_8 = (b_5c_8)b_3$	$x_{15}c_8 = b_3 + \bar{z}_3 + 2x_6 + 3b_9 + 5x_{15}$
$(b_3b_5)x_9 = (b_5x_9)b_3$	$x_{15}x_9 = b_3 + \bar{z}_3 + 2x_6 + 3b_9 + 6x_{15}$
$(b_3b_5)d_3 = (b_5d_3)b_3$	$x_{15}d_3 = \bar{x}_6 + 2\bar{x}_{15} + \bar{b}_9$
$(b_3b_5)\bar{d}_3 = (b_5\bar{d}_3)b_3$	$x_{15}d_3 = s_6 + d_9 + 2t_{15}$
$(b_3b_5)c_9 = (b_5c_9)b_3$	$x_{15}c_9 = r_3 + y_3 + 2s_6 + 6t_{15} + 3d_9$
$(b_3b_5)\bar{c}_9 = (b_5\bar{c}_9)b_3$	$x_{15}\bar{b}_9 = \bar{b}_3 + z_3 + 2\bar{x}_6 + 3\bar{b}_9 + 6\bar{x}_{15}$
$(b_3b_5)\bar{b}_6 = (b_5\bar{b}_6)b_3$	$x_{15}\bar{b}_6 = \bar{b}_3 + z_3 + 2\bar{b}_9 + 4\bar{x}_{15} + \bar{x}_6$
$(b_3b_5)y_{15} = (b_5y_{15})b_3$	$x_{15}y_{15} = 2y_3 + 2r_3 + 4s_6 + 6d_9 + 9t_{15}$
$(b_3b_6)t_{15} = (b_3t_{15})b_6$	$t_{15}^2 = 1 + 5b_8 + 5c_8 + 3b_5 + 3c_5 + 6x_{10} + 6x_9$
$(b_3b_6)d_9 = (b_3d_9)b_6$	$d_9t_{15} = 2b_5 + 2c_5 + 3b_8 + 3c_8 + 3x_9 + 4x_{10}$
$(b_3b_6)y_3 = (b_3y_3)b_6$	$y_3t_{15} = b_5 + c_5 + b_8 + c_8 + x_9 + x_{10}$
$(b_3b_6)z_3 = (b_3z_3)b_6$	$z_3t_{15} = b_6 + c_9 + 2y_{15}$
$(b_3b_6)x_{10} = (b_6x_{10})b_3$	$t_{15}x_{10} = r_3 + y_3 + 4d_9 + 3s_6 + 6t_{15}$
$(b_3b_6)b_5 = (b_6b_5)b_3$	$t_{15}b_5 = r_3 + y_3 + s_6 + 2d_9 + 3t_{15}$
$(b_3b_6)c_5 = (b_6c_5)b_3$	$t_{15}c_5 = r_3 + y_3 + s_6 + 2d_9 + 3t_{15}$
$(b_3b_6)c_8 = (b_6c_8)b_3$	$t_{15}c_8 = r_3 + y_3 + 2s_6 + 3d_9 + 5t_{15}$
$(b_3b_6)x_9 = (b_6x_9)b_3$	$t_{15}x_9 = r_3 + y_3 + 3d_9 + 2s_6 + 6t_{15}$
$(b_3b_6)d_3 = (b_6d_3)b_3$	$t_{15}d_3 = x_6 + b_9 + 2x_{15}$
$(b_3b_6)c_9 = (b_6c_9)b_3$	$t_{15}c_9 = \bar{b}_3 + z_3 + 3\bar{b}_9 + 2\bar{x}_6 + 6\bar{x}_{15}$
$(b_3b_6)b_6 = b_3b_6^2$	$t_{15}b_6 = \bar{b}_3 + z_3 + \bar{x}_6 + 2\bar{b}_9 + 4\bar{x}_{15}$
$(b_3b_6)y_{15} = (b_6y_{15})b_3$	$t_{15}y_{15} = 2\bar{b}_3 + 2z_3 + 4\bar{x}_6 + 6\bar{b}_9 + 9\bar{x}_{15}$
$(b_3\bar{d}_3)d_9 = (b_3d_9)\bar{d}_3$	$d_9^2 = 1 + b_5 + c_5 + 2b_8 + 2c_8 + 2x_9 + 2x_{10}$
$(b_3\bar{d}_3)x_{10} = (b_3x_{10})\bar{d}_3$	$d_9x_{10} = y_3 + r_3 + s_6 + 2d_9 + 4t_{15}$
$(b_3\bar{d}_3)b_5 = (b_3b_5)\bar{d}_3$	$d_9b_5 = s_6 + d_9 + 2t_{15}$
$(b_3\bar{d}_3)c_5 = (b_3c_5)\bar{d}_3$	$d_9c_5 = s_6 + d_9 + 2t_{15}$

(continued on the next page)

Table 2.4 (Continued)

Associativity checked or inner products	New equations
$(b_3\bar{d}_3)c_8 = (b_3c_8)\bar{d}_3$	$d_9c_8 = r_3 + s_6 + 2d_9 + 3t_{15}$
$(b_3\bar{d}_3)d_3 = (b_3d_3)\bar{d}_3$	$d_9d_3 = b_3 + b_9 + x_{15}$
$(b_3\bar{d}_3)c_9 = (b_3c_9)\bar{d}_3$	$d_9c_9 = b_3 + z_3 + 2\bar{x}_6 + 2\bar{b}_9 + 3\bar{x}_{15}$
$(b_3\bar{d}_3)y_{15} = (b_3y_{15})\bar{d}_3$	$d_9y_{15} = \bar{b}_3 + z_3 + 2\bar{x}_6 + 3\bar{b}_9 + 6\bar{x}_{15}$
$(b_3\bar{d}_3)x_9 = (\bar{d}_3x_9)b_3$	$d_9x_9 = r_3 + y_3 + 2s_6 + 2d_9 + 3t_{15}$
$(x_6d_3)z_3 = (z_3x_6)d_3$	$z_3^2 = d_3 + \bar{b}_6$
$(x_6d_3)\bar{z}_3 = (\bar{z}_3x_6)d_3$	$z_3\bar{z}_3 = 1 + b_8$
$(x_6\bar{d}_3)z_3 = (y_3x_6)\bar{d}_3$	$y_3^2 = 1 + c_8$
$(x_6\bar{d}_3)y_3 = (y_3x_6)d_3$	$y_3z_3 = \bar{d}_3 + b_6$
$(x_6d_3)\bar{d}_3 = (d_3\bar{d}_3)x_6$	$z_3\bar{d}_3 = \bar{z}_3 + x_6$
$(x_6d_3)\bar{c}_9 = (d_3\bar{c}_9)x_6$	$z_3\bar{c}_9 = r_3 + d_9 + t_{15}$
$(x_6d_3)\bar{b}_6 = (d_3\bar{b}_6)x_6$	$z_3\bar{b}_6 = y_3 + t_{15}$
$(x_6d_3)b_6 = (d_3b_6)x_6$	$z_3b_6 = \bar{z}_3 + x_{15}$
$(r_3x_{10}, s_6) = (s_6b_8, r_3) = 1$	$r_3s_6 = x_{10} + b_8$
$(d_9y_{15}, \bar{x}_6) = 2, (d_9b_6, \bar{x}_6) = 1, (d_9c_9, \bar{x}_6) = 2$	$x_6d_9 = \bar{b}_6 + 2\bar{y}_{15} + 2\bar{c}_9$
$(x_6y_3, d_3) = (x_6t_{15}, d_3) = 1$	$x_6\bar{d}_3 = y_3 + t_{15}$
$(r_3c_3)\bar{d}_3 = r_3(c_3\bar{d}_3)$	$x_6d_3 = z_3 + \bar{x}_{15}$
$(\bar{x}_6b_6, b_9) = 1, (\bar{b}_9y_{15}, x_6) = (\bar{b}_9c_9, x_6) = 2$	$x_6b_9 = b_6 + 2y_{15} + 2c_9$
$(x_6\bar{b}_9, x_9) = (x_6\bar{x}_{15}, x_9) = 2, (x_6\bar{x}_6, x_9) = 1$	$x_6x_9 = 2b_9 + x_6 + 2x_{15}$
$(b_9\bar{b}_9, c_8) = 2, (b_9\bar{x}_{15}, c_8) = 1, (x_6c_8, b_9) = 1$	$b_9c_8 = 2b_9 + 3x_{15} + x_6 + b_3$
$(b_9\bar{b}_9, b_8) = 2, (b_9\bar{x}_{15}, b_8) = 3,$	
$(b_9z_3, b_8) = (x_6b_8, b_9) = 1$	$b_9b_8 = 2b_9 + 3x_{15} + \bar{z}_3 + x_6$
$(b_9\bar{b}_9, c_5) = (x_6c_5, b_9) = 1, (b_9\bar{x}_{15}, c_5) = 2$	$c_5b_9 = b_9 + 2x_{15} + x_6$
$(x_{15}c_9, y_3) = (x_{15}b_6, y_3) = 1, (x_{15}y_{15}, y_3) = 2$	$x_{15}y_3 = \bar{b}_6 + \bar{c}_9 + 2\bar{y}_{15}$
$(b_3b_5)\bar{z}_3 = (b_3\bar{z}_3)b_5$	$x_{15}\bar{z}_3 = b_6 + c_9 + 2y_{15}$
$(b_3b_5)\bar{d}_3 = (b_5\bar{d}_3)b_3$	$x_{15}\bar{d}_3 = s_6 + 2t_{15} + d_9$
$(b_3b_5)\bar{c}_9 = (b_5\bar{c}_9)b_3$	$x_{15}\bar{c}_9 = \bar{b}_3 + z_3 + 3\bar{b}_9 + 2\bar{x}_6 + 6\bar{x}_{15}$
$(b_3b_5)y_{15} = (b_5y_{15})b_3$	$x_{15}\bar{y}_{15} = 9\bar{x}_{15} + 2z_3 + 2\bar{b}_3 + 4\bar{x}_6 + 6\bar{b}_9$
$(x_6t_{15}, c_3) = (b_9t_{15}, \bar{c}_3) = 1, (x_{15}t_{15}, \bar{c}_3) = 2$	$t_{15}c_3 = \bar{x}_6 + \bar{b}_9 + 2\bar{x}_{15}$
$(d_9b_8, y_3) = (d_9x_{10}, y_3) = (d_9x_9, y_3) = 1$	$d_9y_3 = x_9 + b_8 + x_{10}$
$(d_9y_{15}, z_3) = (d_9c_3, z_3) = (d_9c_9, z_3) = 1$	$d_9z_3 = c_3 + c_9 + y_{15}$
$(b_3\bar{b}_3)d_9 = (b_3d_9)\bar{b}_3$	$d_9b_8 = y_3 + s_6 + 2d_9 + 3t_{15}$
$(b_9d_9, \bar{c}_3) = (d_9z_3, c_3) = (x_{15}d_9, \bar{c}_3) = 1$	$d_9c_3 = \bar{b}_9 + z_3 + \bar{x}_{15}$
$(d_9b_8, y_3) = (t_{15}b_8, y_3) = 1$	$y_3b_8 = d_9 + t_{15}$
$(s_6x_{10}, y_3) = (t_{15}x_{10}, y_3) = (d_9x_{10}, y_3) = 1$	$y_3x_{10} = s_6 + t_{15} + d_9$

(*continued on the next page*)

Table 2.4 (Continued)

Associativity checked or inner products	New equations
$(t_{15}y_3, b_5) = 1$	$y_3b_5 = t_{15}$
$(t_{15}y_3, c_5) = 1$	$y_3c_5 = t_{15}$
$(t_{15}y_3, c_8) = (s_6y_3, c_8) = (y_3c_8, y_3) = 1$	$y_3c_8 = s_6 + t_{15} + y_3$
$(t_{15}y_3, x_9) = (d_9y_3, x_9) = (r_3y_3, x_9) = 1$	$y_3x_9 = t_{15} + d_9 + r_3$
$(b_9y_3, \bar{c}_3) = 1$	$y_3c_3 = \bar{b}_9$
$(b_9y_3, \bar{d}_3) = 1$	$y_3d_3 = \bar{b}_9$
$(b_3y_3, c_9) = (b_9c_9, y_3) = (x_{15}c_9, y_3) = 1$	$y_3c_9 = \bar{b}_3 + \bar{b}_9 + \bar{x}_{15}$
$(x_{15}y_3, b_6) = (y_3z_3, b_6) = 1$	$y_3b_6 = \bar{x}_{15} + z_3$
$(x_6y_3, \bar{y}_{15}) = (b_9y_3, \bar{y}_{15}) = 1, (x_{15}y_3, \bar{y}_{15}) = 2$	$y_3y_{15} = \bar{x}_6 + \bar{b}_9 + 2\bar{x}_{15}$
$(x_6x_{10}, z_3) = (b_9x_{10}, z_3) = (x_{15}x_{10}, z_3) = 1$	$z_3x_{10} = \bar{x}_6 + \bar{b}_9 + \bar{x}_{15}$
$(x_{15}b_5, z_3) = 1$	$z_3b_5 = \bar{x}_{15}$
$(x_{15}c_5, z_3) = 1$	$z_3c_5 = \bar{x}_{15}$
$(x_6c_8, z_3) = (x_{15}c_8, z_3) = (z_3\bar{z}_3, c_8) = 1$	$z_3c_8 = \bar{x}_6 + \bar{x}_{15} + z_3$
$(b_3x_9, z_3) = (b_9x_9, z_3) = (x_{15}x_9, z_3) = 1$	$z_3x_9 = \bar{b}_3 + \bar{b}_9 + \bar{x}_{15}$
$(b_9c_3, z_3) = 1$	$z_3c_3 = b_9$
$(s_6\bar{d}_3, z_3) = (d_3y_3, \bar{z}_3) = 1$	$z_3d_3 = s_6 + y_3$
$(\bar{b}_3c_9, \bar{z}_3) = (\bar{b}_9c_9, \bar{z}_3) = (x_{15}\bar{c}_9, \bar{z}_3) = 1$	$z_3y_{15} = b_3 + b_9 + x_{15}$
$(x_6\bar{y}_{15}, z_3) = (b_9\bar{y}_{15}, z_3) = 1, (x_{15}\bar{y}_{15}, z_3) = 2$	$z_3y_{15} = x_6 + b_9 + 2x_{15}$
$(s_6y_{15}, z_3) = (d_9y_{15}, z_3) = 1, (t_{15}y_{15}, z_3) = 2$	$z_3\bar{y}_{15} = s_6 + 2t_{15} + d_9$

2.8 Structure of NITA Generated by b_3 and Satisfying $b_3^2 = c_3 + b_6$, $c_3 \neq b_3$, $\bar{b}_3$, $(b_3b_8, b_3b_8) = 3$ and c_3 Real

In this section we discuss NITA satisfying the following hypothesis.

Hypothesis 2.4 Let (A, B) be a NITA generated by a nonreal element $b_3 \in B$ of degree 3 and without non-identity basis element of degree 1 or 2, such that:

$$\bar{b}_3b_3 = 1 + b_8$$

and

$$b_3^2 = c_3 + b_6, \quad c_3 = \bar{c}_3$$

where $b_i, c_i, d_i \in B$ are of degree i.

Lemma 2.25 Let (A, B) satisfy Hypothesis 2.4, then

1) $c_3^2 = 1 + b_8$,
2) $\bar{b}_3c_3 = b_3 + x_6, x_6 \neq \bar{x}_6 \in B$,
3) $b_8b_3 = x_6 + b_6\bar{b}_3$,
4) $b_8 + b_8^2 = c_3\bar{b}_6 + b_6c_3 + b_6\bar{b}_6$,

5) $\bar{b}_6 c_3 = x_6 \bar{b}_3$,

6) $c_3 b_8 = b_6 + x_6 b_3$,

7) $(b_8 b_3, b_8 b_3) = 3, 4$.

Proof By Hypothesis 2.4, we have $(\bar{b}_3 \bar{c}_3, b_3) = (c_3, b_3^2) = 1$. So $(\bar{b}_3 b_3, c_3^2) = (\bar{b}_3 c_3, \bar{b}_3 c_3) \geq 2$, hence

$$c_3^2 = 1 + b_8. \tag{1}$$

Furthermore, we have

$$\bar{b}_3 c_3 = b_3 + x_6. \tag{2}$$

If $x_6 = \bar{x}_6$, then $1 = (\bar{b}_3 c_3, b_3 c_3) = (c_3^2, b_3^2)$, a contradiction to Hypothesis 2.4. By the associative law $(\bar{b}_3 b_3) b_3 = \bar{b}_3 b_3^2$ and Hypothesis 2.4, one concludes that $b_3 + b_8 b_3 = c_3 \bar{b}_3 + b_6 \bar{b}_3$. So $b_8 b_3 = x_6 + b_6 \bar{b}_3$ holds by (2). By the associative law $(\bar{b}_3 b_3)^2 = \bar{b}_3^2 b_3^2$ and Hypothesis 2.4 and (1), we see that

$$1 + 2b_8 + b_8^2 = (c_3 + \bar{b}_6)(c_3 + b_6),$$

$$1 + 2b_8 + b_8^2 = c_3^2 + c_3 b_6 + \bar{b}_6 c_3 + \bar{b}_6 b_6,$$

$$b_8^2 + b_8 = c_3 b_6 + \bar{b}_6 c_3 + \bar{b}_6 b_6.$$

By the associative law $(\bar{b}_6 b_3) b_3 = \bar{b}_3 b_3^2$ and Hypothesis 2.4 and (2), it holds that $b_3 + b_8 b_3 = c_3 \bar{b}_3 + b_6 \bar{b}_3$; hence

$$b_8 b_3 = x_6 + b_6 \bar{b}_3. \tag{3}$$

By Hypothesis 2.4, $(b_6 \bar{b}_3, b_3) = (b_6, b_3^2) = 1$, so by (3)

$$(b_8 b_3, b_8 b_3) \geq 3,$$

we shall show that $(b_8 b_3, b_8 b_3) = 3, 4$. By the associative law and Hypothesis 2.4 and (1) and (2), we have that $(b_3^2) c_3 = \bar{b}_3 (\bar{b}_3 c_3)$, $c_3^2 + \bar{b}_6 c_3 = b_3 \bar{b}_3 + x_6 \bar{b}_3$; thus

$$\bar{b}_6 c_3 = x_6 \bar{b}_3. \tag{4}$$

By (3), $(x_6 \bar{b}_3, b_8) = (x_6, b_3 b_8) = 1$ holds. So $(x_6 \bar{b}_3, x_6 \bar{b}_3) \geq 2$. We shall show that $(x_6 \bar{b}_3, x_6 \bar{b}_3) = 2, 3$. If $(x_6 \bar{b}_3, x_6 \bar{b}_3) = 4$, then $x_6 \bar{b}_3 = b_8 + g_3 + w_3 + v_4$; hence

$$b_3 g_3 = x_6 + w_3, \qquad b_3 w_3 = x_6 + k_3, \qquad b_3 v_4 = x_6 + z_6.$$

Therefore $1 \leq (b_3 g_3, b_3 w_3) = (b_3 \bar{b}_3, \bar{g}_3 w_3)$. We obtain $w_3 = g_3$ by Hypothesis 2.4, hence $2 = (x_6 \bar{b}_3, w_3) = (x_6, b_3 w_3)$, so that $b_3 w_3 = 2x_6$, a contradiction by (4). Now we can state that $(\bar{b}_6 c_3, \bar{b}_6 c_3) = (\bar{b}_6 b_6, c_3^2) = 2, 3$. By (1), it follows that $(\bar{b}_6 b_6, b_3 \bar{b}_3) = (b_6 \bar{b}_3, b_6 \bar{b}_3) = 2$ or 3. Therefore $(b_8 b_3, b_8 b_3) = 3, 4$ by (3).

By the associative law $(\bar{b}_3 c_3) b_3 = c_3 (\bar{b}_3 b_3)$ and Hypothesis 2.4, we get $b_3^2 + x_6 b_3 = c_3 + c_3 b_8$ and $c_3 b_8 = b_6 + x_6 b_3$. □

Lemma 2.26 *Let (A, B) satisfy Hypothesis 2.4 and $(b_8 b_3, b_8 b_3) = 3$, then we get*:

1) $b_6 \bar{b}_3 = b_3 + x_{15}$,
2) $b_8 b_3 = x_6 + b_3 + x_{15}$, $x_6 \neq \bar{x}_6$,
3) $\bar{x}_6 b_3 = b_8 + x_{10} = b_6 c_3$, $x_6 \neq b_6$, $x_{10} = \bar{x}_{10}$,
4) $x_6 c_3 = \bar{b}_3 + \bar{x}_{15}$,
5) $b_3 x_6 = c_3 + y_{15}$,
6) $b_8 c_3 = b_6 + c_3 + y_{15}$, $b_6 = \bar{b}_6$, $y_{15} = \bar{y}_{15}$.

Proof By our assumption in this lemma and part 3) in Lemma 2.25, and Hypothesis 2.4:

$$b_6 \bar{b}_3 = b_3 + x_{15}, \tag{1}$$

then

$$b_8 b_3 = x_6 + b_3 + x_{15}$$

by (1),

$$(b_6 \bar{b}_6, b_3 \bar{b}_3) = (b_6 \bar{b}_3, b_6 \bar{b}_3) = 2.$$

Then by 1) in Lemma 2.25,

$$2 = (b_6 \bar{b}_6, c_3^2) = (\bar{b}_6 c_3, \bar{b}_6 c_3). \tag{2}$$

By 6) in Lemma 2.25,

$$(\bar{b}_6 c_3, b_8) = (c_3 b_8, b_6) = 1,$$

so by (2)

$$\bar{b}_6 c_3 = b_8 + x_{10}. \tag{3}$$

So by 5) in Lemma 2.25

$$x_6 \bar{b}_3 = b_8 + x_{10}; \tag{4}$$

so by (1) we get that $x_6 \neq b_6$. By the associative law $(c_3^2)\bar{b}_3 = (c_3 \bar{b}_3)c_3$ and 1), 2) in Lemma 2.25 and (2), it follows that

$$\bar{b}_3 + \bar{b}_3 b_8 = b_3 c_3 + x_6 c_3,$$
$$\bar{b}_3 + \bar{x}_6 + \bar{b}_3 + \bar{x}_{15} = \bar{b}_3 + \bar{x}_6 + x_6 c_3,$$

so

$$x_6 c_3 = \bar{b}_3 + \bar{x}_{15}. \tag{5}$$

By (4) and 2) in Lemma 2.25:

$$2 = (\bar{x}_6 b_3, x_6 b_3) = (x_6 b_3, x_6 b_3), \qquad (b_3 x_6, c_3) = (x_6, \bar{b}_3 c_3) = 1.$$

Therefore

$$b_3 x_6 = c_3 + y_{15}. \tag{6}$$

By the associative law $(b_3\bar{b}_3)c_3 = (\bar{b}_3c_3)b_3$ and 2) in Lemma 2.25 and Hypothesis 2.4, one has

$$c_3 + b_8c_3 = b_3^2 + x_6b_3,$$

$$c_3 + b_8c_3 = c_3 + b_6 + c_3 + y_{15},$$

$$b_8c_3 = b_6 + c_3 + y_{15}.$$

b_8, c_3 are reals, then

$$b_6 = \bar{b}_6, \qquad y_{15} = \bar{y}_{15}.$$

By (3), we have $x_{10} = \bar{x}_{10}$. □

Lemma 2.27 *Let (A, B) satisfy Hypothesis 2.4 and $(b_8b_3, b_8b_3) = 3$, then Lemma 2.26 holds and we get:*

1) $x_{10}b_3 = x_{15} + x_6 + x_9, x_{10} = \bar{x}_{10},$
2) $x_6b_8 = 2x_{15} + b_3 + x_6 + x_9,$
3) $\bar{x}_{15}c_3 = 2x_{15} + x_6 + x_9,$
4) $y_{15}\bar{b}_3 = 2x_{15} + x_6 + x_9,$
5) $x_6b_6 = 2\bar{x}_6 + \bar{x}_{15} + \bar{x}_9,$
6) $c_3x_{10} = b_6 + y_{15} + y_9,$
7) $x_6^2 = 2b_6 + y_{15} + y_9,$
8) $x_{15}b_3 = 2y_{15} + b_6 + y_9,$
9) $b_8b_6 = 2y_{15} + b_6 + c_3 + y_9, y_9 = \bar{y}_9,$
10) $b_8x_{15} = 4x_{15} + 2x_6 + 2x_9 + b_3 + y_9\bar{b}_3.$

Proof By the associative law $(b_3^2)b_6 = (b_3b_6)b_3$ and Hypothesis 2.4 and 1), 3) in Lemma 2.26, we get

$$c_3b_6 + b_6^2 = \bar{b}_3b_3 + \bar{x}_{15}b_3,$$

$$b_8 + x_{10} + b_6^2 = 1 + b_8 + \bar{x}_{15}b_3.$$

Hence

$$x_{10} + b_6^2 = 1 + \bar{x}_{15}b_3. \tag{1}$$

By 3) in Lemma 2.26,

$$(b_3x_{10}, x_6) = (x_{10}, \bar{b}_3x_6) = 1,$$

hence by (1),

$$(x_{10}b_3, x_{15}) = (b_3\bar{x}_{15}, x_{10}) = 1.$$

So

$$x_{10}b_3 = x_{15} + x_6 + x_9,$$

where x_9 is a linear combination of B. We shall show that $x_9 \in B$. If $x_{10}b_3$ has an element of degree 3, then

$$1 = (x_{10}b_3, y_3) = (x_{10}; \bar{b}_3 y_3),$$

hence $\bar{b}_3 y_3 = x_{10}$, a contradiction. So

$$x_{10}b_3 = x_{15} + x_6 + x_9, \tag{2}$$

where $x_9 \in B$. By the associative law and Hypothesis 2.4 and 2), 3) in Lemma 2.26, we get

$$(b_3\bar{x}_6)\bar{b}_3 = \bar{x}_6(b_3\bar{b}_3),$$

$$b_8\bar{b}_3 + x_{10}\bar{b}_3 = \bar{x}_6 + \bar{x}_6 b_8,$$

$$\bar{x}_6 + \bar{x}_6 b_8 = \bar{x}_6 + \bar{b}_3 + \bar{x}_{15} + \bar{x}_{15} + \bar{x}_6 + \bar{x}_9.$$

Hence

$$\bar{x}_6 b_8 = 2\bar{x}_{15} + \bar{b}_3 + \bar{x}_6 + \bar{x}_9. \tag{3}$$

By the associative law $(\bar{b}_3 c_3)b_8 = \bar{b}_3(b_8 c_3)$ and 1), 2), 6) in Lemma 2.26 and 2) in Lemma 2.25, we obtain

$$b_3 b_8 + x_6 b_8 = b_6 \bar{b}_3 + c_3 \bar{b}_3 + y_{15}\bar{b}_3,$$

$$x_6 + b_3 + x_{15} + 2x_{15} + b_3 + x_6 + x_9 = b_3 + x_{15} + b_3 + x_6 + y_{15}\bar{b}_3.$$

So

$$y_{15}\bar{b}_3 = 2x_{15} + x_6 + x_9. \tag{4}$$

By the associative law $(\bar{b}_3 c_3)b_8 = (\bar{b}_3 b_8)c_3$ and 1), 2), 6) in Lemma 2.26 and in Lemma 2.25 and (3), the following equations hold:

$$b_3 b_8 + x_6 b_8 = \bar{x}_6 c_3 + \bar{b}_3 c_3 + \bar{x}_{15}c_3,$$

$$x_6 + b_3 + x_{15} + 2x_{15} + b_3 + x_6 + x_9 = b_3 + x_{15} + b_3 + x_6 + \bar{x}_{15}c_3.$$

So

$$\bar{x}_{15}c_3 = 2x_{15} + x_6 + x_9. \tag{5}$$

By the associative law $(\bar{b}_3 c_3)b_6 = c_3(\bar{b}_3 b_6)$ and 2) in Lemma 2.25 and 1) in Lemma 2.26 and (5), one should have

$$b_3 b_6 + x_6 b_6 = b_3 c_3 + x_{15}c_3,$$

$$\bar{b}_3 + \bar{x}_{15} + x_6 b_6 = \bar{b}_3 + \bar{x}_6 + 2\bar{x}_{15} + \bar{x}_6 + \bar{x}_9,$$

$$x_6 b_6 = 2\bar{x}_6 + \bar{x}_{15} + \bar{x}_9. \tag{6}$$

By the associative law and 5), 4) in Lemma 2.26 and 1) in Lemma 2.25, we get $(b_3x_6)c_3 = b_3(x_6c_3)$, $c_3^2 + y_{15}c_3 = b_3\bar{b}_3 + b_3\bar{x}_{15}$ and $y_{15}c_3 = b_3\bar{x}_{15}$. Now by (2) it follows that $(\bar{x}_{15}b_5, x_{10}) = (b_3x_{10}, x_{15}) = 1$. So

$$1 = (y_{15}c_3, x_{10}) = (y_{15}, c_3x_{10}). \tag{7}$$

By 3) in Lemma 2.26,

$$(c_3x_{10}, b_6) = (c_3b_6, x_{10}) = 1. \tag{8}$$

By (2) and 1) in Lemma 2.25, we have $(c_3x_{10}, c_3x_{10}) = (c_3^2, x_{10}^2) = (b_3\bar{b}_3, x_{10}^2) = (b_3x_{10}, b_3x_{10}) = 3$. Therefore by (7) and (8), the following equation holds:

$$c_3x_{10} = b_6 + y_{15} + y_9, \tag{9}$$

where $y_9 \in B$. By the associative law and 3), 5), 6) in Lemma 2.26 and 2) in Lemma 2.25 and (9), we have that $(\bar{x}_6b_3)c_3 = \bar{x}_6(b_3c_3)$ and

$$b_8c_3 + x_{10}c_3 = \bar{x}_6\bar{b}_3 + \bar{x}_6^2,$$
$$b_6 + c_3 + y_{15} + b_6 + y_{15} + y_9 = c_3 + y_{15} + \bar{x}_6^2,$$

so

$$\bar{x}_6^2 = 2b_6 + y_{15} + y_9. \tag{10}$$

By the associative law and 3), 4), 6) in Lemma 2.26 and Hypothesis 2.4 and (9), we have $(\bar{x}_6b_3)c_3 = (\bar{x}_6c_3)b_3$. Hence

$$b_8c_3 + x_{10}c_3 = b_3^2 + x_{15}b_3,$$
$$b_6 + c_3 + y_{15} + b_6 + y_{15} + y_9 = c_3 + b_6 + x_{15}b_3,$$

so

$$x_{15}b_3 = 2y_{15} + b_6 + y_9. \tag{11}$$

By the associative law and 2), 5), 6) in Lemma 2.26 and (11) and the Hypothesis 2.4, $b_3(b_8b_3) = b_8(b_3^2)$ holds, from which we conclude

$$x_6b_3 + b_3^2 + x_{15}b_3 = b_8c_3 + b_8b_6,$$
$$c_3 + y_{15} + c_3 + b_6 + 2y_{15} + b_6 + y_9 = b_6 + c_3 + y_{15} + b_8b_6,$$

so

$$b_8b_6 = 2y_{12} + b_6 + c_3 + y_9. \tag{12}$$

By the associative law and 1), 2) in Lemma 2.26 and (4) and 2) in Lemma 2.25:

$$(b_6\bar{b}_3)b_8 = \bar{b}_3(b_6b_8), \qquad b_3b_8 + x_{15}b_8 = 2y_{12}\bar{b}_3 + b_6\bar{b}_3 + c_3\bar{b}_3 + y_9\bar{b}_3.$$

$\square$

Lemma 2.28 *Let (A, B) satisfy Hypothesis 2.4 and $(b_8 b_3, b_3 b_8) = 3$, then Lemma 2.27 holds and $x_9 c_3 = \bar{x}_{15} + d_3 + d_9$, $d_3 \neq c_3, b_3, \bar{b}_3$.*

Proof By 3) in Lemma 2.27,

$$(c_3 x_9, \bar{x}_{15}) = (x_9, c_3 \bar{x}_{15}) = 1,$$

so

$$c_3 x_9 = \bar{x}_{15} + \alpha, \tag{1}$$

where α is a linear combination at B. We shall show that α has neither degree 4 nor two elements of degree 3. If $x_4 \in \alpha$, then $1 \leq (c_3 x_9, x_4) = (x_9, c_3 x_4)$, hence $1 = (x_9, c_3 x_4)$. So

$$c_3 x_4 = x_9 + m_3. \tag{2}$$

So

$$(c_3 m_3, x_4) = (m_3, c_3 x_4) = 1,$$

hence

$$c_3 m_3 = y_4 + k_5. \tag{3}$$

Hence

$$2 = (c_3 m_3, c_3 m_3) = (c_3^2, \bar{m}_3 m_3),$$

therefore $\bar{m}_3 m_3 = 1 + b_8$. So by the associative law and (3) and 6) in Lemma 2.26, we get $(\bar{m}_3 m_3) c_3 = \bar{m}_3 (m_3 c_3)$, $c_3 + c_3 b_8 = y_4 \bar{m}_3 + k_5 \bar{m}_3$. Hence

$$2 c_3 + b_6 + y_{15} = y_4 \bar{m}_3 + k_5 \bar{m}_3. \tag{4}$$

By (3)

$$(y_4 \bar{m}_3, c_3) = (y_4, m_3 c_3) = 1, \qquad (k_5 \bar{m}_3, c_3) = (k_5, m_3 c_3) = 1;$$

hence by (4), we get a contradiction. If $x_3, y_3 \in \alpha$, then by (1),

$$1 = (c_3 x_9, x_3) = (x_9, c_3 x_3),$$
$$1 = (c_3 x_9, y_3) = (x_9, c_3 y_3).$$

So

$$c_3 x_3 = x_9, \qquad c_3 y_3 = x_9,$$

hence

$$1 = (c_3 x_3, c_3 y_3) = (c_3^2, \bar{x}_3 y_3).$$

Therefore by 1) in Lemma 2.25, $\bar{x}_3 y_3 = b_8 + 1$, so $x_3 = y_3$; then by (1) $2 = (c_3 x_9, x_3) = (x_9, c_3 x_3)$, a contradiction. Therefore by (1) one of the following holds: a) $\alpha = d_3 + d_9$ or b) $\alpha = d_5 + d_7$ or c) $\alpha = d_6 + k_6$ or d) $\alpha = 2 d_6$ or e) $\alpha = d_{12}$.

We shall show that a) is the only possibility. Assume b) holds, then by (1) we get $c_3 x_9 = \bar{x}_{15} + d_5 + d_7$. So $(c_3 d_5, x_9) = (d_5, c_3 x_9) = 1$. Therefore $c_3 d_5 = x_9 + \beta$, where $\beta = 2x_3$ or $\beta = x_3 + y_3$ or $\beta = m_6$. We shall show that these possibilities are impossible if $\beta = 2x_3$, then $c_3 d_5 = x_9 + 2x_3$. Therefore $(c_3 x_3, d_5) = (x_3, c_3 d_5) = 2$, so $c_3 x_3 = 2d_5$, a contradiction. If $\beta = x_3 + y_3$, then $c_3 d_5 = x_9 + x_3 + y_3$. Hence $(c_3 x_3, d_5) = (x_3, c_3 d_5) = 1$, $(c_3 y_3, d_5) = (y_3, c_3 d_5) = 1$, so $1 \leq (c_3 x_3, c_3 y_3) = (c_3^2, \bar{x}_3 y_3)$. Then by 1) in Lemma 2.25 $x_3 = y_3$ a contradiction. If $\beta = m_6$, then $c_3 d_5 = x_9 + m_6$, so by (1) and (6), we have $(c_3^2) d_5 = c_3(c_3 d_5), d_5 + b_8 d_5 = x_9 c_3 + m_6 c_3$ and $b_8 d_5 = \bar{x}_{15} + d_7 + m_6 c_3$. So $(b_8 x_{15}, d_5) = (b_8 d_5, \bar{x}_{15}) = 1$, hence by 10) in Lemma 2.27: $(y_9 \bar{b}_3, d_5) = 1$ and by 8) in Lemma 2.27: $(y_9 \bar{b}_3, x_{15}) = (y_9, b_3 x_{15}) = 1$, so $y_9 \bar{b}_3 = x_{15} + \bar{d}_5 + k_7$, where k_7 is a linear combination. When $k_7 = e_3 + m_4$, then $y_9 \bar{b}_3 = x_{15} + \bar{d}_5 + e_3 + t_3$, hence $(b_3 m_4, y_9) = (m_4, \bar{b}_3 y_9) = 1$, so $b_3 m_4 = y_9 + t_3$, hence $(t_3 \bar{b}_3, m_4) = (t_3, b_3 m_4) = 1$, therefore

$$t_3 \bar{b}_3 = m_4 + t_5, \tag{5}$$

so $2 = (t_3 \bar{b}_3, t_3 \bar{b}_3) = (t_3 \bar{t}_3, b_3 \bar{b}_3)$, so by Hypothesis 2.4 $\bar{t}_3 t_3 = 1 + b_8$. Hence by the associative law and (5) we obtain:

$$(\bar{t}_3 t_3) b_3 = (\bar{t}_3 b_3) t_3, \qquad b_3 + b_8 b_3 = \bar{m}_4 t_3 + \bar{t}_5 t_3.$$

By 2) in Lemma 2.26, it follows that $2b_3 + x_6 + x_{15} = \bar{m}_4 t_3 + \bar{t}_5 t_3$, so $\bar{m}_4 t_3 = 2b_3 + x_6$. Hence $5 = (\bar{m}_4 t_3, \bar{m}_4 t_3) = (\bar{m}_4 m_4, \bar{t}_5 t_3)$, and then $\bar{m}_4 m_4 = 1 + 4b_8$, a contradiction. If

$$y_9 \bar{b}_3 = x_{15} + \bar{d}_5 + k_7, \tag{6}$$

where $k_7 \in B$, then $(b_3 \bar{d}_5, y_9) = (\bar{d}_5, \bar{b}_3 y_9) = 1$; hence $b_3 \bar{d}_5 = y_9 + \gamma$, where $\gamma = v_6$ or $\gamma = v_3 + w_3$. When $b_3 \bar{d}_5 = y_9 + v_3 + w_3$, then $(b_3 \bar{v}_3, d_5) = (b_3 \bar{d}_5, v_3) = 1$. Also $(b_3 \bar{m}_3, d_5) = 1$, so $1 \leq (b_3 \bar{v}_3, b_3 \bar{m}_3) = (b_3 \bar{b}_3, v_3 \bar{m}_3)$, by Hypothesis 2.4, $v_3 = w_3$, then

$$2 = (b_3 \bar{d}_5, v_3) = (\bar{d}_5, \bar{b}_3 v_3),$$

so

$$\bar{b}_3 v_3 = 2\bar{d}_5,$$

a contradiction. When $b_3 \bar{d}_5 = y_9 + v_6$, then by the associative law and (6) and our assumption:

$$(b_3^2) \bar{d}_5 = b_3(b_3 \bar{d}_5),$$

$$\bar{d}_5 c_3 + \bar{d}_5 b_6 = y_9 b_3 + v_6 b_3,$$

$$\bar{x}_9 + \bar{m}_6 + \bar{d}_5 b_6 = \bar{x}_{15} + d_5 + \bar{k}_7 + v_6 b_3.$$

Hence

$$v_6 b_3 = \bar{x}_9 + \bar{m}_6 + t_3, \tag{7}$$

so $(t_3\bar{b}_3, v_6) = (t_3, b_3v_6) = 1$, hence $t_3\bar{b}_3 = v_6 + p_3$. Therefore $2 = (\bar{b}_3t_3, \bar{b}_3t_3) = (t_3\bar{t}_3, b_3\bar{b}_3)$, Now by the Hypothesis 2.4 we get

$$t_3\bar{t}_3 = 1 + b_8. \tag{8}$$

Again by the associative law:

$$b_3(t_3\bar{t}_3) = t_3(b_3\bar{t}_3), \qquad b_3 + b_3b_8 = t_3\bar{v}_6 + t_3\bar{p}_3;$$

and by 2) in Lemma 2.26, we have $2b_3 + x_6 + x_{15} = t_3\bar{v}_6 + t_3\bar{p}_3$, but $(t_3\bar{v}_6, b_3) = (t_3\bar{b}_3, v_6) = 1$, $(t_3\bar{p}_3, b_3) = (t_3\bar{b}_3, p_3) = 1$. This means $\bar{v}_6t_3 = b_3 + x_{15}$, $t_3\bar{p}_3 = b_3 + x_6$. Then by (7) and (8), it follows that $2 = (\bar{v}_6t_3, \bar{v}_6t_3) = (\bar{v}_6v_6, \bar{t}_3t_3) = (\bar{v}_6v_6, \bar{b}_3b_3) = (v_6b_3, v_6b_5) = 3$, a contradiction.

Assume c) holds. Then by (1) one has that $c_3x_9 = \bar{x}_{15} + d_6 + k_6$, so $(c_3d_6, x_9) = (d_6, c_3x_9) = 1$; hence $c_3d_6 = x_9 + \gamma$, where $\gamma = m_9$ or $\gamma = m_3 + v_3 + k_3$ or $\gamma = v_3 + y_6$ or $\gamma = v_4 + v_5$ or $\gamma = 2v_3 + m_3$ or $\gamma = 3v_3$. If $\gamma = m_9$, then $c_3d_6 = x_9 + m_9$, so by the associative law and 1) in Lemma 2.26:

$$(c_3^2)d_6 = (c_3d_6)c_3,$$

$$d_6 + d_6b_8 = x_9c_3 + m_9c_3,$$

$$d_6 + d_6b_8 = \bar{x}_{15} + d_6 + k_6 + m_9c_3,$$

$$d_6b_8 = \bar{x}_{15} + k_6 + m_9c_3. \tag{9}$$

We shall prove that $d_6 \neq x_6, \bar{x}_6, k_6 \neq x_6, \bar{x}_6$. By 4) in Lemma 2.26 and our assumption if $d_6 = x_6$,

$$1 = (c_3x_9, x_6) = (x_9, c_3x_6) = 0, \qquad 1 = (c_3x_9, \bar{x}_6) = (c_3x_6, \bar{x}_9) = 0,$$

a contradiction, so $d_6 \neq x_6, \bar{x}_6$. In the same way $k_6 \neq x_6, \bar{x}_6$. By (9), $(b_8x_{15}, \bar{d}_6) = (b_8d_6, \bar{x}_{15}) = 1$, in the same way $(b_8x_{15}, \bar{k}_6) = (b_8k_6, \bar{x}_{15}) = 1$, then by 8), 10) in Lemma 2.27, $(y_9\bar{b}_3, x_{15}) = (y_9, b_3x_{15}) = 1$, $(y_9\bar{b}_3, d_6) = 1$, $(y_9\bar{b}_3, k_6) = 1$. So

$$y_9\bar{b}_3 = x_{15} + d_6 + k_6. \tag{10}$$

So by the associative law and 2) in Lemma 2.25:

$$(\bar{b}_3c_3)y_9 = c_3(\bar{b}_3y_9),$$

$$b_3y_9 + y_9x_6 = x_{15}c_3 + \bar{d}_6c_3 + \bar{k}_6c_3,$$

$$\bar{x}_{15} + d_6 + k_6 + y_9x_6 = 2\bar{x}_{15} + \bar{x}_6 + \bar{x}_9 + \bar{x}_9 + \bar{m}_9 + \bar{k}_6c_3,$$

$$d_6 + k_6 + y_9x_6 = \bar{x}_{15} + \bar{x}_6 + 2\bar{x}_9 + \bar{m}_9 + \bar{k}_6c_3.$$

And by 3) in Lemma 2.27, (10) and the fact that $c_3d_6 = x_9 + m_9$, since $k_6, d_6 \neq x_6$, then $1 \leq (\bar{k}_6c_3, d_6) = (\bar{k}_6, c_3d_6)$, a contradiction to the assumption that $c_3d_6 = x_9 + m_9$. If $\gamma = m_3 + v_3 + k_3$, then $c_3d_3 = x_9 + m_3 + v_3 + k_3$, so $(c_3m_3, d_3) =$

$(m_3, c_3d_3) = 1, (c_3v_3, d_3) = (v_3, c_3d_3) = 1$. Hence $1 \le (c_3m_3, c_3v_3) = (m_3\bar{v}_3, c_3^2)$, then by 1) in Lemma 2.25 $m_3 = v_3$, a contradiction. If $\gamma = v_3 + y_6$, then $c_3d_6 = x_9 + v_3 + y_6$, so $(c_3v_3, d_6) = (v_3, c_3d_6) = 1$. Hence $c_3v_3 = d_6 + t_3$, so by the associative law and 1) in Lemma 2.25, we obtain equations $c_3^2v_3 = c_3(c_3v_3)$ and

$$v_3 + v_3b_8 = d_6c_3 + t_3c_3,$$

$$v_3 + v_3b_8 = x_9 + v_3 + y_9 + t_3c_3.$$

Hence $v_3b_8 = x_9 + y_6 + t_3c_3$. Since $(t_3c_3, v_3) = (t_3, c_3v_3) = 1$, then $(v_3b_8, v_3b_8) \ge 4$. On the other hand, $2 = (c_3v_3, c_3v_3) = (c_3^2, \bar{v}_3v_3)$, so by 1) in Lemma 2.25, $\bar{v}_3v_3 = 1 + b_8 = \bar{b}_3b_3$. Therefore $(v_3b_8, v_3b_8) = (b_8^2, v_3\bar{v}_3) = (b_8^2, \bar{b}_3b_3) = (b_8b_3, b_8b_3)$, by 2) in Lemma 2.26, $(b_8b_3, b_8b_3) = 3$, a contradiction. If $\gamma = v_4 + v_5$, then $c_3d_6 = x_4 + v_4 + v_5$, so by the associative law and 1) in Lemma 2.25:

$$c_3^2d_6 = c_3(c_3d_6),$$

$$d_6 + d_6b_8 = x_9c_3 + v_4c_3 + v_5c_3.$$

Hence by our assumption we have $d_6 + d_6b_8 = \bar{x}_{15} + d_6 + k_6 + v_4c_3 + v_5c_3$, thus

$$d_6b_8 = \bar{x}_{15} + k_6 + v_4c_3 + v_5c_3. \tag{11}$$

We shall show that $d_6 \ne x_6, \bar{x}_6, k_6 \ne x_6, \bar{x}_6$. By 4) in Lemma 2.26, if $d_6 = x_6$, or $d_6 = \bar{x}_6$, then $1 = (c_3x_9, x_6) = (x_9, c_3x_6) = 0, 1 = (c_3x_9, \bar{x}_6) = (c_3x_6, \bar{x}_9) = 0$, a contradiction. In the same way $k_6 \ne x_6, \bar{x}_6$. By (12) and 10) in Lemma 2.27,

$$(b_8x_{15}, \bar{d}_6) = (b_8d_6, \bar{x}_{15}) = 1.$$

In the same way $(b_8x_{15}, \bar{k}_6) = 1$ and by 10), 8) in Lemma 2.27 $(y_9\bar{b}_3, \bar{k}_6) = 1$, $(y_9\bar{b}_3, \bar{d}_6) = 1$, $(y_9\bar{b}_3, x_{15}) = 1$, so $y_9\bar{b}_3 = x_{15} + \bar{k}_6 + \bar{d}_6$. Hence by the associative law an 2) in Lemma 2.25:

$$(\bar{b}_3c_3)y_9 = (\bar{b}_3y_9)c_3,$$

$$b_3y_9 + x_6y_9 = x_{15}c_3 + \bar{k}_6c_3 + \bar{d}_6c_3,$$

so by 3) in Lemma 2.27 and our assumption:

$$\bar{x}_{15} + k_6 + d_6 + x_6y_9 = 2\bar{x}_{15} + \bar{x}_6 + \bar{x}_9 + \bar{k}_6c_3 + \bar{d}_6c_3,$$

$$k_6 + d_6 + x_6y_9 = \bar{x}_{15} + \bar{x}_6 + \bar{x}_9 + \bar{v}_4 + \bar{v}_5 + \bar{x}_9 + \bar{k}_6c_3.$$

Hence $1 = (\bar{k}_6c_3, d_6) = (\bar{k}_6, c_3d_6)$, a contradiction to our assumption. If $\gamma = 2v_3 + m_3$, or $\gamma = 3v_3$, then $c_3d_6 = x_9 + 2v_3 + m_3$ or $c_3d_6 = x_9 + 3v_3$. In two cases, it always follows that $(c_3v_3, d_6) = (v_3, c_3d_6) \ge 2$, a contradiction. Assume then by (1) $c_3x_9 = \bar{x}_{15} + 2d_6$. Hence $(c_3d_6, x_9) = (d_6, c_3x_9) = 2$, therefore $c_3d_6 = 2x_9$. So by the associative law and 1) in Lemma 2.25, we have that $(c_3^2)d_6 = c_3(c_3d_6)$

$$d_6 + d_6 b_8 = 2x_9 c_3$$
$$= 2\bar{x}_{15} + 4d_6,$$
$$d_6 b_8 = 2\bar{x}_{15} + 3d_6.$$

So

$$(b_8 x_{15}, \bar{d}_6) = (b_8 d_6, \bar{x}_{15}) = 2. \tag{12}$$

By 4) in Lemma 2.26, if $x_6 = d_6$, then $2 = (c_3 x_9, x_6) = (x_9, c_3 x_6) = 0$, then $d_6 \neq x_6$, also $d_6 \neq \bar{x}_6$. Then by (12) and 8), 10) in Lemma 2.27: $(y_9 \bar{b}_3, x_{15}) = (y_9, b_3 x_{15}) = 1$, $(y_9 \bar{b}_3, \bar{d}_6) = 2$. So $y_9 \bar{b}_3 = x_{15} + 2\bar{d}_6$. Hence by the associative law and 2) in Lemma 2.25:

$$y_9(\bar{b}_3 c_3) = (y_9 \bar{b}_3)c_3,$$
$$y_9 b_3 + y_9 x_6 = x_{15} c_3 + 2\bar{d}_6 c_3,$$

then by 3) in Lemma 2.27, $\bar{x}_{15} + 2d_6 + y_9 x_6 = 2\bar{x}_{15} + \bar{x}_6 + \bar{x}_9 + 2\bar{x}_9$, a contradiction. Assume e) then by 1), $c_3 x_9 = \bar{x}_{15} + d_{12}$. Then by 1) in Lemma 2.25, $2 = (x_9 c_3, x_9 c_3) = (x_9 \bar{x}_9, c_3^2) = (x_9 \bar{x}_9, b_3 \bar{b}_3) = (x_9 \bar{b}_3, x_9 \bar{b}_3)$. So by 1) in Lemma 2.27, $(x_9 \bar{b}_3, x_{10}) = (x_9, b_3 x_{10}) = 1$, hence $x_9 \bar{b}_3 = x_{10} + x_{17}$(13). By the associative law and 1) in Lemma 2.25: $(c_3^2)x_9 = (b_3 \bar{b}_3)x_9 = b_3(\bar{b}_3 x_9) = (c_3 x_9)c_3$, $x_{10} b_3 + b_3 x_{17} = \bar{x}_{15} c_3 + d_{12} c_3$. By 1), 3) and Lemma 2.27, $x_{15} + x_6 + x_9 + b_3 x_{17} = 2x_{15} + x_6 + x_9 + d_{12} c_3$, $b_3 x_{17} = x_{15} + d_{12} c_3$. Then $(x_{15} b_3, x_{17}) = (x_{15}, b_3 x_{17}) = 1$ and by 2) in Lemma 2.26 and 1), 8) in Lemma 2.27:

$$(\bar{b}_3 x_{15}, b_8) = (x_{15}, b_3 b_8) = 1,$$
$$(\bar{b}_3 x_{15}, x_{10}) = (x_{15}, b_3 x_{10}) = 1,$$
$$(\bar{b}_3 x_{15}, \bar{b}_3 x_{15}) = (x_{15} b_3, x_{15} b_3) = 6.$$

Therefore the only possibility is $\bar{b}_3 x_{15} = b_8 + x_{10} + x_{17} + m_3 + z_3 + z_4$. So $(b_3 m_3, x_{15}) = (m_3, \bar{b}_3 x_{15}) = 1$, a contradiction. Hence the only possibility is that $\alpha = d_3 + d_9$, hence by 1) $x_9 c_3 = \bar{x}_{15} + d_3 + d_9$. If $d_3 = c_3$, then $(c_3^2, x_9) = (c_3, c_3 x_9) = 1$, a contradiction to 1) in Lemma 2.25. If $d_3 = b_3$, then $(c_3 b_3, x_9) = (b_3, c_3 x_9) = 1$ a contradiction to 2) in Lemma 2.25 in the same way. If $d_3 = \bar{b}_3$, we get a contradiction so $d_3 \neq c_3, b_3, \bar{b}_3$. ☐

Lemma 2.29 Let (A, B) satisfy Hypothesis 2.4 and $(b_8 b_3, b_8 b_3) = 3$, then Lemma 2.28 holds and we obtain the following:

1) $d_3 c_3 = x_9$, $x_9 \neq \bar{x}_9$,
2) $d_3 b_8 = \bar{x}_{15} + \bar{x}_9$,
3) $b_3 \bar{d}_3 = y_9$, $y_9 = \bar{y}_9$, $y_9 \neq x_9$,
4) $x_9 c_3 = \bar{x}_{15} + d_3 + \bar{x}_9$,
5) $y_9 \bar{b}_3 = \bar{d}_3 + x_9 + x_{15}$,
6) $\bar{d}_3 b_6 = d_3 + \bar{x}_{15}$,

7) $b_8x_{15} = 5x_{15} + 3x_9 + 2x_6 + b_3 + \bar{d}_3$,

8) $x_9b_8 = 3x_{15} + 2x_9 + x_6 + \bar{d}_3$,

9) $y_{15}x_6 = 4\bar{x}_{15} + 2\bar{x}_9 + \bar{x}_6 + b_3 + d_3$,

10) $b_6x_{15} = 4\bar{x}_{15} + \bar{x}_6 + 2\bar{x}_9 + \bar{b}_3 + d_3$,

11) $d_3y_{15} = 2x_{15} + x_6 + x_9$,

12) $x_9x_{15} = 6\bar{x}_{15} + 2\bar{x}_6 + 3\bar{x}_9 + \bar{b}_3 + d_3$,

13) $d_3b_8 = \bar{x}_{15} + \bar{x}_9$,

14) $\bar{d}_3d_3 = 1 + c_8$, $c_8 \neq b_8$,

15) $d_3^2 = b_6 + y_3$, $d_3 \neq c_3, b_3, \bar{b}_3, y_3 \neq b_3, c_3, \bar{b}_3$,

16) $b_3d_3 = z_9$, $z_9 \neq y_9, x_9, \bar{x}_9, z_9 = \bar{z}_9$,

17) $y_3^2 = 1 + c_8$.

Proof   By Lemma 2.28, $(c_3d_3, x_9) = (d_3, c_3x_9) = 1$, hence

$$d_3c_3 = x_9. \tag{1}$$

By the associative law and 1) in Lemma 2.28, we have $d_3(c_3^2) = x_9c_3$ and $d_3 + d_3b_8 = \bar{x}_{15} + d_3 + d_9$, which leads to

$$d_3b_8 = \bar{x}_{15} + d_9. \tag{2}$$

So $(b_8x_{15}, \bar{d}_3) = (b_8d_3, \bar{x}_{15}) = 1$ and by 10) in Lemma 2.27, $(y_9\bar{b}_3, \bar{d}_3) = 1 = (y_9, b_3\bar{d}_3)$. So

$$b_3\bar{d}_3 = y_9. \tag{3}$$

By the associative law $(\bar{b}_3b_3)\bar{d}_3 = \bar{b}_3(b_3\bar{d}_3)$ and Hypothesis 2.4, one has $\bar{d}_3 + \bar{d}_3b_8 = y_9\bar{b}_3$. By (2), we get

$$y_9\bar{b}_3 = \bar{d}_3 + \bar{d}_9 + x_{15}. \tag{4}$$

By the associative law and Hypothesis 2.4 and (3) we have $(b_3^2)\bar{d}_3 = (b_3\bar{d}_3)b_3 = y_9b_3, \bar{d}_3c_3 + \bar{d}_3b_6 = y_9b_3 = d_3 + d_9 + \bar{x}_{15}$. So

$$\bar{d}_3c_3 = d_9, \tag{5}$$

$$\bar{d}_3b_6 = d_3 + \bar{x}_{15}. \tag{6}$$

Hence by (1) we obtain $\bar{d}_9 = x_9$. By Lemma 2.28, we may assume

$$x_9c_3 = \bar{x}_{15} + d_3 + \bar{x}_9.$$

By 10) in Lemma 2.27 and (4) we get

$$b_8x_{15} = 5x_{15} + 3x_9 + 2x_6 + b_3 + \bar{d}_3. \tag{7}$$

By the associative law and 1) in Lemma 2.25 and Lemma 2.28:

$$(c_3^2)x_9 = (c_3x_9)c_3,$$

$$x_9 + x_9 b_8 = \bar{x}_{15} c_3 + d_3 c_3 + \bar{x}_9 c_3.$$

By (1) and 3) in Lemma 2.27 and Lemma 2.28, we obtain

$$x_9 + x_9 b_8 = 2x_{15} + x_6 + x_9 + x_{15} + \bar{d}_3 + x_9,$$

so

$$x_9 b_8 = 3x_{15} + 2x_9 + x_6 + \bar{d}_3. \tag{8}$$

By the associative law and 4) in Lemma 2.27 and 2) in Lemma 2.25, we have that $y_{15}(\bar{b}_3 c_3) = (y_{15}\bar{b}_3)c_3$, $y_{15}b_3 + y_{15}x_6 = 2x_{15}c_3 + x_6 c_3 + x_9 c_3$. Moreover by 4) in Lemma 2.26 and 3), 4) in Lemma 2.27 and Lemma 2.28, we get

$$y_{15}b_3 + y_{15}x_6 = 4\bar{x}_{15} + 2\bar{x}_6 + 2\bar{x}_9 + \bar{b}_3 + \bar{x}_{15} + \bar{x}_{15} + d_3 + \bar{x}_9,$$

then

$$y_{15}x_6 = 4\bar{x}_{15} + 2\bar{x}_9 + \bar{x}_6 + \bar{b}_3 + d_3. \tag{9}$$

By the associative law and 8) in Lemma 2.27 and Hypothesis 2.4, we obtain $(b_3^2)x_{15} = (b_3 x_{15})b_3$, $x_{15}c_3 + b_6 x_{15} = 2y_{15}b_3 + b_6 b_3 + y_9 b_3$, By 3), 4) in Lemma 2.27 and 10) in Lemma 2.26 and (4), we have $2\bar{x}_{15} + \bar{x}_6 + \bar{x}_9 + b_6 x_{15} = 4\bar{x}_{15} + 2\bar{x}_6 + 2\bar{x}_9 + \bar{b}_3 + \bar{x}_{15} + d_3 + \bar{x}_9 + \bar{x}_{15}$, so $b_6 x_{15} = 4\bar{x}_{15} + \bar{x}_6 + 2\bar{x}_9 + \bar{b}_3 + d_3$. By the associative law and 6) in Lemma 2.26 and (1), (6) and (8), we have $(c_3 b_8)d_3 = (c_3 d_3)b_8 = x_9 b_8$. Hence

$$b_6 d_3 + c_3 d_3 + y_{15}d_3 = 3x_{15} + 2x_9 + x_6 + \bar{d}_3,$$

$$\bar{d}_3 + x_{15} + x_9 + y_{15}d_3 = 3x_{15} + 2x_9 + x_6 + \bar{d}_3,$$

$$y_{15}d_3 = 2x_{15} + x_6 + x_9.$$

So by the associative law and (1), it follows that $(d_3 c_3)y_{15} = c_3(d_3 y_{15})$ and $x_9 y_{15} = 2x_{15}c_3 + x_9 c_3 + x_6 c_3$. Hence by 3) in Lemma 2.27 and 4) in Lemma 2.26, we obtain that

$$x_9 c_3 = \bar{x}_{15} + d_3 + \bar{x}_9, \qquad x_9 x_{15} = 6\bar{x}_{15} + 2\bar{x}_6 + 3\bar{x}_9 + \bar{b}_3 + d_3.$$

By the associative law 1) in Lemma 2.25 and Lemma 2.28 and (1), one has that $d_3(c_3^2) = (d_3 c_3)c_3$, $d_3 + d_3 b_8 = c_3 x_9 = \bar{x}_{15} + d_3 + \bar{x}_9$, and $d_3 b_8 = \bar{x}_{15} + \bar{x}_9$. Then we have by (6) that $(d_3^2, b_6) = (d_3, \bar{d}_3 b_6) = 1$, so

$$d_3^2 = b_6 + y_3, \tag{10}$$

hence $(d_3 \bar{d}_3, d_3 \bar{d}_3) = (d_3^2, d_3^2) = 2$. Now by (1) and 1) in Lemma 2.25, we get $(d_3 c_3, d_3 c_3) = (\bar{d}_3 d_3, c_3^2) = 1$. Hence

$$d_3 \bar{d}_3 = 1 + c_8, \tag{11}$$

where $c_8 \neq b_8$ by the main theorem in [CA], so $d_3 \neq c_3, b_3, \bar{b}_3$. Also $y_3 \neq c_3, b_3, \bar{b}_3$, since if $y_3 = c_3$, then $1 = (d_3^2, c_3) = (\bar{d}_3 c_3, d_3)$; therefore $(\bar{d}_3 d_3, c_3 \bar{d}_3) = (\bar{d}_3 c_3, \bar{d}_3 c_3) \geq 2$, a contradiction. In the same way if $y_3 = b_3, \bar{b}_3$. If $x_9 = \bar{x}_9$, then by (1) we have $1 = (c_3 d_3, c_3 \bar{d}_3) = (d_3^2, c_3^2)$, a contradiction to 1) in Lemma 2.25 and that $d_3^2 = b_6 + y_3$, so $x_9 \neq \bar{x}_9$. By (11) and Hypothesis 2.4, it holds that $1 = (\bar{d}_3 d_3, \bar{b}_3 b_3) = (d_3 b_3, d_3 b_3)$, hence $d_3 b_3 = z_9$. By (10) and Hypothesis 2.4, we obtain $1 = (d_3^2, \bar{b}_3^2) = (d_3 b_3, \bar{d}_3 \bar{b}_3)$, so $z_9 = \bar{z}_9$, hence $z_9 \neq x_9, \bar{x}_9$. If $z_9 = y_9$, then by (3) we have $1 = (d_3 b_3, b_3 \bar{d}_3) = (d_3^2, \bar{b}_3 b_3)$, a contradiction to (10) and Hypothesis 2.4; hence $z_9 \neq y_9$. By the associative law and (11), (6) and (10):

$$(d_3 b_6)\bar{d}_3 = b_6(d_3 \bar{d}_3),$$
$$\bar{d}_3^2 + x_{15}\bar{d}_3 = b_6 + c_8 b_6,$$

then

$$c_8 b_6 = \bar{y}_3 + x_{15}\bar{d}_3$$

since c_8, b_6 are reals and

$$(x_{15}\bar{d}_3, y_3) = (x_{15}, d_3 y_3) = 0.$$

So $y_3 = \bar{y}_3$, by (10), $(\bar{d}_3 y_3, d_3) = 1$, hence

$$2 \leq (\bar{d}_3 y_3, \bar{d}_3 y_3) = (d_3 \bar{d}_3, y_3^2),$$

so $y_3^2 = 1 + c_8$. $\qquad\square$

Lemma 2.30 *Let (A, B) satisfy Hypothesis 2.4 and $(b_8 b_3, b_8 b_3) = 3$, then Lemma 2.29 holds and we obtain the following:*

1) $b_8^2 = 1 + 2b_8 + 2x_{10} + z_4 + c_8 + b_5 + c_5$,
2) $\bar{x}_{15}b_3 = x_{10} + b_8 + c_8 + z_9 + b_5 + c_5$, *either b_5, c_5 are reals or $b_5 = \bar{c}_5$,*
3) $b_6^2 = 1 + b_8 + z_9 + c_8 + b_5 + c_5$,
4) $\bar{x}_6 x_6 = 1 + b_8 + c_8 + z_9 + b_5 + c_5$,
5) $x_{10}^2 = 1 + 2b_8 + 2x_{10} + 2c_8 + 2b_5 + 2c_5 + 3z_9$,
6) $x_{10}\bar{x}_6 = \bar{b}_3 + 3\bar{x}_{15} + d_3 + \bar{x}_9$,
7) $b_8 x_{10} = 2b_8 + x_{10} + \bar{x}_9 b_3 + c_8 + z_9 + b_5 + c_5$,
8) $c_3 y_{15} = x_{10} + b_8 + c_8 + z_9 + b_5 + c_5$,
9) $c_3 b_5 = y_{15}$,
10) $c_3 c_5 = y_{15}$.

Proof First we note that either b_5 and c_5 are reals or $b_5 = \bar{c}_5$ holds. But in these two cases, it always follows that $\overline{b_5 + c_5} = \bar{b}_5 + \bar{c}_5 = b_5 + c_5$. By 2) in Lemma 2.26 and 1), 8) in Lemma 2.27, we have $(\bar{x}_{15}b_3, \bar{x}_{15}b_3) = 6$, $(\bar{x}_{15}b_3, x_{10}) = (b_3 x_{10}, x_{15}) = 1$, $(\bar{x}_{15}b_3, b_8) = (b_3 b_8, x_{15})$, then

$$\bar{x}_{15}b_3 = x_{10} + b_8 + x + y + z + w, \qquad (1)$$

where $|x + y + z + w| = 27$. Here the case $\bar{x}_{15}b_3 = x_{10} + b_8 + 2x$ is impossible since $|2x| = 27$ implies that $|x| = 13.5$. But x is an integer, a contradiction. By the associative law $(\bar{b}_3 b_8)b_3 = b_8(\bar{b}_3 b_3)$ and Hypothesis 2.4 and 2), 3) in Lemma 2.26, we get

$$\bar{x}_6 b_3 + \bar{b}_3 b_3 + \bar{x}_{15}b_3 = b_8 + b_8^2, \qquad b_8 + x_{10} + 1 + b_8 + \bar{x}_{15}b_3 = b_8 + b_8^2,$$

$$1 + x_{10} + b_8 + \bar{x}_{15}b_3 = b_8^2.$$

Thus

$$b_8^2 = 1 + 2b_8 + 2x_{10} + x + y + z + w. \tag{2}$$

So by 4) in Lemma 2.25 and 3) in Lemma 2.26,

$$b_6^2 = 1 + b_8 + x + y + z + w. \tag{3}$$

By the associative law $(\bar{b}_3 c_3)(b_3 c_3) = (\bar{b}_3 b_3)c_3^2$ and 1), 2) in Lemma 2.25 and Hypothesis 2.4 and 3) in Lemma 2.26:

$$(b_3 + x_6)(\bar{b}_3 + \bar{x}_6) = (1 + b_8)^2 = 1 + 2b_8 + b_8^2,$$

$$b_3 \bar{b}_3 + b_3 \bar{x}_6 + \bar{b}_3 x_6 + x_6 \bar{x}_6 = 1 + 2b_8 + b_8^2.$$

So by (2) and (1) and 3) in Lemma 2.26:

$$x_6 \bar{x}_6 = 1 + b_8 + x + y + z + w. \tag{4}$$

By the associative law $(\bar{x}_6 b_8)b_3 = (\bar{x}_6 b_3)b_8$ and 3) in Lemma 2.26 and 2) in Lemma 2.27:

$$2\bar{x}_{15}b_3 + \bar{b}_3 b_3 + \bar{x}_6 b_3 + \bar{x}_9 b_3 = b_8^2 + x_{10}b_8,$$

then by 3) in Lemma 2.26 and (1) and (2) we have

$$2x_{10} + 2b_8 + 2x + 2y + 2z + 2w + 1 + b_8 + b_8 + x_{10} + \bar{x}_9 b_3$$
$$= 1 + 2b_8 + 2x_{10} + x + y + z + w + x_{10}b_8,$$

so

$$2b_8 + x_{10} + \bar{x}_9 b_3 + x + y + z + w = x_{10}b_8. \tag{5}$$

By the associative law $(\bar{x}_6^2)b_3 = \bar{x}_6(\bar{x}_6 b_3)$ and 7) in Lemma 2.27 and 3) in Lemma 2.26, we get $2b_6 b_3 + y_{15}b_3 + y_9 b_3 = \bar{x}_6 b_8 + \bar{x}_6 x_{10}$. Then by 1) in Lemma 2.26 and 2), 4) in Lemma 2.27 and 5) in Lemma 2.29, we have the equation:

$$2\bar{b}_3 + 2\bar{x}_{15} + 2\bar{x}_{15} + \bar{x}_6 + \bar{x}_9 + d_3 + \bar{x}_{15} + \bar{x}_9 = 2\bar{x}_{15} + \bar{b}_3 + \bar{x}_6 + \bar{x}_9 + x_{10}\bar{x}_6,$$

so

$$\bar{b}_3 + 3\bar{x}_{15} + d_3 + \bar{x}_9 = x_{10}\bar{x}_6. \tag{6}$$

By the associative $(\bar{x}_6 b_3)x_{10} = b_3(\bar{x}_6 x_{10})$ and 3) in Lemma 2.26, we get $b_8 x_{10} + x_{10}^2 = \bar{b}_3 b_3 + 3\bar{x}_{15} b_3 + d_3 b_3 + \bar{x}_9 b_3$. Then by (3) and Hypothesis 2.4 and (1) and 5) in Lemma 2.29, it follows that

$$2b_8 + x_{10} + \bar{x}_9 b_3 + x + y + z + w + x_{10}^2$$
$$= 1 + b_8 + 3x_{10} + 3b_8 + 3x + 3y + 3z + 3w + z_9 + \bar{x}_9 b_3.$$

Hence

$$x_{10}^2 = 1 + 2b_8 + 2x_{10} + z_9 + 2x + 2y + 2z + 2w. \qquad (7)$$

By 13) and 14) in Lemma 2.29, one has that $(d_3 \bar{d}_3, b_8^2) = (d_3 b_8, d_3 b_8) = 2$, hence we obtain by (2) that $(b_8^2, c_8) = 1$. Without loss of generality, say $x = c_8$. By (4) and (7), it holds that $11 \le (x_{10}^2, x_6 \bar{x}_6) = (x_{10} \bar{x}_6, x_{10} \bar{x}_6)$. But by (6), $(x_{10} \bar{x}_6, x_{10} \bar{x}_6) = 12$ holds, thus we may assume $y = z_9$. By (1), $(\bar{b}_3 \alpha, \bar{x}_{15}) = (\alpha, b_3 \bar{x}_{15}) = 1$ holds, where $\alpha \in \{x, y, z, w\}$. So $|\alpha| \ge 5$. Hence $z = b_5$ and $w = c_5$. Hence by (1), (3), (4), (5), (6) and (7), we obtain

$$\bar{x}_{15} \bar{b}_3 = x_{10} + b_8 + c_8 + z_9 + b_5 + c_5,$$
$$b_8^2 = 1 + 2b_8 + 2x_{10} + c_8 + z_9 + b_5 + c_5,$$
$$b_6^2 = 1 + b_8 + c_8 + z_9 + b_5 + c_5,$$
$$x_6 \bar{x}_6 = 1 + b_8 + c_8 + z_9 + b_5 + c_5,$$
$$x_{10}^2 = 1 + 2b_8 + 2x_{10} + 3z_9 + 2c_8 + 2b_5 + 2c_5,$$
$$b_8 x_{10} = 2b_8 + x_{10} + \bar{x}_9 b_3 + c_8 + z_9 + b_5 + c_5.$$

So by the associative law $(\bar{b}_3 \bar{x}_6)c_3 = \bar{x}_6(\bar{b}_3 c_3)$ and 1), 2) in Lemma 2.25 and 3), 5) in Lemma 2.26:

$$c_3^2 + c_3 y_{15} = b_3 \bar{x}_6 + x_6 \bar{x}_6,$$
$$1 + b_8 + c_3 y_{15} = b_8 + x_{10} + 1 + b_8 + c_8 + z_9 + b_5 + c_5,$$
$$c_3 y_{15} = x_{10} + b_8 + c_8 + z_9 + b_5 + c_5.$$

So $(c_3 b_5, y_{15}) = (b_5, c_3 y_{15}) = 1$, $(c_3 c_5, y_{15}) = (y_{15} c_3, c_5) = 1$, hence $c_3 b_5 = y_{15}$, $c_3 c_5 = y_{15}$. □

Lemma 2.31 *Let (A, B) satisfy Hypothesis 2.4 and $(b_8 b_3, b_8 b_3) = 3$, then Lemma 2.30 holds and we obtain the following:*

1) $b_3 z_9 = x_9 + \bar{d}_3 + x_{15}$,
2) $\bar{x}_9 b_3 = z_9 + x_{10} + c_8$,
3) $b_8 x_{10} = 2b_8 + 2x_{10} + 2z_9 + 2c_8 + b_5 + c_5$,
4) $b_3 c_8 = x_9 + x_{15}$,
5) $d_3 x_6 = x_{10} + c_8$,

6) $d_3b_6 = \bar{d}_3 + x_{15}$,
7) $d_3c_8 = \bar{x}_{15} + \bar{x}_6 + d_3$,
8) $y_3\bar{d}_3 = \bar{x}_6 + d_3$,
9) $d_3\bar{x}_6 = y_{15} + y_3$.

Proof By the associative law $(b_3^2)d_3 = b_3(b_3d_3)$ and 6), 4) in Lemma 2.29 and Hypothesis 2.4, we get $c_3d_3 + b_6d_3 = b_3z_9$. Then by 1), 6) in Lemma 2.29 it follows that

$$x_9 + \bar{d}_3 + x_{15} = b_3z_9. \tag{1}$$

Then $(\bar{b}_3x_9, z_9) = (b_3z_9, x_9) = 1$. And by 1) in Lemma 2.27, $(\bar{x}_9b_3, x_{10}) = (b_3x_{10}, x_9) = 1$ holds. It follows by 4) in Lemma 2.29 and 1) in Lemma 2.25 and Hypothesis 2.4 that $(\bar{x}_9b_3, \bar{x}_9b_3) = (\bar{x}_9x_9, \bar{b}_3b_3) = (\bar{x}_9x_9, c_3^2) = (x_9c_3, x_9c_3) = 3$. Hence

$$\bar{x}_9b_3 = z_9 + x_{10} + d_8. \tag{2}$$

By 1) and 5) in Lemma 2.30, we get

$$(b_8^2, x_{10}^2) = 18 = (b_8x_{10}, b_8x_{10}). \tag{3}$$

By 7) in Lemma 2.30 and (2), we have $b_8x_{10} = 2b_8 + 2x_{10} + 2z_9 + c_8 + b_5 + c_5 + d_8$, then by (3) we obtain that $c_8 = d_8$, so

$$b_8x_{10} = 2b_8 + 2x_{10} + 2z_9 + 2c_8 + b_5 + c_5. \tag{4}$$

Now we have $(b_3c_8, x_9) = (b_3\bar{x}_9, c_8) = 1$ by (2) and $(b_3c_8, x_{15}) = (b_3\bar{x}_{15}, c_8) = 1$ by 2) in Lemma 2.30, so

$$b_3c_8 = x_9 + x_{15}. \tag{5}$$

By the associative law $d_3(c_3\bar{b}_3) = (d_3c_3)\bar{b}_3$ and 1) in Lemma 2.29 and 2) in Lemma 2.25 and (2), we see that $d_3(c_3\bar{b}_3) = (d_3c_3)\bar{b}_3$, $d_3b_3 + d_3x_6 = x_9\bar{b}_3 = z_9 + x_{10} + c_8$. Hence by 14) in Lemma 2.29 we have

$$d_3x_6 = x_{10} + c_8. \tag{6}$$

By the associative law $(b_3^2)d_3 = (b_3d_3)b_3$ and 14) in Lemma 2.29 and (1) and Hypothesis 2.4, it follows $d_3c_3 + d_3b_6 = z_9b_3 = x_9 + \bar{d}_3 + x_{15}$. Hence by 1) in Lemma 2.29 it holds that:

$$d_3b_6 = \bar{d}_3 + x_{15}. \tag{7}$$

By the associative law $(\bar{d}_3d_3)d_3 = \bar{d}_3(d_3^2)$ and 12) and 13) in Lemma 2.29, we have $d_3 + d_3c_8 = b_6\bar{d}_3 + y_3\bar{d}_3$, and get $d_3c_8 = \bar{x}_{15} + y_3\bar{d}_3$ by (7) and $(d_3c_8, \bar{x}_6) = (d_3x_6, c_8) = 1$ by (6). By 12) in Lemma 2.29, $(d_3c_8, d_3) = (c_8, \bar{d}_3d_3) = 1$ holds, hence

$$d_3c_8 = \bar{x}_{15} + \bar{x}_6 + d_3. \tag{8}$$

Therefore

$$y_3\bar{d}_3 = \bar{x}_6 + d_3. \tag{9}$$

Moreover $(d_3\bar{x}_6, y_3) = (\bar{x}_6, \bar{d}_3 y_3) = 1$, and by 11) in Lemma 2.29, it follows that $(d_3\bar{x}_6, y_{15}) = 1$, so $d_3\bar{x}_6 = y_{15} + y_3$. □

Lemma 2.32 *Let (A, B) satisfy Hypothesis 2.4 and $(b_8 b_3, b_8 b_3) = 3$, then Lemma 2.31 holds and we obtain the following:*

1) $x_9 b_3 = y_9 + y_{15} + y_3$,
2) $x_9\bar{d}_3 = y_{15} + c_3 + y_9$,
3) $x_{15}\bar{d}_3 = y_{15} + y_6 + y_9$,
4) $c_8 b_6 = 2y_{15} + y_3 + y_9 + b_6$,
5) $z_9 b_6 = 2y_{15} + 2y_9 + b_6$,
6) $x_{10} b_6 = 3y_{15} + y_3 + c_3 + y_9$,
7) $y_3 b_6 = c_8 + x_{10}$,
8) $d_3 x_{15} = b_8 + z_9 + b_5 + c_5 + x_{10} + c_8$,
9) $y_{15} b_6 = 2b_8 + 3x_{10} + 2z_9 + 2c_8 + b_5 + c_5$,
10) $y_9 b_6 = b_8 + 2z_9 + b_5 + c_5 + x_{10} + c_8$,
11) $\bar{x}_{15} x_{15} = 5b_8 + 6x_{10} + 6z_9 + 5c_8 + 3b_5 + 3c_5 + 1$,
12) $b_6 b_5 = y_{15} + y_9 + b_6$,
13) $b_3 y_3 = \bar{x}_9$.

Proof By the associative law $d_3(c_3 b_3) = (d_3 c_3)b_3 = x_9 b_3$ and 1) in Lemma 2.29 and 2) in Lemma 2.25, we have that $d_3\bar{b}_3 + d_3\bar{x}_6 = x_9 b_3$ and by 3) in Lemma 2.29 that $y_9 + d_3\bar{x}_6 = x_9 b_3$. Finally, we obtain

$$x_9 b_3 = y_9 + y_{15} + y_3, \tag{1}$$

by 9) in Lemma 2.31.

By the associative law $b_3(c_8\bar{d}_3) = \bar{d}_3(b_3 c_8)$ and 4), 7) in Lemma 2.31, we get $x_{15} b_3 + b_3 x_6 + b_3\bar{d}_3 = x_9\bar{d}_3 + x_{15}\bar{d}_3$, and by 5) in Lemma 2.26 and 3) in Lemma 2.29 and 8) in Lemma 2.27, we get $2y_{15} + b_6 + y_9 + c_3 + y_{15} = x_9\bar{d}_3 + x_{15}\bar{d}_3$, $x_9\bar{d}_3 + x_{15}\bar{d}_3 = 3y_{15} + 2y_9 + b_6 + c_3$. Now from 10), 11), 1) in Lemma 2.29, we see that $(x_9\bar{d}_9, c_3) = (x_9, d_3 c_3) = 1$ and $(x_9\bar{d}_3, y_{15}) = (x_9, d_3 y_{15}) = 1$ and $(x_9\bar{d}_3, b_6) = (x_9, d_3 b_6) = 0$, hence

$$x_9\bar{d}_3 = y_{15} + c_3 + y_9, \tag{2}$$

which gives

$$x_{15}\bar{d}_3 = 2y_{15} + y_9 + b_6. \tag{3}$$

By the associative law $(\bar{d}_3 d_3)b_6 = \bar{d}_3(d_3 b_6)$ and 6), 12) in Lemma 2.29, one has that $b_6 + c_8 b_6 = \bar{d}_3^2 + \bar{d}_3 x_{15}$. And by (3) and 13) in Lemma 2.29, we obtain $b_6 + c_8 b_6 = b_6 + y_3 + 2y_{15} + y_9 + b_6$, hence

$$c_8 b_6 = 2y_{15} + y_3 + y_9 + b_6. \tag{4}$$

By the associative law $(d_3b_3)b_6 = d_3(b_3b_6)$ and 1) in Lemma 2.26 and 3) and 16) in Lemma 2.29, we obtain $z_9b_6 = d_3\bar{b}_3 + d_3\bar{x}_{15}$, and hence by (3) and 3) in Lemma 2.29, we have

$$z_9b_6 = 2y_{15} + 2y_9 + b_6. \tag{5}$$

By the associative law $(x_6b_6)d_3 = b_6(x_6d_3)$ and 5) in Lemma 2.27 and 5) in Lemma 2.31, we have $2\bar{x}_6d_3 + \bar{x}_{15}d_3 + \bar{x}_9d_3 = x_{10}b_6 + c_8b_6$. Then by 9) in Lemma 2.31 and (2), (3) and (4), it follows that

$$2y_{15} + 2y_3 + 2y_{15} + y_9 + b_6 + y_{15} + c_3 + y_9 = x_{10}b_6 + 2y_{15} + y_3 + y_9 + b_6.$$

Hence

$$x_{10}b_6 = 3y_{15} + y_3 + c_3 + y_8. \tag{6}$$

Now we have by (4) $(y_3b_6, c_8) = (y_3, b_6c_8) = 1$, $(y_3b_6, x_{10}) = (y_3, b_6x_{10}) = 1$ by (4) and (6), from which we get

$$y_3b_6 = c_8 + x_{10}. \tag{7}$$

By the associative law $(d_3^2)b_6 = d_3(d_3b_6)$ and 15) in Lemma 2.29 and 6) in Lemma 2.31, we obtain $b_6^2 + y_3b_6 = d_3\bar{d}_3 + x_{15}d_3$. Moreover by 14) in Lemma 2.29 and 3) in Lemma 2.30 and (7), we have

$$1 + b_8 + z_9 + c_8 + b_5 + c_5 + x_{10} + c_8 = 1 + c_8 + x_{15}d_3,$$
$$b_8 + z_9 + b_5 + c_5 + x_{10} + c_8 = x_{15}d_3. \tag{8}$$

By the associative law and 5) in Lemma 2.26 and 5) in Lemma 2.27:

$$b_3(x_6b_6) = b_6(b_3x_6),$$
$$2\bar{x}_6b_3 + \bar{x}_{15}b_3 + \bar{x}_9b_3 = b_6c_3 + y_{15}b_6.$$

By 2) in Lemma 2.31 and 2) in Lemma 2.30 and 3) in Lemma 2.26:

$$2b_8 + 2x_{10} + x_{10} + b_8 + z_9 + c_8 + b_5 + c_5 + x_{10} + z_9 + c_8 = b_8 + x_{10} + y_{15}b_6,$$
$$y_{15}b_6 = 2b_8 + 3x_{10} + 2z_9 + 2c_8 + b_5 + c_5. \tag{9}$$

By 3) in Lemma 2.29 and 1) in Lemma 2.26:

$$(b_3b_6)\bar{d}_3 = (b_3\bar{d}_3)b_6 = y_9b_6,$$
$$\bar{b}_3\bar{d}_3 + \bar{x}_{15}\bar{d}_3 = y_9b_6,$$

by 16) in Lemma 2.29 and (8)

$$b_8 + 2z_9 + b_5 + c_5 + x_{10} + c_8 = y_9b_6. \tag{10}$$

By the associative law and 8) in Lemma 2.27 and 1) in Lemma 2.26:

$$(b_3 b_6) x_{15} = b_6 (b_3 x_{15}),$$

$$\bar{b}_3 x_{15} + \bar{x}_{15} x_{15} = 2 y_{15} b_6 + b_6^2 + y_9 b_6,$$

so by (10), (9) and 2), 3) in Lemma 2.30

$$x_{10} + b_8 + z_9 + c_8 + b_5 + c_5 + \bar{x}_{15} x_{15} = 4 b_8 + 6 x_{10} + 4 z_9 + 4 c_8 + 2 c_5 + 1 + b_8 + z_9$$
$$+ c_8 + b_5 + c_5 + b_8 + 2 z_9 + c_8 + x_{10} + b_5 + c_5,$$

$$\bar{x}_{15} x_{15} = 1 + 5 b_8 + 6 x_{10} + 6 z_9 + 5 c_8 + 3 b_5 + 4 c_5.$$

By (9) and (10), we get

$$(b_6 b_5, y_9) = (b_6 y_9, b_5) = 1, \qquad (b_6 b_5, y_{15}) = (b_5, y_{15} b_6) = 1$$

and by (3)

$$(b_5 b_6, b_6) = (b_5, b_6^2) = 1,$$

so

$$b_5 b_6 = y_9 + y_{15} + b_6.$$

By (1)

$$(b_3 y_3, \bar{x}_9) = (b_3 x_9, y_3) = 1,$$

then $b_3 y_3 = \bar{x}_9$.   □

Lemma 2.33 *Let (A, B) satisfy Hypothesis 2.4 and $(b_8 b_3, b_8 b_3) = 3$, then Lemma 2.32 holds and we obtain the following:*

1) $y_9 d_3 = b_3 + x_{15} + x_9,$
2) $z_9 d_3 = \bar{b}_3 + \bar{x}_{15} + \bar{x}_9,$
3) $z_9 y_9 = 3 y_{15} + 2 b_6 + 2 y_9 + c_3 + y_3,$
4) $c_8 b_8 = 2 x_{10} + 2 z_9 + b_8 + c_8 + b_5 + c_5,$
5) $y_9 c_3 = z_9 + x_{10} + c_8,$
6) $\bar{x}_{15} x_6 = 2 b_8 + 3 x_{10} + 2 z_9 + 2 c_8 + b_5 + c_5,$
7) $x_{10} d_3 = \bar{x}_{15} + \bar{x}_6 + \bar{x}_9,$
8) $d_3 x_9 = x_{10} + b_8 + z_9,$
9) $y_3 y_9 = z_9 + b_8 + x_{10},$
10) $b_6 y_9 = 2 z_9 + b_8 + x_{10} + b_5 + c_5 + c_8,$
11) $x_{15}^2 = 9 y_{15} + 4 b_6 + 6 y_9 + 2 c_3 + 2 y_3,$
12) $x_{15} x_6 = c_3 + 4 y_{15} + b_6 + 2 y_9 + y_3.$

Proof By the associative law and 13), 3) in Lemma 2.29:

$$(b_3 \bar{d}_3) d_3 = (\bar{b}_3 d_3) d_3 = \bar{b}_3 (d_3^2), \qquad y_9 d_3 = b_6 \bar{b}_3 + y_3 \bar{b}_3,$$

then by 1) in Lemma 2.26 and 13) in Lemma 2.32,

$$y_9 d_3 = b_3 + x_{15} + x_9. \tag{1}$$

By the associative law and 16), 15) in Lemma 2.29:

$$(d_3^2) b_3 = (d_3 b_3) d_3,$$
$$b_6 b_3 + y_3 b_3 = z_9 d_3,$$

then by 13) in Lemma 2.32 and 1) in Lemma 2.26,

$$z_9 d_3 = \bar{b}_3 + \bar{x}_{15} + \bar{x}_9. \tag{2}$$

By 16) in Lemma 2.29 and (1):

$$(b_3 d_3) y_9 = z_9 y_9 = (d_3 y_9) b_3 = b_3^2 + x_{15} b_3 + x_9 b_3,$$

by 8) in Lemma 2.27 and Hypothesis 2.4 and 1) in Lemma 2.32:

$$c_3 + b_6 + 2 y_{15} + b_6 + y_9 + y_9 + y_{15} + y_3 = z_9 y_9,$$
$$3 y_{15} + 2 b_6 + 2 y_9 + c_3 + y_3 = z_9 y_9. \tag{3}$$

By the associative law and Hypothesis 2.4 and 4) in Lemma 2.31:

$$(b_3 \bar{b}_3) c_8 = \bar{b}_3 (b_3 c_8),$$
$$c_8 + c_8 b_8 = x_9 \bar{b}_3 + x_{15} \bar{b}_3,$$

then by 2) in Lemma 2.31 and 2) in Lemma 2.30:

$$c_8 b_8 = 2 x_{10} + 2 z_9 + b_8 + c_8 + b_5 + c_5. \tag{4}$$

By the associative law and 2) in Lemma 2.25 and 3) in Lemma 2.29:

$$(b_3 c_3) \bar{d}_3 = (b_3 \bar{d}_3) c_3,$$
$$\bar{b}_3 \bar{d}_3 + \bar{x}_6 \bar{d}_3 = y_9 c_3,$$

then by 16) in Lemma 2.29 and 5) in Lemma 2.31:

$$y_9 c_3 = z_9 + x_{10} + c_8. \tag{5}$$

By the associative law and 7) in Lemma 2.27 and 4) in Lemma 2.26:

$$(x_6^2) c_3 = x_6 (x_6 c_3),$$
$$2 b_6 c_3 + y_{15} c_3 + y_9 c_3 = \bar{b}_3 x_6 + \bar{x}_{15} x_6,$$

then by (5) and 3) in Lemma 2.26 and 8) in Lemma 2.30:

$$2 b_8 + 2 x_{10} + x_{10} + b_8 + z_9 + c_8 + b_5 + c_5 + z_9 + x_{10} + c_8 = b_8 + x_{10} + \bar{x}_{15} x_6,$$
$$\bar{x}_{15} x_6 = 2 b_8 + 3 x_{10} + 2 z_9 + 2 c_8 + b_5 + c_5. \tag{6}$$

By the associative law, 9) in Lemma 2.31 and 3) in Lemma 2.26:

$$(\bar{x}_6 b_3)d_3 = (\bar{x}_6 d_3)b_3,$$

$$b_8 d_3 + x_{10}d_3 = y_{15}b_3 + y_3 b_3,$$

then by 4) in Lemma 2.27 and 13) in Lemma 2.32

$$b_8 d_3 + x_{10}d_3 = 2\bar{x}_{15} + \bar{x}_6 + \bar{x}_9 + \bar{x}_9,$$

by 13) in Lemma 2.29

$$x_{10}d_3 = \bar{x}_{15} + \bar{x}_6 + \bar{x}_9. \tag{7}$$

So

$$(d_3 x_9, x_{10}) = (d_3 x_{10}, \bar{x}_9) = 1,$$

and by 13) in Lemma 2.29,

$$(d_3 x_9, b_8) = (d_3 b_8, \bar{x}_9) = 1;$$

and by (2)

$$(d_3 x_9, z_9) = (d_3 z_9, \bar{x}_9) = 1,$$

so

$$d_3 x_9 = x_{10} + b_8 + z_9. \tag{8}$$

By the associative law and 15) in Lemma 2.29 and (1):

$$(d_3^2)y_9 = d_3(d_3 y_9),$$

$$b_6 y_9 + y_3 y_9 = b_3 d_3 + x_{15}d_3 + x_9 d_3,$$

then by 16) in Lemma 2.29 and (8) and 8) in Lemma 2.32:

$$b_6 y_9 + y_3 y_9 = z_9 + b_8 + z_9 + b_5 + c_5 + x_{10} + c_8 + x_{10} + b_8 + z_9,$$

$$b_6 y_9 + y_3 y_9 = 3z_9 + 2b_8 + b_5 + c_5 + c_8 + 2x_{10}.$$

By 5), 6) in Lemma 2.32 and 9) in Lemma 2.27:

$$(b_6 y_9, b_8) = (y_9, b_6 b_8) = 1,$$

$$(b_6 y_9, x_{10}) = (y_9, b_6 x_{10}) = 1,$$

$$(b_6 y_9, z_9) = (y_9, z_9 b_6) = 2;$$

hence

$$y_3 y_9 = z_9 + b_8 + x_{10}. \tag{9}$$

Therefore

$$b_6 y_9 = 2z_9 + b_8 + x_{10} + b_5 + c_5 + c_8. \tag{10}$$

By the associative law and 10) in Lemma 2.29 and 1) in Lemma 2.26:

$$(b_6\bar{b}_3)x_{15} = \bar{b}_3(b_6x_{15}), \qquad b_3x_{15} + x_{15}^2 = 4\bar{x}_{15}\bar{b}_3 + \bar{x}_6\bar{b}_3 + 2\bar{x}_9\bar{b}_3 + \bar{b}_3^2 + d_3\bar{b}_3,$$

by 8) in Lemma 2.27 and 5) in Lemma 2.26 and 1) in Lemma 2.32 and Hypothesis 2.4 and 3) in Lemma 2.29:

$$x_{15}^2 = 6y_{15} + 3b_6 + 3y_9 + c_3 + y_{15} + 2y_9 + 2y_{15} + 2y_3 + c_3 + b_6 + y_9,$$

$$x_{15}^2 = 9y_{15} + 4b_6 + 6y_9 + 2c_3 + 2y_3.$$

By the associative law and 5) in Lemma 2.27 and 1) in Lemma 2.26:

$$(\bar{x}_6 b_6)b_3 = \bar{x}_6(b_6 b_3),$$

$$2x_6 b_3 + x_{15}b_3 + x_9 b_3 = \bar{b}_3\bar{x}_6 + \bar{x}_{15}\bar{x}_6,$$

then by 5) in Lemma 2.26 and 8) in Lemma 2.27 and 1) in Lemma 2.32:

$$2c_3 + 2y_{15} + 2y_{15} + b_6 + y_9 + y_{15} + y_3 = c_3 + y_{15} + \bar{x}_{15}\bar{x}_6,$$

$$x_{15}x_6 = c_3 + 4y_{15} + b_6 + 2y_9 + y_3.$$

□

Lemma 2.34 Let (A, B) satisfy Hypothesis 2.4 and $(b_8 b_3, b_8 b_3) = 3$, then Lemma 2.33 holds, and we obtain the following:

1) $y_3 c_3 = z_9$,
2) $z_9 c_3 = y_9 + y_{15} + y_3$,
3) $y_3 b_8 = y_9 + y_{15}$,
4) $z_9 b_8 = 2z_9 + 2x_{10} + 2c_8 + b_8 + b_5 + c_5$,
5) $z_9^2 = 1 + 2x_{10} + 2z_9 + 2b_8 + 2c_8 + b_5 + c_5$,
6) $c_8^2 = 1 + 2c_8 + 2x_{10} + b_8 + z_9 + b_5 = c_5$,
7) $c_8 c_3 = y_{15} + y_9$,
8) $x_9^2 = 2b_6 + 3y_{15} + 2y_9 + y_3 + c_3$,
9) $c_8 y_3 = b_6 + y_3 + y_{15}$,
10) $x_{10}y_3 = b_6 + y_3 + y_{15}$,
11) $d_3 y_3 = \bar{d}_3 + x_6$,
12) $\bar{x}_6 y_3 = d_3 + x_{15}$,
13) $x_{15}y_3 = 2\bar{x}_{15} + \bar{x}_6 + \bar{x}_9$.

Proof By the associative law and Hypothesis 2.4 and 13) in Lemma 2.32:

$$y_3(b_3^2) = \bar{x}_9 b_3,$$

$$y_3 c_3 + y_3 b_6 = \bar{x}_9 b_3,$$

then by 2) in Lemma 2.31

$$y_3 c_3 + y_3 b_6 = z_9 + x_{10} + c_8,$$

then by 7) in Lemma 2.32

$$y_3c_3 = z_9. \tag{1}$$

By 1), 16) in Lemma 2.29 and the associative law and 1) in Lemma 2.32:

$$b_3(c_3d_3) = (b_3c_3)d_3 = (b_3d_3)c_3 = z_9c_3,$$

$$b_3x_9 = z_9c_3,$$

then

$$z_9c_3 = y_9 + y_{15} + y_3. \tag{2}$$

So by (1) and 1) in Lemma 2.25 and the associative law:

$$y_3(c_3^2) = z_9c_3,$$

$$y_3 + y_3b_8 = y_9 + y_{15} + y_3.$$

Thus

$$y_3b_8 = y_9 + y_{15}. \tag{3}$$

By (2) and the associative law and 1) in Lemma 2.25:

$$z_9(c_3^2) = y_9c_3 + y_{15}c_3 + y_3c_3,$$

$$z_9 + z_9b_8 = y_9c_3 + y_{15}c_3 + y_3c_3,$$

so by (1) and 8) in Lemma 2.30 and 5) in Lemma 2.33:

$$z_9 + z_9b_8 = z_9 + x_{10} + c_8 + x_{10} + b_8 + c_8 + z_9 + b_5 + c_5 + z_9,$$

$$z_9b_8 = 2z_9 + 2x_{10} + 2c_8 + b_8 + b_5 + c_5. \tag{4}$$

By 17) in Lemma 2.29 and 1) in Lemma 2.25 and (1):

$$y_3^2c_3^2 = z_9^2,$$

$$(1 + c_8)(1 + b_8) = z_9^2,$$

$$1 + b_8 + c_8 + c_8b_8 = z_9^2,$$

by 4) in Lemma 2.33

$$z_9^2 = 1 + 2x_{10} + 2z_9 + 2b_8 + 2c_8 + b_5 + c_5. \tag{5}$$

By the associative law and 14) in Lemma 2.29 and 7) in Lemma 2.31:

$$(\bar{d}_3d_3)c_8 = \bar{d}_3(d_3c_8),$$

$$c_8 + c_8^2 = \bar{d}_3\bar{x}_{15} + \bar{x}_6\bar{d}_3 + d_3\bar{d}_3,$$

then by 5) in Lemma 2.31 and 8) in Lemma 2.32:

$$c_8 + c_8^2 = b_8 + z_9 + b_5 + c_5 + x_{10} + c_8 + x_{10} + c_8 + 1 + c_8,$$
$$c_8^2 = 1 + 2c_8 + 2x_{10} + b_8 + z_9 + b_5 + c_5. \tag{6}$$

By 1), 12) in Lemma 2.29 and the associative law:

$$\bar{d}_3(d_3 c_3) = c_3 + c_8 c_3,$$
$$\bar{d}_3 x_9 = c_3 + c_8 c_3,$$

then by 2) in Lemma 2.32

$$c_8 c_3 = y_{15} + y_9. \tag{7}$$

By the associative law and 13) in Lemma 2.32 and 17) in Lemma 2.29 and Hypothesis 2.4:

$$b_3^2 y_3^2 = \bar{x}_9^2,$$
$$(c_3 + b_6)(1 + c_8) = \bar{x}_9^2,$$
$$c_3 + b_6 + c_3 c_8 + b_6 c_8 = \bar{x}_9^2,$$

then by (7) and 4) in Lemma 2.32:

$$c_3 + b_6 + y_{15} + y_9 + 2y_{15} + y_3 + y_9 + b_6 = \bar{x}_9^2,$$
$$x_9^2 = 2b_6 + 3y_{15} + 2y_9 + y_3 + c_3.$$

By the associative law and 4), 7) in Lemma 2.32 and 17) in Lemma 2.29:

$$(y_3^2)b_6 = y_3(y_3 b_6),$$
$$b_6 + c_8 b_6 = c_8 y_3 + x_{10} y_3,$$
$$2y_{15} + y_3 + y_9 + 2b_6 = c_8 y_3 + x_{10} y_3,$$

by 9) in Lemma 2.33, 4) in Lemma 2.32 and 15) in Lemma 2.29: $(c_8 y_3, b_6) = (y_3, c_8 b_6) = 1, (c_8 y_3, y_3) = (c_8, y_3^2) = 1, (c_8 y_3, y_9) = (c_8, y_3 y_9) = 0.$ So

$$c_8 y_3 = b_6 + y_3 + y_{15}, \tag{8}$$

then

$$x_{10} y_3 = b_6 + y_9 + y_{15}. \tag{9}$$

By the associative law and 8) in Lemma 2.31 and 17) in Lemma 2.29:

$$(y_3^2)\bar{d}_3 = y_3(y_3 \bar{d}_3),$$
$$\bar{d}_3 + \bar{d}_3 c_8 = \bar{x}_6 y_3 + d_3 y_3,$$

by 7) in Lemma 2.31, $2\bar{d}_3 + x_{15} + x_6 = \bar{x}_6 y_3 + d_3 y_3$, by 15) in Lemma 2.29, $(d_3 y_3, \bar{d}_3) = (d_3^2, y_3) = 1$. So

$$d_3 y_3 = \bar{d}_3 + x_6, \tag{10}$$

then

$$\bar{x}_6 y_3 = \bar{d}_3 + x_{15}. \tag{11}$$

By the associative law and 6) in Lemma 2.31 and 7) in Lemma 2.32:

$$(d_3 b_6) y_3 = d_3 (y_3 b_6),$$
$$\bar{d}_3 y_3 + x_{15} y_3 = c_8 d_3 + x_{10} d_3,$$

then by 7) in Lemma 2.33 and 7) in Lemma 2.31 and (10):

$$d_3 + \bar{x}_6 + x_{15} y_3 = \bar{x}_{15} + \bar{x}_6 + d_3 + \bar{x}_{15} + \bar{x}_6 + \bar{x}_9,$$
$$y_{15} y_3 = 2\bar{x}_{15} + \bar{x}_6 + \bar{x}_9.$$

$\square$

Lemma 2.35   *Let* (A, B) *satisfy Hypothesis* 2.4 *and* $(b_8 b_3, b_8 b_3) = 3$, *then Lemma* 2.34 *holds and we obtain the following*:

(1) $z_9 x_{10} = 3x_{10} + 2b_8 + 2z_9 + 2c_8 + b_5 + c_5$,
(2) $z_9 c_8 = 2x_{10} + 2b_8 + 2z_9 + b_5 + c_5 + c_8$,
(3) $c_8 x_{10} = 2x_{10} + 2c_8 + 2b_8 + 2z_9 + b_5 + c_5$,
(4) $y_9^2 = 1 + 2z_9 + 2b_8 + 2c_8 + 2x_{10} + b_5 + c_5$,
(5) $x_9 c_8 = 3x_{15} + x_6 + 2x_9 + b_3$,
(6) $\bar{x}_9 y_3 = b_3 + x_9 + x_{15}$,
(7) $\bar{x}_9 x_9 = 1 + 2b_8 + 2x_{10} + 2c_8 + 2z_9 + b_5 + c_5$,
(8) $x_9 y_9 = d_3 + 3\bar{x}_{15} + 2\bar{x}_9 + \bar{b}_3 + 2\bar{x}_6$,
(9) $z_9 y_3 = y_9 + c_3 + y_{15}$,
(10) $x_9 z_9 = 3x_{15} + 2x_9 + 2x_6 + b_3 + \bar{d}_3$.

Proof   By the associative law and 7) in Lemma 2.33 and 16) in Lemma 2.29:

$$(b_3 d_3) x_{10} = b_3 (d_3 x_{10}),$$
$$z_9 x_{10} = \bar{x}_{15} b_3 + \bar{x}_6 b_3 + \bar{x}_9 b_3.$$

Then by 2) in Lemma 2.30 and 3) in Lemma 2.26 and 2) in Lemma 2.31:

$$z_9 x_{10} = x_{10} + b_8 + c_8 + z_9 + b_5 + c_5 + b_8 + x_{10} + z_9 + x_{10} + c_8,$$
$$z_9 x_{10} = 3x_{10} + 2b_8 + 2z_9 + 2c_8 + b_5 + c_5. \tag{1}$$

By the associative law and 16) in Lemma 2.29 and 7) in Lemma 2.31:

$$(b_3 d_3) c_8 = b_3 (d_3 c_8),$$
$$z_9 c_8 = \bar{x}_{15} b_3 + \bar{x}_6 b_3 + d_3 b_3.$$

Then by 2) in Lemma 2.30 and 3) in Lemma 2.26 and 16) in Lemma 2.29, we get
$z_9 c_8 = x_{10} + b_8 + c_8 + z_9 + b_5 + c_5 + b_8 + x_{10} + z_9$ and

$$z_9 c_8 = 2x_{10} + 2b_8 + 2z_9 + b_5 + c_5 + c_8. \tag{2}$$

By the associative law and 7) in Lemma 2.33 and 14) in Lemma 2.29:

$$(\bar{d}_3 d_3)x_{10} = \bar{d}_3(d_3 x_{10}),$$
$$x_{10} + c_8 x_{10} = \bar{d}_3 \bar{x}_{15} + \bar{d}_3 \bar{x}_6 + \bar{d}_3 \bar{x}_9,$$

then by 5) in Lemma 2.31 and 8) in Lemma 2.32 and 8) in Lemma 2.33, it follows
that $x_{10} + c_8 x_{10} = b_8 + z_9 + b_5 + c_5 + x_{10} + c_8 + x_{10} + c_8 + x_{10} + b_8 + z_9$ and

$$c_8 x_{10} = 2x_{10} + 2c_8 + 2b_8 + 2z_9 + b_5 + c_5. \tag{3}$$

By the associative law and 3), 15) in Lemma 2.29 and Hypothesis 2.4, one has

$$b_3^2 \bar{d}_3^2 = y_9^2,$$
$$(c_3 + b_6)(b_6 + y_3) = y_9^2,$$
$$c_3 b_6 + c_3 y_3 + b_6^2 + b_6 y_3 = y_9^2,$$

then by 3) in Lemma 2.26 and 1) in Lemma 2.34 and 3) in Lemma 2.30 and 7) in
Lemma 2.32, we have $y_9^2 = b_8 + x_{10} + z_9 + 1 + b_8 + z_9 + c_8 + b_5 + c_5 + c_8 + x_{10}$.
So

$$y_9^2 = 1 + 2z_9 + 2b_8 + 2c_8 + 2x_{10} + b_5 + c_5. \tag{4}$$

By the associative law and 13) in Lemma 2.32 and 9) in Lemma 2.34:

$$(b_3 y_3)c_8 = b_3(y_3 c_8),$$
$$\bar{x}_9 c_8 = b_6 b_3 + y_3 b_3 + y_{15} b_3,$$

then by 13) in Lemma 2.32 and 1) in Lemma 2.26 and 4) in Lemma 2.27, $\bar{x}_9 c_8 = \bar{b}_3 + \bar{x}_{15} + \bar{x}_9 + 2\bar{x}_{15} + \bar{x}_6 + \bar{x}_9$ and

$$\bar{x}_9 c_8 = \bar{b}_3 + 3\bar{x}_{15} + 2\bar{x}_9 + \bar{x}_6. \tag{5}$$

By the associative law and 13) in Lemma 2.32 and 17) in Lemma 2.29:

$$(y_3^2)b_3 = y_3(y_3 b_3),$$
$$b_3 + c_8 b_3 = \bar{x}_9 y_3,$$

then 4) in Lemma 2.31

$$\bar{x}_9 y_{13} = b_3 + x_9 + x_{15}. \tag{6}$$

By the associative law and 13) in Lemma 2.32 and (6):

$$(b_3 y_3)x_9 = b_3(y_3 x_9),$$
$$\bar{x}_9 x_9 = b_3 \bar{b}_3 + b_3 \bar{x}_9 + b_3 \bar{x}_{15},$$

then by Hypothesis 2.4 and 2) in Lemma 2.31 and 2) in Lemma 2.30:

$$\bar{x}_9 x_9 = 1 + b_8 + z_9 + x_{10} + c_8 + x_{10} + b_8 + c_8 + z_9 + b_5 + c_5,$$
$$\bar{x}_9 x_9 = 1 + 2b_8 + 2x_{10} + 2x_{10} + 2c_8 + 2z_9 + b_5 + c_5. \tag{7}$$

By the associative law and 13) in Lemma 2.32 and 9) in Lemma 2.33:

$$(b_3 y_3)y_9 = b_3(y_3 y_9),$$
$$\bar{x}_9 y_9 = z_9 b_3 + b_8 b_3 + x_{10} b_3,$$

then by 1) in Lemma 2.31 and 2) in Lemma 2.26 and 1) in Lemma 2.27:

$$\bar{x}_9 y_9 = x_9 + \bar{d}_3 + x_{15} + x_6 + b_3 + x_{15} + x_{15} + x_6 + x_9 = 2x_9 + 3x_{15} + 2x_6 + \bar{d}_3 + b_3. \tag{8}$$

By the associative law and 16) in Lemma 2.29 and 11) in Lemma 2.34:

$$(b_3 d_3)y_3 = (d_3 y_3)b_3,$$
$$z_9 y_3 = \bar{d}_3 b_3 + x_6 b_3,$$

then by 3) in Lemma 2.29 and 5) in Lemma 2.26

$$z_9 y_3 = y_9 + c_3 + y_{15}. \tag{9}$$

By the associative law and 13) in Lemma 2.32 and (9):

$$(z_9 y_3)b_3 = z_9(y_3 b_3),$$
$$y_9 b_3 + c_3 b_3 + y_{15} b_3 = z_9 \bar{x}_9,$$

then by 5) in Lemma 2.29 and by 2) in Lemma 2.35 and 4) in Lemma 2.27:

$$z_9 \bar{x}_9 = d_3 + \bar{x}_9 + \bar{x}_{15} + \bar{b}_3 + \bar{x}_6 + 2\bar{x}_{15} + \bar{x}_6 + \bar{x}_9 = 3\bar{x}_{15} + 2\bar{x}_9 + 2\bar{x}_6 + \bar{b}_3 + d_3. \tag{10}$$

$\square$

Lemma 2.36   *Let (A, B) satisfy Hypothesis 2.4 and $(b_8 b_3, b_8 b_3) = 3$, then Lemma 2.25 holds and we obtain the following:*

(1) $x_{15} c_8 = 5x_{15} + 2x_6 + 3x_9 + b_3 + \bar{d}_3,$
(2) $x_{15} x_9 = 6y_{15} + 3y_9 + 2b_6 + c_3 + y - 3,$
(3) $\bar{x}_9 x_{15} = 3b_8 + 4x_{10} + 3z_9 + 3c_8 + 2b_5 + 2c_5,$
(4) $x_{15} y_{15} = 4\bar{x}_6 + 9\bar{x}_{15} + 6\bar{x}_9 + 2\bar{b}_3 + 2d_3,$

(5) $x_{15}y_9 = 6\bar{x}_{15} + 2\bar{x}_6 + 3\bar{x}_9 + \bar{b}_3 + d_3,$
(6) $\bar{x}_9x_6 = 2z_9 + x_{10} + b_8 + c_8 + b_5 + c_5,$
(7) $x_6x_9 = 2y_9 + 2z_{15} + b_6,$
(8) $x_6y_9 = 2\bar{x}_{15} + 2\bar{x}_9 + \bar{x}_6,$
(9) $x_6z_9 = 2x_{15} + 2x_9 + x_6,$
(10) $x_6c_8 = 2x_{15} + x_6 + x_9 + \bar{d}_3,$
(11) $y_{15}y_9 = 4x_{10} + 3b_8 + 3z_9 + 3c_8 + 2c_5 + 2b_5.$

Proof   By the associative law and 6), 7) in Lemma 2.31:

$$(d_3b_6)c_8 = b_6(d_3c_8),$$

$$\bar{d}_3c_8 + x_{15}c_8 = \bar{x}_{15}b_6 + \bar{x}_6b_6 + d_3b_6,$$

then by 7) in Lemma 2.31 and 6), 10) in Lemma 2.29, 5) in Lemma 2.27:

$$x_{15} + x_6\bar{d}_3 + x_{15}c_8 = 4x_{15} + x_6 + 2x_9 + b_3 + \bar{d}_3 + 2x_6 + x_{15} + x_9 + \bar{d}_3 + x_{15},$$

$$x_{15}c_8 = 5x_{15} + 2x_6 + 3x_9 + b_3 + \bar{d}_3. \tag{1}$$

By the associative law and 8) in Lemma 2.33 and 6) in Lemma 2.31:

$$(d_3b_6)x_9 = b_6(d_3x_9),$$

$$\bar{d}_3x_9 + x_{15}x_9 = x_{10}b_6 + b_8b_6 + z_9b_6,$$

then by and 9) in Lemma 2.27 and 2), 5), 6) in Lemma 2.32:

$$y_{15} + c_3 + y_9 + x_{15}x_9 = 3y_{15} + y_3 + c_3 + y_9 + 2y_{15} + b_6 + c_3 + y_9$$

$$+ 2y_{15} + 2y_9 + b_6,$$

$$x_{15}x_9 = 6y_{15} + 3y_9 + 2b_6 + c_3 + y_3. \tag{2}$$

By 13) in Lemma 2.32 and 13) in Lemma 2.34:

$$b_3(y_3x_{15}) = (b_3y_3)x_{15} = \bar{x}_9x_{15},$$

$$2\bar{x}_{15}b_3 + \bar{x}_6b_3 + \bar{x}_9b_3 = \bar{x}_9x_{15}.$$

and 2) in Lemma 2.31 and 3) in Lemma 2.26 and 2) in Lemma 2.30:

$$2x_{10} + 2b_8 + 2c_8 + 2z_9 + 2b_5 + 2c_5 + b_8 + x_{10} + z_9 + x_{10} + c_8 = \bar{x}_9x_{15},$$

$$\bar{x}_9x_{15} = 3b_8 + 4x_{10} + 3z_9 + 3c_8 + 2b_5 + 2c_5. \tag{3}$$

By 11), 12) in Lemma 2.33 and by (2) and 8) in Lemma 2.27 and 3) in Lemma 2.32:

$$(x_{15}y_{15}, \bar{x}_6) = (x_{15}x_6, y_{15}) = 4, \qquad (x_{15}y_{15}, \bar{x}_{15}) = (x_{15}^2, y_{15}) = 9,$$

$$(x_{15}y_{15}, \bar{x}_9) = (x_{15}x_9, y_{15}) = 6, \qquad (x_{15}y_{15}, \bar{b}_3) = (x_{15}b_3, y_{15}) = 2,$$

$$(x_{15}y_{15}, d_3) = (x_{15}\bar{d}_3, y_{15}) = 2.$$

Then

$$x_{15}y_{15} = 4\bar{x}_6 + 9\bar{x}_{15} + 6\bar{x}_9 + 2\bar{b}_3 + 2d_3. \tag{4}$$

By (2) and 11), 12) in Lemma 2.33 and 8) in Lemma 2.27 and 3) in Lemma 2.32:

$$(x_{15}y_9, \bar{x}_{15}) = (y_9, x_{15}^2) = 6, \qquad (x_{15}y_9, \bar{x}_6) = (x_{15}x_6, y_9) = 2,$$

$$(x_{15}y_9, \bar{x}_9) = (x_{15}x_9, y_9) = 3, \qquad (x_{15}y_9, \bar{b}_3) = (x_{15}b_3, y_9) = 1,$$

$$(x_{15}y_9, d_3) = (x_{15}\bar{d}_3, y_9) = 1.$$

Then

$$x_{15}y_9 = 6\bar{x}_{15} + 2\bar{x}_6 + 3\bar{x}_9 + \bar{b}_3 + d_3. \tag{5}$$

By the associative law and 13) in Lemma 2.32 and 12) in Lemma 2.34:

$$b_3(y_3x_6) = (b_3y_3)x_6 = \bar{x}_9x_6,$$

$$b_3d_3 + b_3\bar{x}_{15} = \bar{x}_9x_6,$$

then by 16) in Lemma 2.29 and 2) in Lemma 2.30

$$2z_9 + x_{10} + b_8 + c_8 + b_5 + c_5 = \bar{x}_9x_6. \tag{6}$$

By the associative law and 13) in Lemma 2.32 and 12) in Lemma 2.34:

$$b_3(y_3\bar{x}_6) = \bar{x}_9\bar{x}_6,$$

$$b_3\bar{d}_3 + x_{15}b_3 = \bar{x}_9\bar{x}_6,$$

so by 3) in Lemma 2.29 and 8) in Lemma 2.27

$$\bar{x}_6\bar{x}_9 = 2y_9 + 2y_{15} + b_6. \tag{7}$$

By the associative law and 3) in Lemma 2.29 and 9) in Lemma 2.31:

$$b_3(\bar{d}_3x_6) = (b_3\bar{d}_3)x_6 = x_6y_9,$$

$$b_3y_{15} + b_3y_3 = x_6y_9. \tag{8}$$

By 16) in Lemma 2.29 and 5) in Lemma 2.31 and the associative law:

$$x_6(b_3d_3) = b_3(x_6d_3),$$

$$x_6z_9 = x_{10}b_3 + c_8b_3,$$

then by 1) in Lemma 2.27 and 4) in Lemma 2.31:

$$x_6z_9 = x_{15} + x_6 + x_9 + x_9 + x_{15} = 2x_{15} + 2x_9 + x_6. \tag{9}$$

By (1) and 4) in Lemma 2.30 and 5) in Lemma 2.35 and 7) in Lemma 2.31:

$$(x_6 c_8, x_{15}) = (x_6, c_8 x_{15}) = 2, \qquad (x_6 c_8, x_6) = (x_6 \bar{x}_6, c_8) = 1,$$
$$(x_6 c_8, x_9) = (x_6, c_8 x_9) = 1, \qquad (x_6 c_8, \bar{d}_3) = (c_8 d_3, \bar{x}_6) = 1,$$

then

$$x_6 c_8 = 2x_{15} + x_6 + x_9 + \bar{d}_3. \tag{10}$$

By the associative law and 9) in Lemma 2.31 and 1) in Lemma 2.33:

$$(d_3 \bar{x}_6) y_9 = (d_3 y_9) \bar{x}_6,$$
$$y_{15} y_9 + y_9 y_3 = b_3 \bar{x}_6 + x_{15} \bar{x}_6 + x_9 \bar{x}_6,$$

then by (6) and 6), 9) in Lemma 2.33 and 3) in Lemma 2.26:

$$y_{15} y_9 + z_9 + b_8 + x_{10} = b_8 + x_{10} + 2b_8 + 3x_{10} + 2z_9 + 2c_8$$
$$+ b_5 + c_5 + 2z_9 + x_{10} + b_8 + c_8 + b_5 + c_5,$$
$$y_{15} y_9 = 4x_{10} + 3b_8 + 3z_9 + 3c_8 + 2c_5 + 2b_5.$$

□

Lemma 2.37   *Let (A, B) satisfy Hypothesis* 2.4 *and* $(b_8 b_3, b_8 b_3) = 3$, *then Lemma* 2.36 *holds and we obtain the following*:

1) $y_3 y_{15} = x_{10} + b_8 + b_5 + c_5 + c_8 + z_9$,
2) $y_{15}^2 = 1 + 3b_5 + 6x_{10} + 6z_9 + 5b_8 + 3c_5 + 5c_8$,
3) $x_{10} y_{15} = 6y_{15} + 3b_6 + 4y_9 + c_3 + y_3$,
4) $z_9 x_{10} = 3x_{10} + b_5 + c_5 + 2b_8 + 2z_9 + 2c_8$,
5) $y_9 x_{10} = 4y_{15} + 2y_9 + y_3 + b_6 + c_3$,
6) $y_9 c_8 = 3y_{15} + 2y_9 + b_6 + c_3$,
7) $z_9 y_{15} = 6y_{15} + 2b_6 + 3y_9 + c_3 + y_3$,
8) $x_{10} x_{15} = 3x_6 + 6x_{15} + b_3 + \bar{d}_3 + 4x_9$,
9) $x_{10} x_9 = 4x_{15} + x_6 + 2x_9 + b_3 + \bar{d}_3$,
10) $b_6 x_9 = 2\bar{x}_{15} + 2\bar{x}_9 + \bar{x}_6$,
11) $b_8 y_9 = 3y_{15} + y_3$,
12) $\bar{x}_6 d_3 = y_{15} + y_3$,
13) $b_8 y_{15} = 5y_{15} + 2b_6 + 3y_9 + c_3 + y_3$,
14) $x_{15} z_9 = 2x_6 + 6x_{15} + b_3 + \bar{d}_3 + 3x_9$,
15) $x_9 y_{15} = 6\bar{x}_{15} + 2\bar{x}_6 + 3\bar{x}_9 + \bar{b}_3 + d_3$,
16) $y_{15} c_8 = 2b_6 + 5y_{15} + c_3 + y_3 + 3y_9$.

Proof   By the associative law and 9) in Lemma 2.31 and 15) in Lemma 2.29 and 7) in Lemma 2.27:

$$d_3^2 \bar{x}_6^2 = y_{15}^2 + 2y_{15}y_3 + y_3^2,$$

$$(b_6 + y_3)(2b_6 + y_{15} + y_9) = y_{15}^2 + 2y_{15}y_3 + y_3^2,$$

$$2b_6^2 + y_{15}b_6 + b_6y_9 + 2b_6y_3 + y_3y_{15} + y_3y_9 = y_{15}^2 + 2y_{15}y_3 + y_3^2,$$

by 3) in Lemma 2.30 and 7), 9), 10) in Lemma 2.32 and 9) in Lemma 2.33 and 17) in Lemma 2.29:

$$y_{15}^2 + 1 + c_8 + y_{15}y_3 = 2 + 2b_8 + 2z_9 + 2c_8 + 2b_5 + 2c_5 + 2b_8 + 3x_{10} + 2z_9$$

$$+ 2c_8 + b_5 + c_5 + b_8 + 2z_9 + b_5 + c_5 + x_{10} + c_8. \quad (1)$$

By the associative law and 11), 15) in Lemma 2.29:

$$(d_3^2)y_{15} = (d_3y_{15})d_3, \qquad b_6y_{15} + y_3y_{15} = 2x_{15}d_3 + x_6d_3 + x_9d_3,$$

so by 8), 9) in Lemma 2.32 and 5) in Lemma 2.31 and 8) in Lemma 2.33:

$$2b_8 + 3x_{10} + 2z_9 + 2c_8 + b_5 + c_5 + y_3y_{15} = 2b_8 + 2z_9 + 2b_5 + 2c_5 + 2x_{10}$$

$$+ 2c_8 + x_{10} + c_8 + x_{10} + b_8 + z_9,$$

$$y_3y_{15} = x_{10} + b_8 + b_5 + c_5 + c_8 + z_9. \quad (2)$$

Then by (1),

$$y_{15}^2 = 1 + 3b_5 + 6x_{10} + 6z_9 + 5b_8 + 3c_5 + 5c_8. \quad (3)$$

By 9) in Lemma 2.32 and 11) in Lemma 2.36 and 8) in Lemma 2.30 and (2):

$$(x_{10}y_{15}, y_{15}) = (x_{10}, y_{15}^2) = 6, \qquad (x_{10}y_{15}, b_6) = (x_{10}, y_{15}b_6) = 3,$$

$$(x_{10}y_{15}, y_9) = (x_{10}, y_{15}y_9) = 4, \qquad (x_{10}y_{15}, c_3) = (x_{10}, y_{15}c_3) = 1,$$

$$(y_{15}x_{10}, y_3) = (x_{10}, y_{15}y_3) = 1.$$

So

$$x_{10}y_{15} = 6y_{15} + 3b_6 + 4y_9 + c_3 + y_3. \quad (4)$$

By 16) in Lemma 2.29 and 7) in Lemma 2.33:

$$b_3(d_3x_{10}) = (b_3d_3)x_{10} = z_9x_{10},$$

$$\bar{x}_{15}b_3 + \bar{x}_6b_3 + \bar{x}_9b_3 = z_9x_{10},$$

then by 2) in Lemma 2.30 and 3) in Lemma 2.26 and 2) in Lemma 2.31:

$$x_{10} + b_8 + c_8 + z_9 + b_5 + c_5 + b_8 + x_{10} + z_9 + x_{10} + c_8 = z_9x_{10},$$

$$z_9x_{10} = 3x_{10} + b_5 + c_5 + 2b_8 + 2z_9 + 2c_8. \quad (5)$$

$\square$

Lemma 2.38 *Let (A, B) satisfy Hypothesis 2.4 and $(b_8 b_3, b_8 b_3) = 3$, then $b_5 \neq \bar{c}_5$.*

Proof   Assume $b_5 = \bar{c}_5$. By 2) in Lemma 2.30, $(\bar{b}_3 b_5, \bar{x}_{15}) = (b_5, b_3 \bar{x}_{15}) = 1$, $(b_3 b_5, x_{15}) = (b_3 \bar{x}_{15}, \bar{b}_5) = 1$, then

$$\bar{b}_3 b_5 = \bar{x}_{15}, \tag{1}$$

$$b_3 b_5 = x_{15}. \tag{2}$$

By 9), 10) in Lemma 2.32 and 3) in Lemma 2.30:

$$(b_6 b_5, y_{15}) = (b_5, b_6 y_{15}) = 1,$$
$$(b_6 b_5, y_9) = (b_5, b_6 y_9) = 1,$$
$$(b_6 b_5, b_6) = (b_5, b_6^2) = 1,$$

then

$$b_6 b_5 = y_{15} + y_9 + b_6. \tag{3}$$

By 4) in Lemma 2.30 and 6) in Lemma 2.33 and 6) in Lemma 2.36:

$$(x_6 b_5, x_6) = (b_5, \bar{x}_6 x_6) = 1,$$
$$(x_6 b_5, x_{15}) = (b_5, \bar{x}_6 x_{15}) = 1,$$
$$(x_6 b_5, x_9) = (b_5, \bar{x}_6 x_9) = 1,$$

then

$$x_6 b_5 = x_6 + x_{15} + x_9. \tag{4}$$

By 6) in Lemma 2.33 and 6) in Lemma 2.36 and 4) in Lemma 2.30:

$$(\bar{x}_6 b_5, \bar{x}_{15}) = (b_5, x_6 \bar{x}_{15}) = 1,$$
$$(\bar{x}_6 b_5, \bar{x}_9) = (\bar{x}_6 x_9, \bar{b}_5) = 1,$$
$$(\bar{x}_6 b_5, \bar{x}_6) = (\bar{x}_6 x_6, \bar{b}_5) = 1$$

then

$$\bar{x}_6 b_5 = \bar{x}_{15} + \bar{x}_9 + \bar{x}_6. \tag{5}$$

By 6) in Lemma 2.33 and 8), 11) in Lemma 2.32 and 2) in Lemma 2.30 and 3) in Lemma 2.36:

$$(x_{15} b_5, x_6) = (b_5, \bar{x}_{15} x_6) = 1,$$
$$(x_{15} b_5, \bar{d}_3) = (x_{15} d_3, \bar{b}_5) = 1,$$
$$(x_{15} b_5, x_{15}) = (b_5, \bar{x}_{15} x_{15}) = 3,$$
$$(x_{15} b_5, x_9) = (b_5, \bar{x}_{15} x_9) = 2,$$
$$(x_{15} b_5, b_3) = (x_{15} \bar{b}_3, \bar{b}_5) = 1,$$

then

$$x_{15}b_5 = x_6 + 3x_{15} + 2x_9 + b_3 + \bar{d}_3. \tag{6}$$

By 6) in Lemma 2.33 and 8), 11) in Lemma 2.32 and 3) in Lemma 2.36 and 2) in Lemma 2.30:

$$(\bar{x}_{15}b_5, \bar{x}_6) = (b_5, x_{15}\bar{x}_6) = 1,$$
$$(\bar{x}_{15}b_5, d_3) = (b_5, x_{15}d_3) = 1,$$
$$(\bar{x}_{15}b_5, \bar{x}_{15}) = (\bar{x}_{15}x_{15}, \bar{b}_5) = 3,$$
$$(\bar{x}_{15}b_5, \bar{x}_9) = (\bar{x}_{15}x_9, \bar{b}_5) = 1,$$
$$(\bar{x}_{15}b_5, \bar{b}_3) = (\bar{x}_{15}b_3, \bar{b}_5) = 1,$$

then

$$\bar{x}_{15}b_5 = \bar{x}_6 + 3\bar{x}_{15} + 2\bar{x}_9 + \bar{b}_3 + d_3. \tag{7}$$

By 3), 6) in Lemma 2.36 and 7) in Lemma 2.35:

$$(x_9b_5, x_{15}) = (x_9\bar{x}_{15}, \bar{b}_5) = 2,$$
$$(x_9b_5, x_6) = (b_5, \bar{x}_9x_6) = 1,$$
$$(x_9b_5, x_9) = (x_9\bar{x}_9, \bar{b}_5) = 1,$$

then

$$x_9b_5 = 2x_{15} + x_6 + x_9. \tag{8}$$

By 3), 6) in Lemma 2.36 and 7) in Lemma 2.35:

$$(\bar{x}_9b_5, \bar{x}_{15}) = (\bar{x}_9x_{15}, \bar{b}_5) = 2,$$
$$(\bar{x}_9b_5, \bar{x}_6) = (b_5, \bar{x}_9x_6) = 1,$$
$$(x_9b_5, x_9) = (x_9\bar{x}_9x_9, \bar{b}_5) = 1,$$

then

$$\bar{x}_9b_5 = 2\bar{x}_{15} + \bar{x}_6 + \bar{x}_9. \tag{9}$$

By 9) in Lemma 2.32 and 1), 2) in Lemma 2.37 and 11) in Lemma 2.36 and 8) in Lemma 2.30

$$(y_{15}b_5, b_6) = (b_5, y_{15}b_6) = 1,$$
$$(y_{15}b_5, y_{15}) = (b_5, y_{15}^2) = 3,$$
$$(y_{15}b_5, y_3) = (y_{15}y_3, \bar{b}_5) = 1,$$
$$(y_{15}b_5, c_3) = (b_5, y_{15}c_3) = 1,$$
$$(y_{15}b_5, y_9) = (b_5, y_{15}y_9) = 2,$$

then

$$y_{15}b_5 = b_6 + 3y_{15} + 2y_9 + c_3 + y_3. \tag{10}$$

By 11) in Lemma 2.36 and 10) in Lemma 2.32 and 4) in Lemma 2.35:

$$(y_9b_5, y_{15}) = (b_5, y_9y_{15}) = 2,$$
$$(y_9b_5, b_6) = (b_5, y_9b_6) = 1,$$
$$(y_9b_5, y_9) = (b_5, y_9^2) = 1,$$

then

$$y_9b_5 = 2y_{15} + b_6 + y_9. \tag{11}$$

By 8) in Lemma 2.32,

$$(d_3b_5, \bar{x}_{15}) = (d_3x_{15}, \bar{b}_5) = 1,$$
$$(\bar{d}_3b_5, x_{15}) = (b_5, d_3x_{15}) = 1,$$

then

$$d_3b_5 = \bar{x}_{15}, \tag{12}$$
$$\bar{d}_3b_5 = x_{15}. \tag{13}$$

Then by 16) in Lemma 2.29 and associative law:

$$b_3(b_5d_3) = (b_3d_3)b_5, \qquad \bar{x}_{15}b_3 = z_9b_5,$$

then by 2) in Lemma 2.30,

$$z_9b_5 = x_{10} + b_8 + c_8 + z_9 + b_5 + c_5. \tag{14}$$

By 1) in Lemma 2.37, $(y_3b_5, y_{15}) = (b_5, y_3y_{15}) = 1$, then

$$y_3b_5 = y_{15}. \tag{15}$$

Then $1 = (y_3b_5, y_3\bar{b}_5) = (b_5^2, y_3^2)$, then by 17) in Lemma 2.29,

$$(b_5^2, c_8) = 1, \tag{16}$$

By (1), (2) we get $1 = (b_3\bar{b}_5, b_3b_5) = (b_3\bar{b}_3, b_5^2)$, then by Hypothesis 2.4,

$$(b_5^2, b_8) = 1. \tag{17}$$

By (2), (12) $1 = (\bar{d}_3\bar{b}_5, b_3b_5) = (\bar{d}_3\bar{b}_3, b_5^2)$, then by 16) in Lemma 2.29 $(b_5^2, z_9) = 1$, so by (16), (17), we obtain

$$b_5^2 = c_8 + b_8 + z_9. \tag{18}$$

By (1) $1 = (\bar{b}_3 b_5, \bar{b}_3 b_5) = (\bar{b}_3 b_3, \bar{b}_5 b_5)$, then by Hypothesis 2.4,

$$(b_5 \bar{b}_5, b_8) = 0. \tag{19}$$

By (2), (4), $1 = (x_6 b_5, b_3 b_5) = (x_6 \bar{b}_3, \bar{b}_5 b_5)$, then by 3) in Lemma 2.26 and (19):

$$(\bar{b}_5 b_5, x_{10}) = 1.$$

By (13), $3 = (b_6 b_5, b_6 b_5) = (b_6^2, \bar{b}_5 b_5)$, so by 3) in Lemma 2.30, $m_5 = b_5$, or $m_5 = \bar{b}_5$, in the two cases $(\bar{b}_5 b_5, b_5) \neq 0$ or $(\bar{b}_5 b_5, \bar{b}_5) \neq 0$, a contradiction to (18). $\qquad\square$

Lemma 2.39 *Let (A, B) satisfy Hypothesis 2.4 and $(b_8 b_3, b_8 b_3) = 3$, then Lemma 2.37 holds and we obtain the following*:

1) $b_3 b_5 = x_{15}$;
2) $b_3 b_5 = x_{15}$;
3) $b_6 b_5 = b_6 + y_{15} + y_9$;
4) $b_6 c_5 = b_6 + y_{15} + y_9$;
5) $b_5 x_6 = x_{15} + x_9 + x_6$;
6) $c_5 x_6 = x_{15} + x_9 + x_6$;
7) $x_{15} b_5 = x_6 + \bar{d}_3 + 3x_{15} + b_3 + 2x_9$;
8) $x_{15} c_5 = x_6 + \bar{d}_3 + 3x_{15} + b_3 + 2x_9$;
9) $x_9 b_5 = 2x_{15} + x_6 + x_9$;
10) $x_9 c_5 = 2x_{15} + x_6 + x_9$;
11) $y_{15} b_5 = b_6 + y_3 + 3y_{15} + 2y_9 + c_3$;
12) $y_{15} c_5 = b_6 + y_3 + 3y_{15} + 2y_9 + c_3$;
13) $y_9 b_5 = 2y_{15} + b_6 + y_9$;
14) $y_9 c_5 = 2y_{15} + b_6 + y_9$;
15) $d_3 b_5 = \bar{x}_{15}$;
16) $d_3 c_5 = \bar{x}_{15}$;
17) $z_9 b_5 = x_{10} + b_8 + c_8 + z_9 + b_5 + c_5$;
18) $z_9 c_5 = x_{10} + b_8 + c_8 + z_9 + b_5 + c_5$;
19) $y_3 b_5 = y_{15}$;
20) $y_3 c_5 = y_{15}$;
21) $c_5 b_5 = c_8 + b_8 + z_9$;
22) $b_8 b_5 = x_{10} + b_8 + c_8 + c_5 + z_9$;
23) $b_8 c_5 = x_{10} + b_8 + c_8 + c_5 + z_9$;
24) $b_5^2 = 1 + b_5 + z_9 + x_{10}$;
25) $c_5^2 = 1 + c_5 + z_9 + x_{10}$;
26) $b_5 x_{10} = 2x_{10} + b_8 + z_9 + b_5 + c_8$;
27) $c_5 x_{10} = 2x_{10} + b_8 + z_9 + c_5 + c_8$;
28) $c_8 b_5 = b_8 + z_9 + c_5 + x_{10} + c_8$;
29) $c_8 c_5 = b_8 + z_9 + x_{10} + c_8 + b_5$.

Proof By 2) in Lemma 2.30, $(b_3 b_5, x_{15}) = (b_3 \bar{x}_{15}, b_5) = 1$, $(b_3 c_5, x_{15}) = (b_3 \bar{x}_{15}, c_5) = 1$, then

$$b_3 b_5 = x_{15}, \tag{1}$$

$$b_3 c_5 = x_{15}. \tag{2}$$

By 9), 10) in Lemma 2.32 and 3) in Lemma 2.30:

$$(b_6 b_5, b_6) = (b_6^2, b_5) = 1, \qquad (b_6 c_5, b_6) = (b_6^2, c_5) = 1,$$
$$(b_6 b_5, y_{15}) = (b_5, b_6 y_{15}) = 1, \qquad (b_6 c_5, y_{15}) = (c_5, b_6 y_{15}) = 1,$$
$$(b_6 b_5, y_9) = (b_5, b_6 y_9) = 1, \qquad (b_6 c_5, y_9) = (c_5, b_6 y_9) = 1,$$

then

$$b_5 b_6 = b_6 + y_{15} + y_9, \tag{3}$$
$$b_6 c_5 = y_{15} + y_9 + b_6. \tag{4}$$

By 4) in Lemma 2.30 and 6) in Lemma 2.33 and 6) in Lemma 2.36:

$$(b_5 x_6, x_6) = (b_5, \bar{x}_6 x_6) = 1, \qquad (c_5 x_6, x_6) = (c_5, \bar{x}_6 x_6) = 1,$$
$$(b_5 x_6, x_{15}) = (x_6 \bar{x}_{15}, b_5) = 1, \qquad (c_5 x_6, x_{15}) = (x_6 \bar{x}_{15}, c_5) = 1,$$
$$(b_5 x_6, x_9) = (x_6 \bar{x}_9, b_5) = 1, \qquad (c_5 x_6, x_9) = (x_6 \bar{x}_9, c_5) = 1,$$

then

$$b_5 x_6 = x_6 + x_{15} + x_9, \tag{5}$$
$$c_5 x_6 = x_{15} + x_9 + x_6. \tag{6}$$

By 6) in Lemma 2.33 and 8), 11) in Lemma 2.32 and 2) in Lemma 2.30 and 3) in Lemma 2.36:

$$(x_{15} b_5, x_6) = (b_5, \bar{x}_{15} x_6) = 1, \qquad (x_{15} c_5, x_6) = (c_5, \bar{x}_{15} x_6) = 1,$$
$$(x_{15} b_5, \bar{d}_3) = (x_{15} d_3, b_5) = 1, \qquad (x_{15} c_5, \bar{d}_3) = (x_{15} d_3, c_5) = 1,$$
$$(x_{15} b_5, x_{15}) = (x_{15} \bar{x}_{15}, b_5) = 3, \qquad (x_{15} c_5, x_{15}) = (x_{15} \bar{x}_{15}, c_5) = 3,$$
$$(x_{15} b_5, b_3) = (b_5, \bar{x}_{15} b_3) = 1, \qquad (x_{15} c_5, b_3) = (c_5, \bar{x}_{15} b_3) = 1,$$
$$(x_{15} b_5, x_9) = (b_5, \bar{x}_{15} x_9) = 2, \qquad (x_{15} c_5, x_9) = (x_{15} \bar{x}_9, c_5) = 2,$$

then

$$x_{15} b_5 = x_6 + \bar{d}_3 + 3x_{15} + b_3 + 2x_9, \tag{7}$$
$$x_{15} c_5 = x_6 + \bar{d}_3 + 3x_{15} + b_3 + 2x_9. \tag{8}$$

By 3), 6) in Lemma 2.36 and 7) in Lemma 2.35:

$$(x_9 b_5, x_{15}) = (b_5, \bar{x}_9 x_{15}) = 2, \qquad (x_9 c_5, x_{15}) = (c_5, \bar{x}_9 x_{15}) = 2,$$
$$(x_9 b_5, x_6) = (b_5, \bar{x}_9 x_6) = 1, \qquad (x_9 c_5, x_6) = (c_5, \bar{x}_9 x_6) = 1,$$
$$(x_9 b_5, x_9) = (b_5, \bar{x}_9 x_9) = 1, \qquad (x_9 c_5, x_9) = (c_5, \bar{x}_9 x_9) = 1,$$

then

$$x_9b_5 = 2x_{15} + x_6 + x_9, \tag{9}$$

$$x_9c_5 = 2x_{15} + x_6 + x_9. \tag{10}$$

By 9) in Lemma 2.32 and 1), 2) in Lemma 2.37 and 11) in Lemma 2.36 and 8) in Lemma 2.30:

$$(y_{15}b_5, b_6) = (y_{15}b_6, b_5) = 1, \qquad (y_{15}c_5, b_6) = (y_{15}b_6, c_5) = 1,$$
$$(y_{15}b_5, y_3) = (y_{15}y_3, b_5) = 1, \qquad (y_{15}c_5, y_3) = (y_{15}y_3, c_5) = 1,$$
$$(y_{15}b_5, y_{15}) = (b_5, y_{15}^2) = 3, \qquad (y_{15}c_5, y_{15}) = (y_{15}^2, c_5) = 3,$$
$$(y_{15}b_6, c_3) = (y_{15}c_3, b_6) = 1, \qquad (y_{15}c_5, c_3) = (y_{15}c_3, c_5) = 1,$$
$$(y_{15}b_5, y_9) = (y_{15}y_9, b_5) = 2, \qquad (y_{15}c_5, y_9) = (c_5, y_{15}y_9) = 2,$$

then

$$y_{15}b_5 = b_6 + y_3 + 3y_{15} + 2y_9 + c_3, \tag{11}$$

$$y_{15}c_5 = b_6 + y_3 + 3y_{15} + 2y_9 + c_3. \tag{12}$$

By 11) in Lemma 2.36 and 10) in Lemma 2.32 and 4) in Lemma 2.35:

$$(y_9b_5, y_{15}) = (b_5, y_9y_{15}) = 2, \qquad (y_9c_5, y_{15}) = (c_5, y_9y_{15}) = 2,$$
$$(y_9b_5, b_6) = (y_9b_6, b_5) = 1, \qquad (y_9c_5, b_6) = (y_9b_6, c_5) = 1,$$
$$(y_9b_5, y_9) = (b_5, y_9^2) = 1, \qquad (y_9c_5, y_9) = (c_5, y_9^2) = 1,$$

then

$$y_9b_5 = 2y_{15} + b_6 + y_9, \tag{13}$$

$$y_9c_5 = 2y_{15} + b_6 + y_9. \tag{14}$$

By 8) in Lemma 2.32, $(d_3b_5, \bar{x}_{15}) = (d_3x_{15}, b_5) = 1$, $(d_3c_5, \bar{x}_{15}) = (d_3x_{15}, c_5) = 1$. Then

$$d_3b_5 = \bar{x}_{15}, \tag{15}$$

$$d_3c_5 = \bar{x}_{15}. \tag{16}$$

So by the associative law and 16) in Lemma 2.29, we get $(b_3d_3)b_5 = b_3(d_3b_5)$, $(b_3d_3)c_5 = b_3(d_3c_5)$, $z_9b_5 = b_3\bar{x}_{15}$, $z_9c_5 = b_3\bar{x}_{15}$, then by 2) in Lemma 2.30,

$$z_9b_5 = x_{10} + b_8 + c_8 + z_9 + b_5 + c_5, \tag{17}$$

$$z_9c_5 = x_{10} + b_8 + c_8 + z_9 + b_5 + c_5. \tag{18}$$

By 1) in Lemma 2.37, $(y_3 b_5, y_{15}) = (b_5, y_3 y_{15}) = 1$, $(y_3 c_5, y_{15}) = (c_5, y_3 y_{15}) = 1$, then

$$y_3 b_5 = \bar{y}_{15}, \tag{19}$$

$$y_3 c_5 = \bar{y}_{15}. \tag{20}$$

So $(y_3^2, b_5^2) = (y_3 b_5, y_3 b_5) = 1$, $(y_3^2, c_5 b_5) = (y_3 b_5, y_3 c_5) = 1$, then by 17) in Lemma 2.29, $(c_5 b_5, c_8) = 1$. By (1), (2), $(\bar{b}_3 b_3, b_5 c_5) = (b_3 b_5, b_3 c_5) = 1$. then by Hypothesis 2.4, $(c_5 b_5, b_8) = 1$. By (17),

$$c_5 b_5 = c_8 + b_8 + z_9. \tag{21}$$

By the associative and Hypothesis 2.4 and (1), (2), it follows that $(b_3 \bar{b}_3) b_5 = (b_3 b_5) \bar{b}_3$, $(b_3 \bar{b}_3) c_5 = (b_3 c_5) \bar{b}_3$, $b_5 + b_8 b_5 = x_{15} \bar{b}_3$, $c_5 + b_8 c_5 = x_{15} \bar{b}_3$, then by 2) in Lemma 2.30,

$$b_8 b_5 = x_{10} + b_8 + c_8 + c_5 + z_9, \tag{22}$$

$$b_8 c_5 = x_{10} + b_8 + c_8 + b_5 + z_9. \tag{23}$$

By 5) in Lemma 2.30, $(x_{10} b_5, x_{10}) = (b_5, x_{10}^2) = 2$, $(x_{10} c_5, x_{10}) = (c_5, x_{10}^2) = 2$. By (22) and (23), $(x_{10} b_5, b_8) = (x_{10}, b_5 b_8) = 1$, $(x_{10} c_5, b_8) = (x_{10}, c_5 b_8) = 1$. By (17), (18), $(x_{10} b_5, z_9) = (x_{10}, z_9 b_5) = 1$, $(x_{10} c_5, b_8) = (x_{10}, c_5 b_8) = 1$, $(b_5^2, z_9) = (b_5, b_5 z_9) = 1$ and $(c_5^2, z_9) = (c_5, c_5 z_9) = 1$.

By (19), (20), we obtain $(y_3^2, b_5^2) = (y_3 b_5, y_3 b_5) = 1$, $(y_3^2, c_5^2) = (y_3 c_5, y_3 c_5) = 1$, then by 17) in Lemma 2.29 $(b_5^2, c_8) = 0$, $(c_5^2, c_8) = 0$.

By (1) and (2), we see that $(\bar{b}_3 b_3, b_5^2) = (b_3 b_5, b_3 b_5) = 1$, $(\bar{b}_3 b_3, c_5^2) = (b_3 c_5, b_3 c_5) = 1$, then by Hypothesis 2.4, $(b_5^2, b_8) = 0$, $(c_5^2, b_8) = 0$ holds.

By (5), (6), (15) and (16), $(b_5^2, d_3 x_6) = (b_5 \bar{d}_3, b_5 x_6) = 1$, $(c_5^2, d_3 x_6) = (c_5 \bar{d}_3, c_5 x_6) = 1$, so by 5) in Lemma 2.31, $(b_5^2, x_{10}) = 1$, $(c_5^2, x_{10}) = 1$.

By (5), (21) and (6) we have the following inner-products: $(b_5^2, \bar{x}_6 x_6) = (b_5 x_6, b_5 x_6) = 3$, $(c_5^2, \bar{x}_6 x_6) = (c_5 x_6, c_5 x_6) = 3$. $(c_5^2, b_5) = (c_5, c_5 b_5) = 0$ and $(b_5^2, c_5) = (b_5, b_5 c_5) = 0$, then by 4) in Lemma 2.30 $(b_5^2, b_5) = 1$, $(c_5^2, c_5) = 1$, hence

$$b_5^2 = 1 + b_5 + z_9 + x_{10}, \qquad c_5^2 = 1 + b_5 + z_9 + x_{10}.$$

So $(b_5 x_{10}, b_5) = (x_{10}, b_5^2) = 1$, $(c_5 x_{10}, c_5) = (x_{10}, c_5^2) = 1$.

By 14) in Lemma 2.29 and (15), (16), one has $b_5(\bar{d}_3 d_3) = \bar{d}_3(b_5 d_3)$, $c_5(\bar{d}_3 d_3) = \bar{d}_3(c_5 d_3)$. Hence $b_5 + c_8 b_5 = \bar{x}_{15} \bar{d}_3$, $c_5 + c_8 c_5 = \bar{d}_3 \bar{x}_{15}$.

Therefore by 8) in Lemma 2.32, we have the equations $c_8 b_5 = b_8 + z_9 + c_5 + x_{10} + c_8$ and $c_8 c_5 = b_8 + z_9 + x_{10} + c_8 + b_5$. Consequently $(b_5 x_{10}, c_8) = (x_{10}, b_5 c_8) = (x_{10}, c_5 c_8) = 1$.

Finally, we get $b_5 x_{10} = c_8 + 2 x_{10} + b_8 + z_9 + b_8 + z_9 + b_5$ and $c_5 x_{10} = 2 x_{10} + b_8 + z_9 + c_5 + c_8$. $\qquad \square$

Theorem 2.10 *Let* (A, B) *be a NITA generated by a nonreal element* $b_3 \in B$ *of degree* 3 *and without non-identity basis element of degree* 1 *or* 2, *such that:* $b_3\bar{b}_3 = 1 + b_8$ *and* $b_3^2 = c_3 + b_6$, $c_3 = \bar{c}_3$, $(b_8 b_3, b_8 b_3) = 3$. *Then* (A, B) *is a Table Algebra of dimension* 22:

$$B = \{b_1, \bar{b}_3, b_3, c_3, b_6, x_6, \bar{x}_6, b_8, x_{15}, \bar{x}_{15}, x_{10}, x_9, \bar{x}_9, y_{15}, y_9, d_3, \bar{d}_3, z_9,$$

$$c_8, b_5, c_5, y_3\}$$

and B *has increasing series of table subsets* $\{b_1\} \subseteq \{b_1, b_8, x_{10}, z_9, c_8, b_5, c_5\} \subseteq B$ *defined by:*

1) $b_3\bar{b}_3 = 1 + b_8$;
2) $b_3^2 = c_3 + b_6$;
3) $b_3 c_3 = \bar{b}_3 + \bar{x}_6$;
4) $b_3 b_6 = \bar{b}_3 + \bar{x}_{15}$;
5) $b_3 x_6 = c_3 + y_{15}$;
6) $b_3 \bar{x}_6 = b_8 + x_{10}$;
7) $b_3 b_8 = x_6 + b_3 + x_{15}$;
8) $b_3 x_{15} = 2y_{15} + b_6 + y_9$;
9) $b_3 x_{10} = x_{15} + x_6 + x_9$;
10) $b_3 \bar{x}_{15} = x_{10} + b_8 + z_9 + c_8 + b_5 + c_5$;
11) $b_3 x_9 = y_9 + y_{15} + y_3$;
12) $b_3 \bar{x}_9 = x_{10} + z_9 + c_8$;
13) $b_3 y_{15} = 2\bar{x}_{15} + \bar{x}_6 + \bar{x}_9$;
14) $b_3 y_9 = d_3 + \bar{x}_{15} + \bar{x}_9$;
15) $b_3 d_3 = z_9$;
16) $b_3 \bar{d}_3 = y_9$;
17) $b_3 z_9 = x_9 + \bar{d}_3 + x_{15}$;
18) $b_3 c_8 = x_{15} + x_9$;
19) $b_3 b_5 = x_{15}$;
20) $b_3 c_5 = x_{15}$;
21) $b_3 y_3 = \bar{x}_9$;
22) $c_3^2 = 1 + b_8$;
23) $c_3 b_6 = b_8 + x_{10}$;
24) $c_3 x_6 = \bar{b}_3 + \bar{x}_{15}$;
25) $c_3 b_8 = b_6 + c_3 + y_{15}$;
26) $c_3 x_{15} = 2\bar{x}_{15} + \bar{x}_6 + \bar{x}_9$;
27) $c_3 x_{10} = b_6 + y_{15} + y_9$;
28) $c_3 x_9 = d_3 + \bar{x}_{15} + \bar{x}_9$;
29) $c_3 y_{15} = x_{10} + b_8 + z_9 + c_8 + b_5 + c_5$;
30) $c_3 y_9 = z_9 + x_{10} + c_8$;
31) $c_3 d_3 = x_9$;
32) $c_3 z_9 = y_3 + y_9 + y_{15}$;
33) $c_3 c_8 = y_9 + y_{15}$;
34) $c_3 b_5 = y_{15}$;
35) $c_3 c_5 = y_{15}$;

36) $c_3 y_3 = z_9$;

37) $b_6^2 = 1 + b_8 + z_9 + c_8 + b_5 + c_5$;

38) $b_6 x_6 = 2\bar{x}_6 + \bar{x}_{15} + \bar{x}_9$;

39) $b_6 b_8 = 2y_{15} + b_6 + y_9 + c_3$;

40) $b_6 x_{15} = 4\bar{x}_{15} + \bar{x}_6 + 2\bar{x}_9 + \bar{b}_3 + d_3$;

41) $b_6 x_{10} = c_3 + 3y_{15} + y_9 + y_3$;

42) $b_6 x_9 = 2\bar{x}_{15} + 2\bar{x}_9 + \bar{x}_6$;

43) $b_6 y_{15} = 2b_8 + 3x_{10} + 2z_9 + 2c_8 + b_5 + c_5$;

44) $b_6 y_9 = b_8 + 2z_9 + c_8 + x_{10} + b_5 + c_5$;

45) $b_6 d_3 = \bar{d}_3 + x_{15}$;

46) $b_6 z_9 = 2y_{15} + 2y_9 + b_6$;

47) $b_6 c_8 = b_6 + y_3 + y_9 + 2y_{15}$;

48) $b_6 b_5 = y_{15} + y_9 + b_6$;

49) $b_6 c_5 = y_{15} + b_6 + y_9$;

50) $b_6 y_3 = x_{10} + c_8$;

51) $x_6^2 = 2b_6 + y_{15} + y_9$;

52) $x_6 \bar{x}_6 = 1 + b_8 + z_9 + c_8 + b_5 + c_5$;

53) $x_6 b_8 = b_3 + 2x_{15} + x_6 + x_9$;

54) $x_6 x_{15} = 4y_{15} + b_6 + 2y_9 + c_3 + y_3$;

55) $x_6 x_{10} = b_3 + 3x_{15} + \bar{d}_3 + x_9$;

56) $x_6 \bar{x}_{15} = 2b_8 + 3x_{10} + 2z_9 + 2c_8 + b_5 + c_5$;

57) $x_6 x_9 = 2y_{15} + 2y_9 + b_6$;

58) $x_6 \bar{x}_9 = x_{10} + b_8 + 2z_9 + c_8 + b_5 + c_5$;

59) $x_6 y_{15} = 4\bar{x}_{15} + 2\bar{x}_9 + \bar{x}_6 + \bar{b}_3 + d_3$;

60) $x_6 y_9 = 2\bar{x}_{15} + 2\bar{x}_9 + \bar{x}_6$;

61) $x_6 d_3 = x_{10} + c_8$;

62) $x_6 \bar{d}_3 = y_{15} + y_3$;

63) $x_6 z_9 = 2x_9 + 2x_{15} + x_6$;

64) $x_6 c_8 = 2x_{15} + x_6 + x_9 + \bar{d}_3$;

65) $x_6 b_5 = x_6 + x_{15} + x_9$;

66) $x_6 c_5 = x_{15} + x_9 + x_6$;

67) $x_6 y_3 = d_3 + \bar{x}_{15}$;

68) $b_8^2 = 1 + 2b_8 + 2x_{10} + z_9 + c_8 + b_5 + c_5$;

69) $b_8 x_{10} = 2x_{10} + 2z_9 + 2c_8 + 2b_8 + b_5 + c_5$;

70) $b_8 x_9 = 3x_{15} + 2x_9 + x_6 + \bar{d}_3$;

71) $b_8 y_{15} = 5y_{15} + 2b_6 + 3y_9 + c_3 + y_3$;

72) $b_8 y_9 = 3y_{15} + 2y_9 + b_6 + y_3$;

73) $b_8 d_3 = \bar{x}_{15} + \bar{x}_9$;

74) $b_8 z_9 = 2x_{10} + b_8 + 2c_8 + 2z_9 + b_5 + c_5$;

75) $b_8 c_8 = 2x_{10} + 2z_9 + b_8 + c_8 + b_5 + c_5$;

76) $b_8 b_5 = b_8 + z_9 + c_5 + c_8 + x_{10}$;

77) $b_8 c_5 = b_8 + x_{10} + c_8 + z_9 + b_5$;

78) $b_8 y_3 = y_9 + y_{15}$;

79) $b_8 x_{15} = 5x_{15} + 2x_6 + 3x_9 + b_3 + \bar{d}_3$;

80) $x_{15}^2 = 9y_{15} + 4b_6 + 6y_9 + 2c_3 + 2y_3$;

81) $x_{15}x_{10} = 3x_6 + 6x_{15} + b_3 + \bar{d}_3 + 4x_9$;

82) $x_{15}\bar{x}_{15} = 5b_8 + 6x_{10} + 6z_9 + 5c_8 + 3b_5 + 3c_5 + 1$;

83) $x_{15}x_9 = 6y_{15} + 3y_9 + 2b_6 + c_3 + y_3$;

84) $x_{15}\bar{x}_9 = 3b_8 + 4x_{10} + 3z_9 + 3c_8 + 2b_5 + 2c_5$;

85) $x_{15}y_{15} = 4\bar{x}_6 + 9\bar{x}_{15} + 6\bar{x}_9 + 2\bar{b}_3 + 2d_3$;

86) $x_{15}y_9 = 6\bar{x}_{15} + 2\bar{x}_6 + 3\bar{x}_9 + \bar{b}_3 + d_3$;

87) $x_{15}d_3 = b_8 + z_9 + b_5 + c_5 + x_{10} + c_8$;

88) $x_{15}\bar{d}_3 = 2y_{15} + y_9 + b_6$;

89) $x_{15}z_9 = 2x_6 + 6x_{15} + b_3 + \bar{d}_3 + 3x_9$;

90) $x_{15}c_8 = 5x_{15} + 2x_6 + 3x_9 + b_3 + \bar{d}_3$;

91) $x_{15}b_5 = x_6 + 3x_{15} + 2x_9 + b_3 + \bar{d}_3$;

92) $x_{15}c_5 = x_6 + 3x_{15} + 2x_9 + b_3 + \bar{d}_3$;

93) $x_{15}y_3 = \bar{x}_6 + 2\bar{x}_{15} + \bar{x}_9$;

94) $x_{10}^2 = 1 + 2b_8 + 2x_{10} + 2c_8 + 2b_5 + 2c_5 + 3z_9$;

95) $x_{10}x_9 = b_3 + 4x_{15} + \bar{d}_3 + 2x_9 + x_6$;

96) $x_{10}y_{15} = 6y_{15} + 3b_6 + 4y_9 + c_3 + y_3$;

97) $x_{10}y_9 = 4y_{15} + 2y_9 + y_3 + b_6 + c_3$;

98) $x_{10}d_3 = \bar{x}_{15} + \bar{x}_6 + \bar{x}_9$;

99) $x_{10}z_9 = 3x_{10} + b_5 + c_5 + 2b_8 + 2z_9 + 2c_8$;

100) $x_{10}c_8 = 2x_{10} + 2c_8 + 2b_8 + 2z_9 + b_5 + c_5$;

101) $x_{10}b_5 = b_8 + 2x_{10} + z_9 + c_8 + b_5$;

102) $x_{10}c_5 = b_8 + z_9 + c_5 + 2x_{10} + c_8$;

103) $x_{10}y_3 = y_{15} + y_9 + b_6$;

104) $x_9^2 = y_3 + c_3 + 2b_6 + 3y_{15} + 2y_9$;

105) $x_9\bar{x}_9 = 1 + 2b_8 + 2x_{10} + 2c_8 + 2z_9 + b_5 + c_5$;

106) $x_9y_{15} = 6\bar{x}_{15} + 2\bar{x}_6 + 3\bar{x}_9 + \bar{b}_3 + d_3$;

107) $x_9y_9 = d_3 + 3\bar{x}_{15} + 2\bar{x}_9 + \bar{b}_3 + 2\bar{x}_6$;

108) $x_9d_3 = x_{10} + b_8 + z_9$;

109) $x_9\bar{d}_3 = y_9 + c_3 + y_{15}$;

110) $x_9z_9 = 3x_{15} + 2x_9 + 2x_6 + b_3 + \bar{d}_3$;

111) $x_9b_5 = 2x_{15} + x_6 + x_9$;

112) $x_9c_5 = 2x_{15} + x_6 + x_9$;

113) $x_9y_3 = \bar{b}_3 + \bar{x}_9 + \bar{x}_{15}$;

114) $y_{15}^2 = 1 + 3b_5 + 6x_{10} + 6z_9 + 5b_8 + 3c_5 + 5c_8$;

115) $y_{15}y_9 = 4x_{10} + 3b_8 + 3z_9 + 3c_8 + 2c_5 + 2b_5$;

116) $y_{15}d_3 = 2x_{15} + x_6 + x_9$;

117) $y_{15}z_9 = 6y_{15} + 2b_6 + 3y_9 + c_3 + y_3$;

118) $y_{15}c_8 = 2b_6 + 5y_{15} + c_3 + y_3 + 3y_9$;

119) $y_{15}b_5 = b_6 + 3y_{15} + 2y_9 + c_3 + y_3$;

120) $y_{15}c_5 = b_6 + c_3 + 3y_{15} + y_3 + 2y_9$;

121) $y_{15}y_3 = b_5 + b_8 + z_9 + c_5 + x_{10} + c_8$;

122) $y_9^2 = 1 + 2z_9 + 2b_8 + 2c_8 + 2x_{10} + b_5 + c_5$;

123) $y_9d_3 = b_3 + x_{15} + x_9$;

124) $y_9z_9 = 3y_{15} + 2b_6 + 2y_9 + c_3 + y_3$;

125) $y_9c_8 = 3y_{15} + 2y_9 + b_6 + c_3$;

126) $y_9 b_5 = 2y_{15} + b_6 + y_9$;

127) $y_9 c_5 = 2y_{15} + y_9 + b_6$;

128) $y_9 y_3 = z_9 + b_8 + x_{10}$;

129) $d_3 \bar{d}_3 = 1 + c_8$;

130) $d_3^2 = b_6 + y_3$;

131) $d_3 z_9 = \bar{b}_3 + \bar{x}_{15} + \bar{x}_9$;

132) $d_3 c_8 = \bar{x}_{15} + d_3 + \bar{x}_6$;

133) $d_3 b_5 = \bar{x}_{15}$;

134) $d_3 c_5 = \bar{x}_{15}$;

135) $d_3 y_3 = \bar{d}_3 + x_6$;

136) $z_9^2 = 1 + 2x_{10} + 2z_9 + 2b_8 + 2c_8 + b_5 + c_5$;

137) $z_9 c_8 = 2x_{10} + 2b_8 + 2z_9 + b_5 + c_5 + c_8$;

138) $z_9 b_5 = x_{10} + b_8 + z_9 + c_8 + b_5 + c_5$;

139) $z_9 c_5 = x_{10} + b_8 + z_9 + c_8 + b_5 + c_5$;

140) $z_9 y_3 = y_9 + c_3 + y_{15}$;

141) $c_8^2 = 1 + 2c_8 + 2x_{10} + b_8 + z_9 + b_5 + c_5$;

142) $c_8 b_5 = b_8 + z_9 + c_5 + x_{10} + c_8$;

143) $c_8 c_5 = b_8 + z_9 + x_{10} + c_8 + b_5$;

144) $c_8 y_3 = b_6 + y_3 + y_{15}$;

145) $b_5^2 = 1 + b_5 + x_{10} + z_9$;

146) $b_5 c_5 = b_8 + z_9 + c_8$;

147) $b_5 y_3 = y_{15}$;

148) $c_5^2 = 1 + z_9 + x_{10} + c_5$;

149) $c_5 y_3 = y_{15}$;

150) $y_3^2 = 1 + c_8$.

Proof The equations hold by Lemmas 2.25–2.37 and 2.39.   □

References

[AB]   Arad, Z., Blau, H.: On table algebras and application to finite group theory. J. Algebra **138**, 137–185 (1991)

[B]   Blau, H.I.: Integral table algebra, affine diagrams, and the analysis of degree two. J. Algebra **178**, 872–918 (1995)

[B1]   Blau, H.I.: Quotient structures in C-algebra. J. Algebra **177**, 297–337 (1995)

[Bl]   Blichfeldt, H.F.: Finite Collineation Groups. University of Chicago Press, Chicago (1917)

[Br]   Brauer, R.: Über endliche lineare Gruppen von Primzahlgrad. Math. Ann. **169**, 73–96 (1967)

[CA]   Chen, G.Y., Arad, Z.: On four normalized table algebras generated by a faithful nonreal element of degree 3. J. Algebra **283**, 457–484 (2005)

[F]   Feit, W.: The current situation in the theory of finite simple groups, Actes du Congrès International des Mathématiciens (Nice, 1970), Tome 1, pp. 55–93. Gauthier-Villars, Paris (1971)

[L]   Lindsey, J.H.: On a projective representation of Hall-Janko group. Bull. Am. Math. Soc. **74**, 1094 (1968)

[W]   Wales, D.B.: Finite linear groups of degree seven I. Can. J. Math. **21**, 1025–1041 (1969)

Chapter 3
A Proof of a Non-existence of Sub-case (2)

3.1 Introduction

In Chap. 3 we shall freely use the definitions and notation used in Chap. 2. The Main Theorem 1 of Chap. 2 left three unsolved problems for the complete classification of the Normalized Integral Table Algebra (NITA) (A, B) generated by a faithful non-real element of degree 3 with $L_1(B) = 1$ and $L_2(B) = \emptyset$. In this chapter we solve the first problem, and prove that the NITA which satisfies case (2) of the Main Theorem 1 of Chap. 2 does not exist. Consequently, we can state the Main Theorem 2 of this chapter as follows.

Main Theorem 2 *Let (A, B) be a NITA that is generated by a non-real element $b_3 \in B$ of degree 3 and has no non-identity basis elements of degrees 1 or 2. Then $b_3\bar{b}_3 = 1 + b_8$, $b_8 \in B$, and one of the following holds:*

(1) *There exists a real element $b_6 \in B$ such that $b_3^2 = \bar{b}_3 + b_6$ and*

$$(A, B) \cong_x \big(CH\,PSL(2, 7), Irr\,PSL(2, 7) \big).$$

(2) *There exist $b_6, b_{10}, b_{15} \in B$, where b_6 is non-real, such that $b_3^2 = \bar{b}_3 + b_6$, $\bar{b}_3 b_6 = b_3 + b_{15}$, $b_3 b_6 = b_8 + b_{10}$, and $(b_3 b_8, b_3 b_8) = 3$. Moreover, if b_{10} is real then $(A, B) \cong_x (CH(3 \cdot A_6), Irr(3 \cdot A_6))$ of dimension 17.*

(3) *There exist $c_3, b_6 \in B$, $c_3 \neq b_3$ or $\bar{b}_3$, such that $b_3^2 = c_3 + b_6$ and $(b_3 b_8, b_3 b_8) = 3$ or 4. If $(b_3 b_8, b_3 b_8) = 3$ and c_3 is non-real, then $(A, B) \cong_x (A(3 \cdot A_6 \cdot 2), B_{32})$ of dimension 32. (See Theorem 2.9 of Chap. 2 for the definition of this specific NITA.) If $(b_3 b_8, b_3 b_8) = 3$ and c_3 is real, then $(A, B) \cong_x (A(7 \cdot 5 \cdot 10), B_{22})$ of dimension 22. (See Theorem 2.10 of Chap. 2 for the definition of this specific NITA.)*

In the above Main Theorem 2, we still have 2 open problems in the cases (2) and (3) that have not been solved. In case (2) we must classify the NITA such that b_{10} is non-real, and in case (3) we must classify the NITA such that $(b_3 b_8, b_3 b_8) = 4$. In Chaps. 4 and 5, the first problem will be almost completely solved.

Z. Arad et al., *On Normalized Integral Table Algebras (Fusion Rings)*,
Algebra and Applications 16, DOI 10.1007/978-0-85729-850-8_3,
© Springer-Verlag London Limited 2011

Let us emphasize that the NITA's of dimensions 7 and 17 in the cases (1) and (2) of the Main Theorem 2 are strictly isomorphic to the NITA's induced from finite groups G via the basis of the irreducible characters of G. However, the NITA's of dimensions 22 and 32 are not induced from finite groups as described in Chap. 2.

The Main Theorem 2 follows from the Main Theorem 1 in Chap. 2 and the following theorem:

Theorem 3.1 *There exists no NITA (A, B) that is generated by a non-real element $b_3 \in B$ of degree 3 and has no non-identity basis elements of degrees 1 or 2 such that $b_3 \bar{b}_3 = 1 + b_8$, $b_8 \in B$, and*

$$b_3^2 = b_4 + b_5, \quad b_4, b_5 \in B, \quad (b_3 b_8, b_3 b_8) = 3, \quad and \quad (b_4^2, b_4^2) = 3.$$

The rest of this chapter is devoted to proving the above Theorem 3.1.

3.2 Preliminary Results

We are going to prove Theorem 3.1, by contradiction. That is, for the rest of this chapter we shall always assume that (A, B) is a NITA that is generated by a non-real element $b_3 \in B$ of degree 3 and has no non-identity basis elements of degrees 1 or 2 such that

$$b_3 \bar{b}_3 = 1 + b_8, \quad b_8 \in B, \tag{3.1}$$

$$b_3^2 = b_4 + b_5, \quad b_4, b_5 \in B, \tag{3.2}$$

and $(b_3 b_8, b_3 b_8) = 3$, $(b_4^2, b_4^2) = 3$. Then we are going to derive a contradiction. From $(b_3 \bar{b}_3) b_3 = b_3^2 \bar{b}_3$ and (3.1), (3.2), we obtain that

$$\bar{b}_3 b_4 = b_3 + x_9, \quad x_9 \in B, \tag{3.3}$$

$$\bar{b}_3 b_5 = b_3 + y_{12}, \quad y_{12} \in B, \tag{3.4}$$

and

$$b_3 b_8 = b_3 + x_9 + y_{12}. \tag{3.5}$$

Now $2 = (\bar{b}_3 b_4, \bar{b}_3 b_4) = (\bar{b}_3 b_3, \bar{b}_4 b_4)$ implies that $(\bar{b}_4 b_4, b_8) = 1$. Hence from $(\bar{b}_4 b_4, \bar{b}_4 b_4) = (b_4^2, b_4^2) = 3$, we see that

$$b_4 \bar{b}_4 = 1 + b_7 + b_8, \quad b_7 \in B \text{ is real.} \tag{3.6}$$

Furthermore, $b_3^2 \bar{b}_4 = (b_3 \bar{b}_4) b_3$ and (3.6), (3.3), (3.1) imply that

$$b_7 + \bar{b}_4 b_5 = b_3 \bar{x}_9. \tag{3.7}$$

Since $(b_3 \bar{x}_9, b_8) = 1$ by (3.5), and b_7, b_8 are real by (3.6), from the above equality we may assume that

$$b_4 \bar{b}_5 = b_8 + \Sigma_{12}, \quad \Sigma_{12} \in \mathbb{N}(B), \tag{3.8}$$

and

$$\bar{b}_3 x_9 = b_7 + b_8 + \Sigma_{12}. \tag{3.9}$$

Thus, from (3.4), (3.2), (3.1), (3.8), and $\bar{b}_3 (\bar{b}_3 b_5) = \bar{b}_3^2 b_5$, we see that

$$1 + \bar{b}_3 y_{12} = \bar{\Sigma}_{12} + b_5 \bar{b}_5. \tag{3.10}$$

Note that $(b_3 b_4, b_3 b_4) = (b_3 \bar{b}_3, b_4 \bar{b}_4) = 2$ by (3.1) and (3.6). So we may set

$$b_3 b_4 = z + w, \quad z, w \in B. \tag{3.11}$$

Thus, $\bar{b}_3 z + \bar{b}_3 w = \bar{b}_3 (b_3 b_4) = (b_3 \bar{b}_3) b_4 = b_4 + b_4 b_8$. But $(b_4 b_8, b_5) = 1$ by (3.8). Thus by (3.8), without loss of generality, we may assume that $(\bar{b}_3 z, b_5) = 1$. Since $(b_3 b_5, b_3 b_5) = (\bar{b}_3 b_5, \bar{b}_3 b_5) = 2$ by (3.4), we can set

$$b_3 b_5 = z + u, \quad u \in B. \tag{3.12}$$

Note that in (3.11) and (3.12), the degrees of z, w, and u are not yet determined. According to the possibilities of their degrees, we will have difference cases to consider. But let us first prove the following proposition.

Proposition 3.1 *Let Σ_{12} be the same as in (3.8). Then $\Sigma_{12} \in B$.*

Proof Toward a contradiction, assume that $\Sigma_{12} \notin B$. Then $(b_4 \bar{b}_5, b_4 \bar{b}_5) \geq 3$ by (3.8), and

$$(b_3 \Sigma_{12}, x_9) \geq 2 \tag{3.13}$$

by (3.9). Note that $b_3 (\bar{b}_3 b_5) = (b_3 \bar{b}_3) b_5$ and (3.4), (3.1), (3.2) imply that $b_5 b_8 = b_4 + b_3 y_{12}$, and $(b_3 y_{12}, b_5) = 1$ by (3.4). So $(b_5 \bar{b}_5, b_8) = (b_5 b_8, b_5) = 1$. Thus,

$$3 \leq (b_4 \bar{b}_5, b_4 \bar{b}_5) = (b_4 \bar{b}_4, b_5 \bar{b}_5) = 2 + (b_7, b_5 \bar{b}_5).$$

Hence $(b_7, b_5 \bar{b}_5) \geq 1$. But $L_2(B) = \emptyset$. So $(b_7, b_5 \bar{b}_5) = 1$. Thus, we must have that $(b_4 \bar{b}_5, b_4 \bar{b}_5) = 3$. Therefore, we may assume that

$$\Sigma_{12} = r + s, \quad r, s \in B, \tag{3.14}$$

and

$$b_5 \bar{b}_5 = 1 + b_7 + b_8 + \Sigma_9, \tag{3.15}$$

where $\Sigma_9 \in \mathbb{N}(B)$ is real. Furthermore, $(\bar{b}_3 y_{12}, b_7) \geq 1$ by (3.10), and hence $(b_3 b_7, y_{12}) \geq 1$. Thus, $(b_3 b_7, y_{12}) = 1$, $(\bar{b}_3 y_{12}, b_7) = 1$, and $b_7 \neq \bar{r}$ or $\bar{s}$. But we also have $(b_3 b_7, x_9) = 1$ by (3.7). Therefore,

$$b_3 b_7 = x_9 + y_{12}. \tag{3.16}$$

Thus, from $b_3(b_4 \bar{b}_4) = b_4(b_3 \bar{b}_4)$, (3.6), (3.3), and (3.5), we obtain that

$$b_4 \bar{x}_9 = b_3 + x_9 + 2 y_{12}. \tag{3.17}$$

From (3.10), (3.14), and (3.15), we see that

$$\bar{b}_3 y_{12} = \bar{r} + \bar{s} + b_7 + b_8 + \Sigma_9. \tag{3.18}$$

Thus, $(\bar{b}_3 b_4)\bar{x}_9 = (b_4 \bar{x}_9)\bar{b}_3$ and (3.3), (3.9), (3.14), (3.17), (3.1), (3.18) yield that

$$x_9 \bar{x}_9 = 1 + 2b_7 + 3b_8 + 2\Sigma_9 + r + s + \bar{r} + \bar{s}. \tag{3.19}$$

From (3.8), (3.5), (3.3), (3.4), and $\bar{b}_3(b_4 \bar{b}_5) = (\bar{b}_3 b_4)\bar{b}_5$, we obtain that

$$\bar{b}_5 x_9 = \bar{x}_9 + \bar{b}_3 r + \bar{b}_3 s. \tag{3.20}$$

Thus by (3.18)

$$(\bar{b}_5 x_9, \bar{y}_{12}) = 2. \tag{3.21}$$

Note that the degrees of both r and s are at least 4 by (3.18). If one of r and s has degree 4, without loss of generality, we may assume that $r = r_4$ and $s = s_8$. Then by (3.18), again, we see that $b_3 \bar{r}_4 = y_{12}$. So $(r_4 \bar{x}_9, b_3 \bar{b}_4) = (b_4 \bar{x}_9, b_3 \bar{r}_4) = 2$ by (3.17). That is, $(r_4 \bar{x}_9, \bar{b}_3 + \bar{x}_9) = 2$. Note that $(b_3 r_4, b_3 r_4) = (b_3 \bar{r}_4, b_3 \bar{r}_4) = 1$. So $(r_4 \bar{x}_9, \bar{b}_3) = (b_3 r_4, x_9) = 0$. Thus, $(r_4 \bar{x}_9, \bar{x}_9) = 2$. On the other hand, $(\bar{b}_5 x_9, \bar{b}_3 r_4) = (\bar{b}_5 x_9, \bar{y}_{12}) = 2$ by (3.21). So $(r_4 \bar{x}_9, b_3 \bar{b}_5) = 2$, and hence $(r_4 \bar{x}_9, \bar{b}_3) = 0$ forces that $(r_4 \bar{x}_9, \bar{y}_{12}) = 2$ by (3.4). Therefore,

$$r_4 \bar{x}_9 = 2\bar{x}_9 + 2\bar{y}_{12} + \cdots,$$

a contradiction. This proves that the degrees of both r and s are at least 5.

Now we claim that $\Sigma_9 \in B$. If $\Sigma_9 \notin B$, then (3.18) forces that $\Sigma_9 = c_4 + d_5$, where $c_4, d_5 \in B$ are real, and

$$b_3 c_4 = y_{12}, \qquad b_3 d_5 = y_{12} + e_3, \qquad e_3 \in B.$$

Thus, c_4 real implies that $(c_4 \bar{x}_9, b_3 \bar{b}_5) = (b_5 \bar{x}_9, b_3 c_4) = 2$ by (3.21). But $(c_4 \bar{x}_9, \bar{b}_3) = (b_3 c_4, x_9) = 0$. So by (3.4),

$$(c_4 \bar{x}_9, \bar{y}_{12}) = 2.$$

From (3.19) we also have that

$$(c_4 \bar{x}_9, \bar{x}_9) = (x_9 \bar{x}_9, c_4) \geq 2.$$

Thus,

$$c_4 \bar{x}_9 = 2\bar{y}_{12} + 2\bar{x}_9 + \cdots,$$

a contradiction. This proves that $\Sigma_9 \in B$. Therefore, $(b_5\bar{b}_5, x_9\bar{x}_9) = 8$ by (3.15) and (3.19). So

$$(b_5\bar{x}_9, b_5\bar{x}_9) = 8. \tag{3.22}$$

Since $(\bar{b}_3 r + \bar{b}_3 s, \bar{y}_{12}) = 2$ by (3.18), (3.20) and (3.22) force that $(\bar{b}_3 r + \bar{b}_3 s, \bar{x}_9) = 0$, and $(\bar{b}_3 r + \bar{b}_3 s, \bar{b}_3 r + \bar{b}_3 s) = 7$. Hence by $\bar{b}_3 r + \bar{b}_3 s = 2\bar{y}_{12}+$ sum of three distinct elements of B. Recall that the degrees of both r and s are at least 5. So without loss of generality, we may assume that $(b_3\bar{r}, b_3\bar{r}) = 2$ and $(b_3\bar{s}, b_3\bar{s}) = 3$. In the following, we show that

$$r = r_5, \quad s = s_7, \quad b_3\bar{r}_5 = y_{12} + c_3, \quad b_3\bar{s}_7 = y_{12} + d_4 + e_5, \quad c_3, d_4, e_5 \in B. \tag{3.23}$$

If (3.23) does not hold, then $(b_3\bar{r}, b_3\bar{r}) = 2$ and $(b_3\bar{s}, b_3\bar{s}) = 3$ force that

$$r = r_6, \quad s = s_6, \quad b_3\bar{r}_5 = y_{12} + c_6,$$

$$b_3\bar{s}_6 = y_{12} + d_3 + e_3, \quad c_6, d_3, e_3 \in B, \quad d_3 \neq e_3.$$

Hence $(\bar{b}_3 d_3, \bar{s}_6) = 1$ and $(\bar{b}_3 e_3, \bar{s}_6) = 1$. So $(\bar{b}_3 d_3, \bar{b}_3 e_3) \geq 1$. Thus, $(b_3\bar{b}_3, d_3\bar{e}_3) \geq 1$. But $d_3 \neq e_3$. So $(1, d_3\bar{e}_3) = 0$ and $(b_8, d_3\bar{e}_3) = 1$, a contradiction to the assumption that $L_1(B) = \{1\}$. This proves (3.23).

Since $(b_3\bar{c}_3, r_5) = 1$ by (3.23), we have that $(c_3\bar{c}_3, b_3\bar{b}_3) = (b_3\bar{c}_3, b_3\bar{c}_3) = 2$. Thus,

$$c_3\bar{c}_3 = 1 + b_8. \tag{3.24}$$

Hence (3.1) and (3.16) imply that

$$(c_3b_7, c_3b_7) = (c_3\bar{c}_3, b_7^2) = (b_3\bar{b}_3, b_7^2) = (b_3b_7, b_3b_7) = 2.$$

Similarly, $(c_3b_8, c_3b_8) = 3$ by (3.1) and (3.5). That is,

$$(c_3b_7, c_3b_7) = 2 \quad \text{and} \quad (c_3b_8, c_3b_8) = 3. \tag{3.25}$$

Furthermore, $(b_3b_4)\bar{b}_4 = b_3(b_4\bar{b}_4)$ and (3.11), (3.6), (3.5) yield that

$$z\bar{b}_4 + w\bar{b}_4 = 2b_3 + 2x_9 + 2y_{12}.$$

Hence the degree of z must be one of 3, 6, and 9. According to the degree of z, we have three cases to consider. In the following, we will derive a contradiction for each case. Note that $(z\bar{b}_4, b_3) = (w\bar{b}_4, b_3) = 1$ by (3.11).

Case A Assume that

$$z = z_3, \quad w = w_9, \quad z_3\bar{b}_4 = b_3 + x_9, \quad w_9\bar{b}_4 = b_3 + x_9 + 2y_{12}. \tag{3.26}$$

By (3.11) and (3.12) we obtain that

$$\bar{b}_3 z_3 = b_4 + b_5. \tag{3.27}$$

Thus, $(z_3 \bar{z}_3, b_3 \bar{b}_3) = (\bar{b}_3 z_3, \bar{b}_3 z_3) = 2$. Hence $z_3 \bar{z}_3 = 1 + b_8$. Therefore, (3.27) and (3.5) yield that

$$\bar{b}_4 z_3 + \bar{b}_5 z_3 = (b_3 \bar{z}_3) z_3 = b_3 (\bar{z}_3 z_3) = 2 b_3 + x_9 + y_{12}.$$

So $\bar{b}_4 z_3 = b_3 + x_9$, and by (3.9),

$$b_3 (\bar{b}_4 \bar{z}_3) = b_3 (\bar{b}_3 + \bar{x}_9) = 1 + b_7 + 2 b_8 + \bar{\Sigma}_{12}.$$

But, on the other hand, from (3.27), (3.6), and (3.8), we see that

$$b_3 (b_4 \bar{z}_3) = b_4 (b_3 \bar{z}_3) = b_4 (\bar{b}_4 + \bar{b}_5) = 1 + b_7 + 2 b_8 + \Sigma_{12}.$$

Thus, $\bar{\Sigma}_{12} = \Sigma_{12}$, and (3.10) implies that $(b_3 \Sigma_{12}, y_{12}) = (\bar{b}_3 y_{12}, \Sigma_{12}) \geq 2$ (because $\Sigma_{12} \notin B$). But $(b_3 \Sigma_{12}, x_9) \geq 2$ by (3.13). Hence

$$b_3 \Sigma_{12} = 2 x_9 + 2 y_{12} + \cdots ,$$

a contradiction.

Case B Assume that

$$z = z_6, \qquad w = w_6, \qquad u = u_9, \qquad z_6 \bar{b}_4 = b_3 + x_9 + y_{12},$$
$$w_6 \bar{b}_4 = b_3 + x_9 + y_{12}. \tag{3.28}$$

Note that (3.12), (3.8), (3.5), (3.28), (3.14), (3.23), and $(b_3 b_5) \bar{b}_4 = (\bar{b}_4 b_5) b_3$ yield that

$$\bar{b}_4 u_9 = 2 y_{12} + c_3 + d_4 + e_5. \tag{3.29}$$

Thus,

$$c_3 b_4 = u_9 + p_3, \qquad p_3 \in B. \tag{3.30}$$

Also $(\bar{b}_4 p_3, c_3) = 1$ by (3.30). So

$$1 < (\bar{b}_4 p_3, \bar{b}_4 p_3) = (p_3 \bar{p}_3, b_4 \bar{b}_4) \leq 2.$$

Hence $(\bar{b}_4 p_3, \bar{b}_4 p_3) = 2$, and $\bar{b}_4 p_3 = c_3 + q_9$, $q_9 \in B$. Therefore from $(c_3 b_4) \bar{b}_4 = c_3 (b_4 \bar{b}_4)$ and (3.30), (3.29), (3.6), we obtain that

$$c_3 b_7 + c_3 b_8 = 2 y_{12} + c_3 + d_4 + e_5 + q_9.$$

Since $(c_3b_7, c_3b_7) = 2$ and $(c_3b_8, c_3b_8) = 3$ by (3.25), the above equality cannot be true, a contradiction.

Case C Assume that

$$z = z_9, \qquad w = w_3, \qquad u = u_6, \qquad z_9\bar{b}_4 = b_3 + x_9 + 2y_{12}, \qquad w_3\bar{b}_4 = b_3 + x_9. \tag{3.31}$$

Note that (3.12), (3.8), (3.5), (3.31), (3.14), (3.23), and $(b_3b_5)\bar{b}_4 = (\bar{b}_4b_5)b_3$ yield that

$$\bar{b}_4u_6 = y_{12} + c_3 + d_4 + e_5. \tag{3.32}$$

Since $(c_3b_4, c_3b_4) = (c_3\bar{c}_3, b_4\bar{b}_4) = 2$ by (3.24) and (3.6), from (3.32) we may set

$$c_3b_4 = u_6 + p_6, \qquad p_6 \in B. \tag{3.33}$$

Thus, $(c_3b_4)\bar{b}_4 = c_3(b_4\bar{b}_4)$ and (3.33), (3.32) yield that

$$c_3b_7 + c_3b_8 = y_{12} + d_4 + e_5 + \bar{b}_4p_6. \tag{3.34}$$

Since $(c_3b_7, c_3b_7) = 2$ and $(c_3b_8, c_3b_8) = 3$ by (3.25), the above equality forces that $(\bar{b}_4p_6, \bar{b}_4p_6) = 2$. But $(\bar{b}_4p_6, c_3) = 1$ by (3.33). So we may assume that $\bar{b}_4p_6 = c_3 + q_{21}, q_{21} \in B$. Hence (3.34) yields that

$$c_3b_7 + c_3b_8 = y_{12} + d_4 + e_5 + c_3 + q_{21},$$

a contradiction to (3.25).

Therefore, we must have that $\Sigma_{12} \in B$.   □

From Proposition 3.1, we may assume that $\Sigma_{12} = b_{12} \in B$. Hence (3.9) implies that

$$\bar{b}_3x_9 = b_7 + b_8 + b_{12}, \tag{3.35}$$

and (3.8) yields that

$$b_4\bar{b}_5 = b_8 + b_{12}. \tag{3.36}$$

Furthermore, by (3.10) we see that

$$1 + \bar{b}_3y_{12} = \bar{b}_{12} + b_5\bar{b}_5. \tag{3.37}$$

Since $(b_3b_7, x_9) = 1$ by (3.35), we may assume that

$$b_3b_7 = x_9 + \Sigma_{12}^*, \qquad \Sigma_{12}^* \in \mathbb{N}(B). \tag{3.38}$$

Thus, from (3.6), (3.3), (3.5), (3.38), and $b_3(b_4\bar{b}_4) = b_4(b_3\bar{b}_4)$, we see that

$$b_4\bar{x}_9 = b_3 + y_{12} + x_9 + \Sigma_{12}^*. \tag{3.39}$$

Table 3.1 Splitting to seven cases

Case 1	$z = z_3$	$w = w_9$	$\bar{b}_3 z_3 = b_4 + b_5,$	$\bar{b}_3 w_9 = b_4 + \sigma + \theta$
Case 2	$z = z_4$	$w = w_8$	$\bar{b}_3 z_4 = b_4 + b_5 + \sigma_3,$	$\bar{b}_4 w_8 = b_4 + \theta_{20}$
Case 3	$z = z_5$	$w = w_7$	$\bar{b}_3 z_5 = b_4 + b_5 + \sigma_6,$	$\bar{b}_3 w_7 = b_4 + \theta_{17}$
Case 4	$z = z_6$	$w = w_6$	$\bar{b}_3 z_6 = b_4 + b_5 + \sigma_9,$	$\bar{b}_3 w_6 = b_4 + \theta_{14}$
Case 5	$z = z_7$	$w = w_5$	$\bar{b}_3 z_7 = b_4 + b_5 + \sigma_{12},$	$\bar{b}_3 w_5 = b_4 + \theta_{11}$
Case 6	$z = z_8$	$w = w_4$	$\bar{b}_3 z_8 = b_4 + b_5 + \sigma_{15},$	$\bar{b}_3 w_4 = b_4 + \theta_8$
Case 7	$z = z_9$	$w = w_3$	$\bar{b}_3 z_9 = b_4 + b_5 + \sigma_{18},$	$\bar{b}_3 w_3 = b_4 + \theta_5$

Therefore, (3.11), (3.6), (3.38), (3.5), and $(b_3 b_4)\bar{b}_4 = (b_4 \bar{b}_4)b_3$ force that

$$z\bar{b}_4 + w\bar{b}_4 = 2b_3 + 2x_9 + y_{12} + \Sigma_{12}^*, \tag{3.40}$$

and (3.12), (3.36), (3.5), and $(b_3 b_5)\bar{b}_4 = b_3(\bar{b}_4 b_5)$ yield that

$$z\bar{b}_4 + u\bar{b}_4 = b_3 + x_9 + y_{12} + b_3\bar{b}_{12}. \tag{3.41}$$

Note that $(b_3 x_9, b_3 x_9) = (\bar{b}_3 x_9, \bar{b}_3 x_9) = 3$ by (3.35), and $(b_3 x_9, b_4) = 1$ by (3.3). So we may assume that

$$b_3 x_9 = b_4 + \sigma + \theta, \quad \sigma, \theta \in B, \quad \text{and} \quad |\sigma| + |\theta| = 23. \tag{3.42}$$

Thus, from (3.11), (3.3), (3.2), and $\bar{b}_3(b_3 b_4) = (\bar{b}_3 b_4)b_3$, we get that

$$\bar{b}_3 z + \bar{b}_3 w = 2b_4 + b_5 + \sigma + \theta. \tag{3.43}$$

Recall that $(\bar{b}_3 z, b_4) = (\bar{b}_3 z, b_5) = 1$ by (3.11) and (3.12), respectively. So from (3.43), we have the cases in Table 3.1.

In the rest of this chapter, we derive a contradiction for each case in Table 3.1. Here let us summarize the identities obtained in this section.

$$\begin{aligned}
b_3\bar{b}_3 &= 1 + b_8, & b_3^2 &= b_4 + b_5, & \bar{b}_3 b_4 &= b_3 + x_9, \\
\bar{b}_3 b_5 &= b_3 + y_{12}, & b_3 b_8 &= b_3 + x_9 + y_{12}, & b_4\bar{b}_4 &= 1 + b_7 + b_8, \\
\bar{b}_3 x_9 &= b_7 + b_8 + b_{12}, & b_4 b_5 &= b_8 + b_{12}, & b_3 x_9 &= b_4 + \sigma + \theta, \\
b_3 b_4 &= z + w, & b_3 b_5 &= z + u, & b_3 b_7 &= x_9 + \Sigma_{12}^*, \\
b_4\bar{x}_9 &= b_3 + y_{12} + x_9 + \Sigma_{12}^*.
\end{aligned}$$

3.3 Case $z = z_3$

In this section we assume that $z = z_3$. Then we have that

$$z = z_3, \quad w = w_9, \quad u = u_{12}, \quad \bar{b}_3 z_3 = b_4 + b_5, \quad \bar{b}_3 w_9 = b_4 + \sigma + \theta. \tag{3.44}$$

Since $(b_3\bar{b}_3, z_3\bar{z}_3) = (\bar{b}_3 z_3, \bar{b}_3 z_3) = 2$ by (3.44), we see that

$$z_3\bar{z}_3 = 1 + b_8. \tag{3.45}$$

Also note that $(b_3 z_3, b_3 z_3) = (\bar{b}_3 z_3, \bar{b}_3 z_3) = 2$ by (3.44). So we may set $b_3 z_3 = \mu + \nu$, $\mu, \nu \in B$ and $|\mu| < |\nu|$. Then from (3.44), (3.11), (3.12) and $(b_3 z_3)\bar{b}_3 = (\bar{b}_3 z_3)b_3$, we obtain that

$$\bar{b}_3\mu + \bar{b}_3\nu = 2z_3 + w_9 + u_{12}. \tag{3.46}$$

But $(\bar{b}_3\mu, z_3) = (\bar{b}_3\nu, z_3) = 1$. So we must have either $\mu = \mu_4$ and $\nu = \nu_5$, or $\mu = 1$ and $\nu = \nu_8$. In the following, we derive a contradiction for each of these two cases.

Subcase 1A $\mu = \mu_4$ and $\nu = \nu_5$.

Thus,

$$b_3 z_3 = \mu_4 + \nu_5, \tag{3.47}$$

and (3.46) implies that

$$\bar{b}_3\mu_4 = z_3 + w_9, \tag{3.48}$$

and

$$\bar{b}_3\nu_5 = z_3 + u_{12}. \tag{3.49}$$

On the other hand, from $(\bar{b}_3 z_3)b_3 = (b_3\bar{b}_3)z_3$, (3.44), (3.1), (3.11), and (3.12), we see that

$$z_3 b_8 = z_3 + w_9 + u_{12}. \tag{3.50}$$

Furthermore, $\bar{b}_3^2\mu_4 = \bar{b}_3(\bar{b}_3\mu_4)$ and (3.44), (3.2), (3.48) imply that

$$\bar{b}_4\mu_4 + \bar{b}_5\mu_4 = 2b_4 + b_5 + \sigma + \theta.$$

If $(\bar{b}_4\mu_4, b_4) \geq 2$, then $(b_4^2, \mu_4) \geq 2$, and hence $(b_4^2, b_4^2) \geq 5$, a contradiction to the assumption that $(b_4^2, b_4^2) = 3$. Thus, $(\bar{b}_4\mu_4, b_4) \leq 1$. So $(\bar{b}_5\mu_4, b_4) \geq 1$, and hence $(b_4 b_5, \mu_4) \geq 1$. But $(b_4 b_5, b_4 b_5) = (\bar{b}_4\bar{b}_5, b_4 b_5) = 2$ by (3.36). So we must have that $(b_4 b_5, \mu_4) = 1$. Thus, we may assume that

$$b_4 b_5 = \mu_4 + n_{16}, \quad n_{16} \in B. \tag{3.51}$$

Hence from (3.12), (3.47), (3.2), (3.51), and $b_3(b_3 b_5) = b_3^2 b_5$, we get that

$$\mu_4 + \nu_5 + b_3 u_{12} = \mu_4 + n_{16} + b_5^2.$$

Since $(b_3 u_{12}, \nu_5) = 1$ by (3.49), there exists $\Sigma_{15} \in \mathbb{N}(B)$ such that

$$b_3 u_{12} = \nu_5 + n_{16} + \Sigma_{15} \quad \text{and} \quad b_5^2 = 2\nu_5 + \Sigma_{15}.$$

Therefore

$$(b_3 u_{12}, b_3 u_{12}) < (b_5^2, b_5^2). \tag{3.52}$$

Note that (3.11), (3.2) and $b_3(b_3 b_4) = b_3^2 b_4$ yield that $b_3 z_3 + b_3 w_9 = b_4^2 + b_4 b_5$, and $z_3(\bar{b}_3 b_4) = (\bar{b}_3 z_3) b_4$ implies that $b_3 z_3 + z_3 x_9 = b_4^2 + b_4 b_5$ by (3.3) and (3.44). Thus,

$$b_3 w_9 = z_3 x_9.$$

However, from $z_3(b_3 b_8) = (z_3 b_8) b_3$, (3.5), and (3.50), we get that

$$b_3 z_3 + z_3 x_9 + z_3 y_{12} = b_3 z_3 + b_3 w_9 + b_3 u_{12}.$$

So $z_3 y_{12} = b_3 u_{12}$, and hence by (3.52),

$$(z_3 y_{12}, z_3 y_{12}) < (b_5^2, b_5^2). \tag{3.53}$$

Since $b_3 \bar{b}_3 = z_3 \bar{z}_3$ by (3.1) and (3.45), we must have that

$$(z_3 y_{12}, z_3 y_{12}) = (z_3 \bar{z}_3, y_{12} \bar{y}_{12}) = (b_3 \bar{b}_3, y_{12} \bar{y}_{12}) = (\bar{b}_3 y_{12}, \bar{b}_3 y_{12}).$$

So $(\bar{b}_3 y_{12}, \bar{b}_3 y_{12}) < (b_5^2, b_5^2)$ by (3.53). But, on the other hand, from (3.37) we also have that

$$(\bar{b}_3 y_{12}, \bar{b}_3 y_{12}) \geq (b_5 \bar{b}_5, b_5 \bar{b}_5) = (b_5^2, b_5^2),$$

a contradiction.

Subcase 1B $\mu = 1$ and $\nu = \nu_8$.

Thus, we must have that $z_3 = \bar{b}_3$, and $\bar{b}_3^2 = b_4 + b_5$ by (3.44). Hence both b_4 and b_5 are real by (3.2), and b_{12} is real by (3.36). Thus, we see that $w_9 = \bar{x}_9$ by (3.3) and (3.11), $u_{12} = \bar{y}_{12}$ by (3.4) and (3.12), and $\bar{b}_3 \bar{x}_9 = b_4 + \sigma + \theta$ by (3.44). So $b_3 x_9 = b_4 + \sigma + \theta$. Since $|\sigma| + |\theta| = 23$, (3.42) implies that both σ and θ are real. Hence, $b_3 x_9 = b_4 + \sigma + \theta$.

Since $(\bar{b}_3 y_{12}, b_8) = 1$ by (3.5), we see that $(b_5^2, b_8) = 1$ by (3.37). So we may assume that

$$b_5^2 = 1 + b_8 + \Sigma_{16}, \quad \Sigma_{16} \in \mathbb{N}(B) \text{ is real.} \tag{3.54}$$

Then by (3.37),

$$\bar{b}_3 y_{12} = b_8 + b_{12} + \Sigma_{16}. \tag{3.55}$$

Hence the degree of any component of Σ_{16} is at least 4 by (3.55). Our goal is to prove that $(\Sigma_{16}, \Sigma_{16}) = 2$. But we need first to prove the next lemma:

Lemma 3.1 *Let* Σ_{16} *be the same as in* (3.54). *Then* $(\Sigma_{16}, \Sigma_{16}) \leq 3$.

Proof Toward a contradiction, assume that $(\Sigma_{16}, \Sigma_{16}) \geq 4$. Then $(\bar{b}_3 y_{12}, \Sigma_{16}) \geq (\Sigma_{16}, \Sigma_{16}) \geq 4$ by (3.55). So $(b_3 \Sigma_{16}, y_{12}) \geq 4$. Hence $(b_3 \Sigma_{16}, y_{12}) = 4$, $b_3 \Sigma_{16} = 4 y_{12}$, and $(\Sigma_{16}, \Sigma_{16}) = 4$. Thus, either Σ_{16} has four distinct components of degree 4, or $\Sigma_{16} = 2 x_8$, $x_8 \in B$. Furthermore, $(b_5 \bar{y}_{12}, \bar{b}_3 b_5) = (\bar{b}_3 y_{12}, b_5^2) \geq 5$ by (3.54) and (3.55). But $(b_5 \bar{y}_{12}, b_3) = 1$ by (3.4). So we must have that

$$b_5 \bar{y}_{12} = b_3 + 4 y_{12} + \cdots . \tag{3.56}$$

If Σ_{16} has four distinct components of degree 4, then for a component x_4 of Σ_{16}, $b_3 \Sigma_{16} = 4 y_{12}$ implies that $b_3 x_4 = y_{12}$. Thus, (3.4) and (3.56) imply that

$$(x_4 y_{12}, b_3 + y_{12}) = (x_4 y_{12}, \bar{b}_3 b_5) = (b_3 x_4, b_5 \bar{y}_{12}) = 4. \tag{3.57}$$

Note that by (3.3), (3.4), and (3.55),

$$(b_4 \bar{y}_{12}, y_{12}) = (b_4 \bar{y}_{12}, b_3 + y_{12}) = (b_4 \bar{y}_{12}, \bar{b}_3 b_5) = (b_4 b_5, \bar{b}_3 y_{12}) = 2.$$

So

$$(x_4 y_{12}, b_3 + x_9) = (x_4 y_{12}, \bar{b}_3 b_4) = (b_3 x_4, b_4 \bar{y}_{12}) = (y_{12}, b_4 \bar{y}_{12}) = 2. \tag{3.58}$$

Hence (3.57) and (3.58) force that $(x_4 y_{12}, b_3) = 1$, and

$$x_4 y_{12} = b_3 + x_9 + 3 y_{12}, \qquad \bar{b}_3 x_4 = \bar{y}_{12}. \tag{3.59}$$

Therefore,

$$(\bar{y}_{12} y_{12}, \bar{x}_4 x_4) = (x_4 y_{12}, x_4 y_{12}) = 11.$$

But from $(\bar{b}_3 x_4) y_{12} = \bar{b}_3 (x_4 y_{12})$, (3.59), (3.1), (3.35), and (3.55), we get that

$$\bar{y}_{12} y_{12} = 1 + b_7 + 5 b_8 + 4 b_{12} + 3 \Sigma_{16}.$$

Since we assume that Σ_{16} has four distinct components of degree 4, and $(\bar{x}_4 x_4, b_8) = 0$ by (3.59), the above equality implies that it is impossible that $(\bar{y}_{12} y_{12}, \bar{x}_4 x_4) = 11$, a contradiction.

If $\Sigma_{16} = 2 x_8$, $x_8 \in B$, then $b_3 \Sigma_{16} = 4 y_{12}$ implies that $b_3 x_8 = 2 y_{12}$, and similar to the proof in the above paragraph, we have that

$$x_8 y_{12} = 2 b_3 + 2 x_9 + 6 y_{12}, \qquad \bar{b}_3 x_8 = 2 \bar{y}_{12}, \qquad (\bar{y}_{12} y_{12}, \bar{x}_8 x_8) = 44. \tag{3.60}$$

Hence from $(\bar{b}_3 x_8) y_{12} = \bar{b}_3 (x_8 y_{12})$, (3.60), (3.1), (3.35), and (3.55), we see that

$$\bar{y}_{12} y_{12} = 1 + b_7 + 5 b_8 + 4 b_{12} + 6 x_8.$$

Since $(\bar{x}_8 x_8, \bar{b}_3 b_3) = (b_3 x_8, b_3 x_8) = 4$, we may set

$$\bar{x}_8 x_8 = 1 + 3 b_8 + \Sigma_{39}, \qquad \Sigma_{39} \in \mathbb{N}(B).$$

But $b_3\bar{x}_8 = 2y_{12}$ by (3.60), and $b_3b_8 = b_3 + x_9 + y_{12}$ by (3.5). Multiplying both sides of the above equation by b_3, we obtain that

$$x_8 y_{12} = \frac{1}{2}(4b_3 + 3x_9 + 3y_{12} + b_3\Sigma_{39}).$$

Recall that $(b_3b_7, x_9) = 1$ by (3.38), and $(b_3b_{12}, x_9) = 1$ by (3.35). So the above equality and (3.60) force that $(\Sigma_{39}, b_7 + b_{12}) \le 1$. Therefore, $(\bar{y}_{12}y_{12}, \bar{x}_8x_8) = 44$ yields that

$$\bar{x}_8 x_8 = 1 + 3b_8 + b_{12} + 4x_8.$$

However, the above equality is impossible, a contradiction.

So Lemma 3.1 holds.   □

Since $(b_3b_8, y_{12}) = 1$ by (3.5), we see that $(\Sigma_{16}, b_8) = 0$ by (3.55). If $(\Sigma_{16}, b_{12}) = 1$, then $(b_3b_{12}, y_{12}) = 2$ by (3.55). But we also have $(b_3b_{12}, x_9) = 1$ by (3.35). Hence $b_3b_{12} = x_9 + 2y_{12} + x_3$ for some $x_3 \in B$, a contradiction. Therefore, we must have that $(\Sigma_{16}, b_{12}) = 0$. Thus, Lemma 3.1 and (3.55) yield that

$$(\bar{b}_3 y_{12}, \bar{b}_3 y_{12}) \le 5.$$

Since $(b_3b_{12}, x_9) = 1$ by (3.35) and $(b_3b_{12}, y_{12}) = 1$ by (3.55), we may set

$$b_3b_{12} = x_9 + y_{12} + \Sigma_{15}, \quad \Sigma_{15} \in \mathbb{N}(B). \tag{3.61}$$

Thus, from (3.35), (3.39), (3.38), (3.5), (3.61), and $\bar{b}_3^2 x_9 = \bar{b}_3(\bar{b}_3 x_9)$, we get that

$$b_5\bar{x}_9 = 2x_9 + y_{12} + \Sigma_{15}. \tag{3.62}$$

Hence $b_3^2 x_9 = b_3(b_3 x_9)$ and (3.42), (3.2), (3.3), (3.39), (3.62) imply that

$$b_3\sigma + b_3\theta = 2\bar{x}_9 + 2\bar{y}_{12} + \bar{\Sigma}_{12}^* + \bar{\Sigma}_{15}. \tag{3.63}$$

Furthermore, from $b_3(b_4b_5) = b_4(b_3b_5)$, (3.4), (3.3), (3.36), (3.5), and (3.61), we obtain that

$$b_4\bar{y}_{12} = x_9 + 2y_{12} + \Sigma_{15}. \tag{3.64}$$

Lemma 3.2 *Let Σ_{16} be the same as in (3.54). Then $(\Sigma_{16}, \Sigma_{16}) = 2$. Assume that $\Sigma_{16} = p + q$, $p, q \in B$. Then $|p| \ge 5$, $|q| \ge 5$, and*

$$\bar{b}_3 y_{12} = b_8 + b_{12} + p + q, \quad b_3 y_{12} = \sigma + \theta + b_5 + x_8, \quad x_8 \in B.$$

Proof Let us first prove that $(b_3\sigma, \bar{y}_{12}) = 1$ and $(b_3\theta, \bar{y}_{12}) = 1$. If $(b_3\sigma, \bar{y}_{12}) \ge 2$, then $(b_3 y_{12}, \sigma) \ge 2$. Since $(b_3 y_{12}, b_5) = 1$ by (3.4), we must have that

$$b_3 y_{12} = 2\sigma + b_5 + \cdots.$$

Thus, $(b_3 y_{12}, b_3 y_{12}) \geq 6$. So $(\bar{b}_3 y_{12}, \bar{b}_3 y_{12}) = (b_3 y_{12}, b_3 y_{12}) \geq 6$. But we have already shown that $(b_3 y_{12} b_3 y_{12}) \leq 5$, a contradiction. This proves that $(b_3 \sigma, \bar{y}_{12}) < 2$. Similarly, we can prove that $(b_3 \theta, \bar{y}_{12}) < 2$. Therefore, (3.63) forces that $(b_3 \sigma, \bar{y}_{12}) = 1$ and $(b_3 \theta, \bar{y}_{12}) = 1$.

Thus, we have that $(b_3 y_{12}, \sigma) = 1$, and $(b_3 y_{12}, \theta) = 1$. So we can set

$$b_3 y_{12} = \sigma + \theta + b_5 + x_8, \quad x_8 \in \mathbb{N}(B). \tag{3.65}$$

Now we prove that $x_8 \in B$. If $x_8 \notin B$, then we must have that $x_8 = s_4 + t_4$, $s_4, t_4 \in B$, and $s_4 \neq t_4$. Hence

$$\bar{b}_3 s_4 = \bar{b}_3 t_4 = y_{12} \quad \text{and} \quad (\Sigma_{16}, \Sigma_{16}) = 3.$$

So from $\bar{b}_3^2 s_4 = b_3 (\bar{b}_3 s_4)$, (3.2), and (3.55), we see that

$$b_4 s_4 + b_5 s_4 = b_8 + b_{12} + \Sigma_{16}.$$

If $b_4 s_4 = \Sigma_{16}$, then $(b_4 s_4, b_5^2) = 3$ by (3.54), and hence $(b_4 b_5, b_5 s_4) = 3$. Thus, $|b_5 s_4| \geq 24$ by (3.36), a contradiction. So $b_4 s_4 \neq \Sigma_{16}$. But from $(b_3 \bar{b}_3) s_4 = b_3 (\bar{b}_3 s_4)$, we see that

$$s_4 b_8 = \sigma + \theta + b_5 + t_4. \tag{3.66}$$

So $(b_5 s_4, b_8) = 1$. Thus, $b_4 s_4 \neq \Sigma_{16}$ forces that $(b_5 s_4, b_{12}) = 0$. So we must have that $(b_4 s_4, b_{12}) = 1$. Therefore, we may assume that

$$\Sigma_{16} = p_4 + q + r, \quad b_4 s_4 = b_{12} + p_4, \quad b_5 s_4 = b_8 + q + r, \quad p_4, q, r \in B.$$

Since $(s_4 \bar{t}_4, b_8) = 1$ by (3.66), $s_4 \neq t_4$, and $L_1(B) = \{1\}$, we see that $(b_4 s_4, b_4 t_4) = (s_4 \bar{t}_4, b_4^2) = 1$ by (3.6). So in a way that is similar to the above discussions, we get that

$$\Sigma_{16} = p_4 + q_4 + r_8, \quad b_4 t_4 = b_{12} + q_4, \quad b_5 t_4 = b_8 + p_4 + r_8,$$

$$p_4, q_4, r_8 \in B, \quad p_4 \neq q_4.$$

Note that $b_3 p_4 = y_{12}$ and $b_3 q_4 = y_{12}$ by (3.55). So $b_3^2 p_4 = b_3 (b_3 p_4)$ yields that

$$b_4 p_4 + b_5 p_4 = \sigma + \theta + b_5 + s_4 + t_4.$$

But $(b_4 p_4, s_4) = 1$, $(b_5 p_4, t_4) = 1$, and $(b_5 p_4, b_5) = 1$. Thus, without loss of generality, we may assume that $\sigma = \sigma_{12}$, $\theta = \theta_{11}$, and

$$b_4 p_4 = s_4 + \sigma_{12}, \qquad b_5 p_4 = t_4 + b_5 + \theta_{11}.$$

Similarly, we also have that

$$b_4 q_4 = s_4 + \sigma_{12}, \qquad b_5 q_4 = t_4 + b_5 + \theta_{11}.$$

Hence $(p_4\bar{q}_4, b_4^2) = (b_4 p_4, b_4 q_4) = 2$, and $(p_4\bar{q}_4, b_5^2) = (b_5 p_4, b_5 q_4) = 3$. However, it follows from $(p_4\bar{q}_4, b_4^2) = 2$, $p_4 \neq q_4$, $L_1(B) = \{1\}$, and $L_2(B) = \emptyset$ that $p_4\bar{q}_4 = 2b_8$. Hence $(p_4\bar{q}_4, b_5^2) = 2$ by (3.54), a contradiction. This proves that $x_8 \in B$.

Thus, $(\bar{b}_3 y_{12}, \bar{b}_3 y_{12}) = (b_3 y_{12}, b_3 y_{12}) = 4$, and hence $(\Sigma_{16}, \Sigma_{16}) = 2$. Assume that $\Sigma_{16} = p + q$, $p, q \in B$. Then by (3.55), $\bar{b}_3 y_{12} = b_8 + b_{12} + p + q$. Hence $|p| \geq 4$ and $|q| \geq 4$. If $|p| = 4$, then $p = p_4$, $b_3 p_4 = y_{12}$, and both p_4 and q are real (because Σ_{16} is real). Thus, (3.2) and $b_3^2 p_4 = b_3(b_3 p_4)$ yield that

$$b_4 p_4 + b_5 p_4 = \sigma + \theta + b_5 + x_8.$$

Since $(b_5 p_4, b_5) = 1$, we see that $(b_5 p_4, b_5 p_4) = 2$ or 3. If $(b_5 p_4, b_5 p_4) = 2$, then $(b_4 p_4, b_4 p_4) = 2$. Since $|\sigma| + |\theta| = 23$, without loss of generality, we may assume that $\sigma = \sigma_8$, and $b_4 p_4 = x_8 + \sigma_8$. Recall that σ_8 is real, $(b_3\sigma_8, \bar{x}_9) = 1$ by (3.42), and $(b_3\sigma_8, \bar{y}_{12}) = 1$ by (3.65). So we have that $b_3\sigma_8 = \bar{x}_9 + \bar{y}_{12} + x_3$, where x_3 is a component of $\bar{\Sigma}_{12}^*$ or $\bar{\Sigma}_{15}$ by (3.63). But any component of $\bar{\Sigma}_{12}^*$ and $\bar{\Sigma}_{15}$ has degree at least 4, a contradiction. If $(b_5 p_4, b_5 p_4) = 3$, then $(b_4 p_4, b_4 p_4) = 1$, and hence we have either $b_4 p_4 = \sigma$ or $b_4 p_4 = \theta$. Thus,

$$(b_4\bar{y}_{12}, y_{12}) = (b_4\bar{y}_{12}, b_3 p_4) = (b_4 p_4, b_3 y_{12}) = 1,$$

a contradiction to (3.64). This proves that $|p| > 4$. Similarly we can prove that $|q| > 4$. So the lemma holds. $\square$

For the rest of this section, we will set $\Sigma_{16} = p+q$, $p, q \in B$. Then $b_5^2 = 1+b_8+p+q$ by (3.54). From $(\bar{b}_3 b_3)b_8 = \bar{b}_3(b_3 b_8)$, (3.1), (3.5), (3.35), and Lemma 3.2, we see that $b_8^2 = 1 + b_7 + 2b_8 + 2b_{12} + p + q$. Thus, we have that

$$b_5^2 = 1 + b_8 + p + q \quad \text{and} \quad b_8^2 = 1 + b_7 + 2b_8 + 2b_{12} + p + q. \tag{3.67}$$

Recall that $b_3 b_7 = x_9 + \Sigma_{12}^*$ by (3.38). In the following, we prove that $\Sigma_{12}^* \in B$. But first we need to prove the next lemma.

Lemma 3.3 *Let Σ_{12}^* be the same as in (3.38). Then $(\Sigma_{12}^*, \Sigma_{12}^*) \leq 2$, and either $(b_3\sigma, \bar{\Sigma}_{12}^*) = 0$ or $(b_3\theta, \bar{\Sigma}_{12}^*) = 0$.*

Proof Let us first show that

$$(\Sigma_{12}^*, \Sigma_{12}^*) \leq 2. \tag{3.68}$$

If (3.68) does not hold, then $(\Sigma_{12}^*, \Sigma_{12}^*) \geq 3$, and hence $(b_3 b_7, b_4\bar{x}_9) \geq 4$ by (3.38) and (3.39). So b_7 real implies that $(b_4 b_7, b_3 x_9) \geq 4$. Also b_4 real implies $(b_4 b_7, b_4) = 1$ by (3.6). But $b_3 x_9 = b_4 + \sigma + \theta$ by (3.42). Hence

$$(b_4 b_7, \sigma + \theta) \geq 3. \tag{3.69}$$

Since $|\sigma| + |\theta| = 23$ by (3.42), and $(b_4 b_7, b_4) = 1$, the above inequality forces that either $(b_4 b_7, \sigma) = 0$ or $(b_4 b_7, \theta) = 0$. Without loss of generality, we may assume that $(b_4b_7, \theta) = 0$. So (3.69) yields that $(b_4b_7, \sigma) \geq 3$, and

$$b_4b_7 = b_4 + 3\sigma + \cdots .$$

Therefore, $(b_4b_7, b_4b_7) \geq 10$. So $(b_4^2, b_7^2) \geq 10$, and hence (3.6) forces that

$$|b_7^2| \geq 1 + 7 \cdot 9 = 64,$$

a contradiction. Thus, (3.68) must hold.

If both $(b_3\sigma, \bar{\Sigma}_{12}^*) \neq 0$ and $(b_3\theta, \bar{\Sigma}_{12}^*) \neq 0$, since any component of Σ_{15} has degree at least 4 by (3.61), we must have that $(\Sigma_{12}^*, \Sigma_{12}^*) = 2$ by (3.68) and (3.63). So we may set $\Sigma_{12}^* = r + s$, $r, s \in B$. Thus, $b_3b_7 = x_9 + r + s$ by (3.38). So $(\bar{b}_3r, \bar{b}_3s) \geq 1$, and hence $(b_3r, b_3s) = (\bar{b}_3r, \bar{b}_3s) \geq 1$. Let g be a common component of b_3r and b_3s. Then $(\bar{b}_3g, r) \geq 1$, and $(\bar{b}_3g, s) \geq 1$. Hence $|r| + |s| = 12$ forces that $|g| \geq 4$. Since $(b_3\sigma, \bar{\Sigma}_{12}^*) \neq 0$ and $(b_3\theta, \bar{\Sigma}_{12}^*) \neq 0$, from (3.63) we see that $(b_3\Sigma_{12}^*, \sigma) = 1$ and $(b_3\Sigma_{12}^*, \theta) = 1$. Therefore, there is $h \in B$ such that $|h| \leq 5$ and

$$b_3\Sigma_{12}^* = \sigma + \theta + 2g + h.$$

Hence it follows from (3.38), (3.2), (3.42), and $b_3^2b_7 = b_3(b_3b_7)$ that

$$b_4b_7 + b_5b_7 = b_4 + 2\sigma + 2\theta + 2g + h. \tag{3.70}$$

Similar to the proof in the above paragraph, we may assume that $b_4b_7 = b_4 + 2\sigma + \cdots$ and $(b_4b_7, \theta) = 0$. Note that $(b_3\bar{b}_3, b_7^2) = (b_3b_7, b_3b_7) = 3$ by (3.38). Hence $(b_7^2, b_8) = 2$ by (3.1). Thus, $(b_4b_7, b_5b_7) = (b_4b_5, b_7^2) \geq 2$ by (3.36), a contradiction to (3.70). This proves that either $(b_3\sigma, \bar{\Sigma}_{12}^*) = 0$ or $(b_3\theta, \bar{\Sigma}_{12}^*) = 0$. □

From Lemma 3.3 we can prove the following

Lemma 3.4 *Let Σ_{12}^* be the same as in (3.38). Then $\Sigma_{12}^* \in B$.*

Proof Toward a contradiction, assume that $\Sigma_{12}^* \notin B$. Then $(\Sigma_{12}^*, \Sigma_{12}^*) = 2$ by (3.68). As in the proof of Lemma 3.3, we may assume that $b_4b_7 = b_4 + 2\sigma + \cdots$. Then $L_2(B) = \emptyset$ forces that either $|\sigma| = 12$ or $|\sigma| \leq 10$. If $(b_3\theta, \bar{\Sigma}_{12}^*) = 0$, then

$$b_3\sigma = \bar{x}_9 + \bar{y}_{12} + \bar{\Sigma}_{12}^* + \cdots$$

by (3.63) and Lemma 3.2. Since any component of Σ_{15} has degree at least 4 by (3.61), the above equality implies that either $|\sigma| = 11$ or $|\sigma| \geq 13$, a contradiction. Thus, we must have that $(b_3\sigma, \bar{\Sigma}_{12}^*) = 0$ by Lemma 3.3. But, on the other hand, from (3.38), (3.2), (3.42), and $b_3^2b_7 = b_3(b_3b_7)$, we obtain that

$$b_4b_7 + b_5b_7 = b_4 + \sigma + \theta + b_3\Sigma_{12}^*.$$

So $(b_4b_7, \sigma) = 2$ and the above equality force that $(b_3\sigma, \bar{\Sigma}_{12}^*) = 1$, a contradiction. This proves that $\Sigma_{12}^* \in B$. □

In the following, we denote $\Sigma_{12}^* = c_{12} \in B$. Then (3.38) and (3.17) yield that

$$b_3 b_7 = x_9 + c_{12} \quad \text{and} \quad b_4 \bar{x}_9 = b_3 + y_{12} + x_9 + c_{12}. \tag{3.71}$$

Lemma 3.5 *Let Σ_{15} be the same as in (3.61). Then $\Sigma_{15} \in B$.*

Proof It follows from $(\bar{b}_3 b_4)\bar{x}_9 = (b_4 \bar{x}_9)\bar{b}_3$, (3.3), (3.35), (3.71), (3.1), (3.42), and Lemma 3.2 that

$$x_9 \bar{x}_9 = 1 + 2b_8 + b_{12} + p + q + \bar{b}_3 c_{12}.$$

Since $(b_3 p + b_3 q, y_{12}) = 2$ by Lemma 3.2, and $|p| + |q| = 16$, we must have that $(b_3 p + b_3 q, c_{12}) \leq 2$. Thus, $(\bar{b}_3 c_{12}, p + q) \leq 2$, and hence $(b_5^2, x_9 \bar{x}_9) \leq 7$ by Lemma 3.2 and (3.54). So $(b_5 \bar{x}_9, b_5 \bar{x}_9) \leq 7$, and (3.62) forces that $(\Sigma_{15}, \Sigma_{15}) \leq 2$ and (3.54). On the other hand, since $(b_5 \bar{x}_9, b_5 \bar{x}_9) \geq 6$ by (3.62), we see that $(b_5^2, x_9 \bar{x}_9) \geq 6$, and hence $(\bar{b}_3 c_{12}, p + q) \geq 1$ by Lemma 3.2. Therefore,

$$(\bar{b}_3 c_{12}, p + q) = \begin{cases} 1, & \text{if } \Sigma_{15} \in B; \\ 2, & \text{otherwise.} \end{cases}$$

Furthermore, since $|p| \geq 5$ and $|q| \geq 5$ by Lemma 3.2, we see that

$$\text{if } \Sigma_{15} \notin B, \quad \text{then} \quad p = p_8, \quad q = q_8, \quad \text{and} \quad b_3 p_8 = b_3 q_8 = y_{12} + c_{12}. \tag{3.72}$$

Toward a contradiction, in the following we assume that $\Sigma_{15} \notin B$. Then from the proof above, $(\Sigma_{15}, \Sigma_{15}) = 2$. Since $p + q$ is real, it follows from Lemma 3.2, (3.72), and $\bar{b}_3^2 y_{12} = \bar{b}_3(\bar{b}_3 y_{12})$ that

$$b_4 y_{12} + b_5 y_{12} = \bar{b}_3 + 2\bar{x}_9 + 4\bar{y}_{12} + 2\bar{c}_{12} + \bar{\Sigma}_{15}.$$

But, on the other hand, from $b_3^2 y_{12} = b_3(b_3 y_{12})$, (3.63), and Lemma 3.2, we see that

$$b_4 y_{12} + b_5 y_{12} = \bar{b}_3 + 2\bar{x}_9 + 3\bar{y}_{12} + \bar{c}_{12} + \bar{\Sigma}_{15} + b_3 x_8.$$

Thus,

$$b_3 x_8 = \bar{y}_{12} + \bar{c}_{12}.$$

Hence $b_3^2 x_8 = b_3(b_3 x_8)$ and Lemma 3.2 yield that

$$b_4 x_8 + b_5 x_8 = b_8 + b_{12} + p_8 + q_8 + b_3 \bar{c}_{12}.$$

Since $(b_3 \bar{c}_{12}, b_{12}) = 0$ by (3.61), the above equation implies that

$$(b_4 x_8, b_{12}) \leq 1. \tag{3.73}$$

Now from (3.42), (3.71), (3.6), (3.1), (3.35), Lemma 3.2, and $b_4(b_3 x_9) = b_3(b_4 x_9)$, we get that

$$b_4 \sigma + b_4 \theta = 2b_8 + 2b_{12} + p + q + b_3 \bar{c}_{12}.$$

Since $b_4(b_3 y_{12}) = b_4\sigma + b_4\theta + b_4 b_5 + b_4 x_8$ by Lemma 3.2, and $b_3(b_4 y_{12}) = b_3\bar{x}_9 + 2b_3\bar{y}_{12} + b_3\bar{\Sigma}_{15}$ by (3.64), it follows that

$$b_3\bar{c}_{12} + b_4 x_8 = b_7 + p + q + b_3\bar{\Sigma}_{15}.$$

Since $(b_3\bar{c}_{12}, b_{12}) = 0$ and $(b_3\bar{\Sigma}_{15}, b_{12}) = 2$ by (3.61), the above equation forces that $(b_4 x_8, b_{12}) = 2$, a contradiction to (3.73). This proves that Σ_{15} must be an element in B. $\square$

In the following, we denote $\Sigma_{15} = c_{15} \in B$. Then from (3.61), (3.62), and (3.64), we obtain that

$$b_3 b_{12} = x_9 + y_{12} + c_{15}, \qquad b_5\bar{x}_9 = 2x_9 + y_{12} + c_{15}, \qquad b_4\bar{y}_{12} = x_9 + 2y_{12} + c_{15}.$$
$$(3.74)$$

Lemma 3.6 *Assume that $|\sigma| < |\theta|$. Then $\sigma = \sigma_{11}$, $\theta = \theta_{12}$, and*

$$b_3\sigma_{11} = \bar{x}_9 + \bar{y}_{12} + \bar{c}_{12}, \qquad b_3\theta_{12} = \bar{x}_9 + \bar{y}_{12} + \bar{c}_{15}.$$

Proof Since $|\sigma| < |\theta|$, we must have that $(b_3\sigma, \bar{c}_{15}) = 0$. In the following, we show that $(b_3\theta, \bar{c}_{12}) = 0$ by (3.42), (3.63), and Lemma 3.2. Toward a contradiction, assume that $(b_3\theta, \bar{c}_{12}) \neq 0$. Then from (3.42), (3.63), and Lemma 3.2, we must have that $\sigma = \sigma_7$, $\theta = \theta_{16}$, and

$$b_3\sigma_7 = \bar{x}_9 + \bar{y}_{12}, \qquad b_3\theta_{16} = \bar{x}_9 + \bar{y}_{12} + \bar{c}_{12} + \bar{c}_{15}. \qquad (3.75)$$

Thus, $b_3^2\sigma_7 = b_3(b_3\sigma_7)$ yields that

$$b_4\sigma_7 + b_5\sigma_7 = b_7 + 2b_8 + 2b_{12} + p + q.$$

Note that $(b_3\bar{b}_3)b_4 = b_3(\bar{b}_3 b_4)$ and (3.1), (3.2), (3.3, (3.42) yield that $(b_4 b_8, \sigma_7) = 1$. Hence $(b_4\sigma_7, b_8) = 1$ and similarly, $(b_5\sigma_7, b_8) = 1$. But the degrees of p and q are at least 5, and the above equality forces that $p = p_8$, $q = q_8$. Without loss of generality, we may assume that

$$b_4\sigma_7 = b_8 + b_{12} + p_8, \qquad b_5\sigma_7 = b_7 + b_8 + b_{12} + q_8.$$

Note that from $b_4(b_3\sigma_7) = (b_3 b_4)\sigma_7$, we get that

$$\sigma_7\bar{x}_9 = b_3 + x_9 + 2y_{12} + c_{12} + c_{15}.$$

Hence $b_3(b_4\sigma_7) = (b_3 b_4)\sigma_7$ implies that $b_3 p_8 = y_{12} + c_{12}$. So we may set $b_3 q_8 = y_{12} + \varepsilon_{12}$, $\varepsilon_{12} \in \mathbb{N}(B)$. Thus, from $b_3(b_5\sigma_7) = (b_3 b_5)\sigma_7$, we have that

$$\sigma_7\bar{y}_{12} = b_3 + 2x_9 + 2y_{12} + c_{12} + c_{15} + \varepsilon_{12}.$$

Note that $(b_3\bar{b}_3, \sigma_7^2) = 2$, $(b_4^2, \sigma_7^2) = 3$, $(b_5^2, \sigma_7^2) = 4$, and $(b_4 b_5, \sigma_7^2) = 2$. So there is $x_5 \in B$ such that σ_7^2 is one of the following:

$$1 + b_7 + b_8 + p_8 + q_8 + b_{12} + x_5,$$

$$1 + b_7 + b_8 + 2p_8 + b_{12} + x_5,$$
$$1 + b_7 + b_8 + 2q_8 + b_{12} + x_5.$$

Hence from $(b_3\sigma_7)\sigma_7 = b_3\sigma_7^2$, we must have that

$$\sigma_7^2 = 1 + b_7 + b_8 + p_8 + q_8 + b_{12} + x_5 \quad \text{and} \quad b_3 x_5 = c_{15}.$$

Thus,

$$(b_5 x_5, b_4 + \sigma_7 + \theta_{16}) = (b_5 x_5, \bar{b}_3 \bar{x}_9) = (b_5 \bar{x}_9, b_3 x_5) = 1.$$

So $(b_5 x_5, \theta_{16}) = 1$. On the other hand,

$$(b_4 x_5, b_5 + x_8 + \sigma_7 + \theta_{16}) = (b_4 x_5, \bar{b}_3 \bar{y}_{12}) = (b_3 x_5, b_4 \bar{y}_{12}) = 1.$$

Thus, $(b_4 x_5, x_8 + \theta_{16}) = 1$. But $b_4 x_5 + b_5 x_5 = b_3^2 x_5 = b_3 c_{15}$, and $(b_3 c_{15}, \theta_{16}) = 1$ by (3.75). So we must have that $(b_4 x_5, \theta_{16}) = 0$, and hence $(b_4 x_5, x_8) = 1$. Thus, $(b_4 x_8, x_5) = 1$. Since $b_3^2 x_8 = b_3(b_3 x_8)$ and $b_3 x_8 = y_{12} + \varepsilon_{12}$, we see that

$$b_4 x_8 + b_5 x_8 = b_5 + x_8 + \sigma_7 + \theta_{16} + b_3 \varepsilon_{12}.$$

Hence $(b_3 \varepsilon_{12}, x_5) \geq 1$. So $(\bar{b}_3 x_5, \varepsilon_{12}) \geq 1$. However, $(\bar{b}_3 x_5, \bar{b}_3 x_5) = (b_3 x_5, b_3 x_5) = 1$, a contradiction.

This proves that (3.75) is not true. So we must have that $(b_3 \theta, \bar{c}_{12}) = 0$, and the lemma holds by (3.42), (3.63), and Lemma 3.2.   □

For the rest of this section, we will assume that $\sigma = \sigma_{11}$ and $\theta = \theta_{12}$. Note that $b_3^2 b_8 = b_3(b_3 b_8)$ yields that

$$b_4 b_8 + b_5 b_8 = 2b_4 + 2b_5 + x_8 + 2\sigma_{11} + 2\theta_{12}.$$

Since $(b_4 b_8, b_4) = 1$ and $(b_4 b_8, b_5) = 1$, the above equality forces that

$$b_4 b_8 = b_4 + b_5 + \sigma_{11} + \theta_{12} \quad \text{and} \quad b_5 b_8 = b_4 + b_5 + \sigma_{11} + \theta_{12} + x_8. \tag{3.76}$$

Lemma 3.7 *Let p and q be the same as in Lemma 3.2. Then $p = p_8$ and $q = q_8$.*

Proof Toward a contradiction, we may assume that $|p| > |q|$. From $\bar{b}_3^2 y_{12} = \bar{b}_3(\bar{b}_3 y_{12})$, Lemma 3.2, and (3.74), we obtain that

$$b_5 y_{12} = \bar{b}_3 + \bar{x}_9 + \bar{b}_3 p + \bar{b}_3 q.$$

On the other hand, from $b_3^2 y_{12} = b_3(b_3 y_{12})$, Lemma 3.2, and (3.74), we see that

$$b_5 y_{12} = \bar{b}_3 + \bar{x}_9 + \bar{c}_{12} + \bar{y}_{12} + b_3 x_8.$$

Thus, the components of $\bar{b}_3 p$ and $\bar{b}_3 q$ have degrees at least 4. Since $|p| > |q|$, p and q are real, $(\bar{b}_3 c_{12}, p + q) = 1$ from the proof of Lemma 3.5, and $(\bar{b}_3 y_{12}, p) = (\bar{b}_3 y_{12}, p) = 1$ by Lemma 3.2, we must have that $p = p_{10}$, $q = q_6$, and

$$\bar{b}_3 p_{10} = \bar{y}_{12} + \bar{c}_{12} + x_6, \quad \bar{b}_3 q_6 = \bar{y}_{12} + y_6, \quad b_3 x_8 = \bar{y}_{12} + x_6 + y_6, \quad x_6, y_6 \in B.$$

Thus, $\bar{b}_3^2 q_6 = \bar{b}_3(\bar{b}_3 q_6)$ and Lemma 3.2 yield that

$$b_4 q_6 + b_5 q_6 = b_5 + x_8 + \sigma_{11} + \theta_{12} + \bar{b}_3 y_6.$$

Since $(b_5 q_6, b_5) = 1$ by (3.67), the above equality forces that either $(b_5 q_6, \sigma_{11}) = 0$ or $(b_5 q_6, \theta_{12}) = 0$. In the following, we derive a contradiction for each of these two cases.

First assume that $(b_5 q_6, \sigma_{11}) = 0$. Since $(\bar{b}_3 y_6, x_8) = 1$, we must have that

$$b_4 q_6 = \sigma_{11} + x_8 + d_5, \quad b_5 q_6 = b_5 + \theta_{12} + x_8 + e_5,$$
$$\bar{b}_3 y_6 = x_8 + d_5 + e_5, \quad d_5, e_5 \in B.$$

Hence it follows from $b_5(\bar{b}_3 q_6) = b_3(b_5 q_6)$ that $\bar{c}_{12} + b_5 \bar{y}_6 = \bar{c}_{15} + \bar{y}_{12} + b_3 e_5$. Thus,

$$b_3 e_5 = \bar{c}_{12} + x_3, \quad b_5 \bar{y}_6 = \bar{c}_{15} + \bar{y}_{12} + x_3, \quad x_3 \in B.$$

On the other hand, $b_4(\bar{b}_3 q_6) = b_3(b_4 q_6)$ implies that $\bar{c}_{15} + b_4 \bar{y}_6 = \bar{c}_{12} + x_6 + y_6 + b_3 d_5$. Thus,

$$b_3 d_5 = \bar{c}_{15} \quad \text{and} \quad b_4 \bar{y}_6 = \bar{c}_{12} + x_6 + y_6.$$

Therefore, $(b_5 \bar{y}_6, b_4 \bar{y}_6) = 0$, and hence $(y_6 \bar{y}_6, b_4 b_5) = 0$. So $(y_6 \bar{y}_6, b_8) = 0$ by (3.36). But $\bar{b}_3 y_6 = x_8 + d_5 + e_5$ forces that $(y_6 \bar{y}_6, b_8) = 2$, a contradiction.

Now assume that $(b_5 q_6, \theta_{12}) = 0$. Then $(\bar{b}_3 y_6, x_8) = 1$ forces that

$$b_4 q_6 = \theta_{12} + x_8 + d_4, \quad b_5 q_6 = b_5 + \sigma_{11} + x_8 + \Sigma_6,$$
$$\bar{b}_3 y_6 = x_8 + d_4 + \Sigma_6, \quad d_4 \in B, \ \Sigma_6 \in \mathbb{N}(B).$$

Hence from $b_4(\bar{b}_3 q_6) = b_3(b_4 q_6)$ we see that

$$b_4 \bar{y}_6 = x_6 + y_6 + b_3 d_4.$$

But $(b_3 d_4, y_6) = 1$. So $(b_4 \bar{y}_6, b_4 \bar{y}_6) \geq 6$, and hence $(y_6 \bar{y}_6, b_4^2) \geq 6$. Note that $\bar{b}_3 y_6 = x_8 + d_4 + \Sigma_6$ implies that $(y_6 \bar{y}_6, b_8) \geq 2$. Thus, by (3.6) we get that

$$|y_6 \bar{y}_6| \geq 1 + 2 \cdot 8 + 3 \cdot 7 = 38,$$

a contradiction.

This proves that $|p| = |q| = 8$, and the lemma holds. $\square$

Since $(\bar{b}_3 c_{12}, p_8 + q_8) = 1$ from the proof of Lemma 3.5, and $(\bar{b}_3 y_{12}, p_8) = (\bar{b}_3 y_{12}, p_8) = 1$ by Lemma 3.2, without loss of generality, we may set

$$\bar{b}_3 p_8 = \bar{y}_{12} + \bar{c}_{12}, \qquad \bar{b}_3 q_8 = \bar{y}_{12} + \varepsilon_{12}, \quad \varepsilon_{12} \in \mathbb{N}(B). \tag{3.77}$$

Thus, from the proof of Lemma 3.7 we see that

$$b_3 x_8 = \bar{y}_{12} + \varepsilon_{12}, \quad \text{and} \quad b_5 y_{12} = \bar{b}_3 + \bar{x}_9 + \bar{c}_{12} + 2\bar{y}_{12} + \varepsilon_{12}. \tag{3.78}$$

Lemma 3.8 *There exists real elements* $t_9, t_{12} \in B$ *such that*

$$b_3 \bar{c}_{12} = b_7 + p_8 + t_9 + t_{12}, \qquad b_4 \sigma_{11} = b_7 + b_8 + b_{12} + t_9 + p_8$$

and

$$b_5 \sigma_{11} = b_7 + b_8 + b_{12} + p_8 + q_8 + t_{12}, \qquad b_4 \theta_{12} = b_8 + b_{12} + t_{12} + p_8 + q_8.$$

Proof From $b_4(b_3 x_9) = b_3(b_4 x_9)$, (3.42), and (3.71), we obtain that

$$b_4 \sigma_{11} + b_4 \theta_{12} = 2b_8 + 2b_{12} + p_8 + q_8 + b_3 \bar{c}_{12}. \tag{3.79}$$

From $b_4(b_3 y_{12}) = b_3(b_4 y_{12})$, Lemma 3.2, and (3.64), we see that

$$b_4 \sigma_{11} + b_4 \theta_{12} + b_4 x_8 = b_7 + 2b_8 + 2b_{12} + 2p_8 + 2q_8 + b_3 \bar{c}_{15}. \tag{3.80}$$

Thus, if τ is a component of $b_3 \bar{c}_{12}$ and $\tau \neq b_7$ or p_8, then τ is also a component of $b_3 \bar{c}_{15}$. Hence both $\bar{c}_{12}$ and $\bar{c}_{15}$ are components of $\bar{b}_3 \tau$. So the degree of τ is at least 9. But from (3.71), (3.77), (3.79), and (3.80), $b_3 \bar{c}_{12} = b_7 + p_8 +$ the linear combination of some common components of $b_3 \bar{c}_{12}$ and $b_3 \bar{c}_{15}$. Therefore

$$(b_3 \bar{c}_{12}, b_3 \bar{c}_{12}) = 3 \text{ or } 4.$$

It follows from $b_4^2 b_8 = b_4(b_4 b_8)$, (3.6), (3.67), and (3.36) that

$$b_8 + b_7 b_8 + b_{12} + p_8 + q_8 = b_4 \sigma_{11} + b_4 \theta_{12}.$$

Thus,

$$b_7 b_8 = b_8 + b_{12} + b_3 \bar{c}_{12}.$$

Hence $(b_7 b_8, b_7 b_8) = 5$ or 6. So

$$(b_7^2, b_8^2) = 5 \text{ or } 6. \tag{3.81}$$

In the following, we show that $(b_4 b_7, \sigma_{11}) = 1$. If $(b_4 b_7, \sigma_{11}) \neq 1$, since $(b_4 b_7, b_4) = 1$ by (3.6) and $L_2(B) = \emptyset$, we must have that $(b_4 b_7, \sigma_{11}) = 0$. Note that $b_3^2 b_7 = b_3(b_3 b_7)$ implies that $b_4 b_7 + b_5 b_7 = b_4 + \sigma_{11} + \theta_{12} + b_3 c_{12}$. So $(b_3 c_{12}, \sigma_{11}) = 1$ forces that

$$b_4 b_7 = b_4 + \theta_{12} + \cdots \quad \text{and} \quad b_5 b_7 = 2\sigma_{11} + \cdots .$$

Thus, $(b_4b_7, b_4b_7) \geq 3$, and $(b_5b_7, b_5b_7) \geq 5$. So $(b_3b_7, b_3b_7) = 2$ implies that

$$(b_7^2, b_8) = 1, \qquad (b_7^2, b_7) \geq 1, \qquad (b_7^2, p_8 + q_8) \geq 3.$$

Therefore, $(b_7^2, b_8^2) \geq 7$ by (3.67), a contradiction to (3.81). This proves that $(b_4b_7, \sigma_{11}) = 1$. Thus,

$$b_4b_7 = b_4 + \sigma_{11} + \cdots \quad \text{and} \quad b_5b_7 = \sigma_{11} + \theta_{12} + \cdots.$$

Note that $b_3^2\theta_{12} = b_3(b_3\theta_{12})$, (3.2), and Lemma 3.6 imply that

$$b_4\theta_{12} + b_5\theta_{12} = b_7 + 2b_8 + 2b_{12} + p_8 + q_8 + b_3\bar{c}_{15}.$$

Hence $(b_4\theta_{12}, p_8) \leq 1$. So $(b_4\sigma_{11}, p_8) \geq 1$ by (3.79). It follows from $b_4^2b_5 = b_4(b_4b_5)$, (3.6), (3.36), and (3.76) that

$$b_4b_{12} = b_5 + x_8 + b_5b_7 = b_5 + x_8 + \sigma_{11} + \theta_{12} + \cdots.$$

Thus, $b_4\sigma_{11} = b_7 + b_8 + b_{12} + p_8 + \cdots$. So (3.79) forces that

$$b_3\bar{c}_{12} = b_7 + p_8 + t_9 + t_{12}, \qquad t_9, t_{12} \in B,$$

and

$$b_4\sigma_{11} = b_7 + b_8 + b_{12} + t_9 + p_8, \qquad b_4\theta_{12} = b_8 + b_{12} + t_{12} + p_8 + q_8.$$

Clearly t_9 and t_{12} are real by the above equalities. Furthermore, $b_3^2\sigma_{11} = b_3(b_3\sigma_{11})$ implies that $b_5\sigma_{11} = b_7 + b_8 + b_{12} + p_8 + q_8 + t_{12}$. So the lemma holds. □

Lemma 3.9 *There exists $d_{12} \in B$ such that*

$$b_3\bar{c}_{15} = t_9 + t_{12} + b_{12} + d_{12}.$$

Furthermore, $(b_3\bar{c}_{15}, b_3\bar{c}_{15}) = 4$.

Proof It follows from Lemma 3.8 and (3.80) that

$$t_9 + t_{12} + b_4x_8 = q_8 + b_3\bar{c}_{15}.$$

Thus, q_8 is a component of b_4x_8. Furthermore, since $(b_3\bar{c}_{15}, p_8) = 0$ by (3.77), the above equation forces that $(b_4x_8, p_8) = 0$. Note that $b_3^2x_8 = b_3(b_3x_8)$ yields that

$$b_4x_8 + b_5x_8 = b_8 + b_{12} + p_8 + q_8 + b_3\varepsilon_{12}$$

by (3.78) and Lemma 3.2. Since $(b_3\bar{c}_{15}, b_{12}) = 1$ by (3.74), we see that $(b_4x_8, b_{12}) = 1$. Also $(b_4x_8, b_8) = (b_4b_8, x_8) = 0$ by (3.76). Thus, if τ is a component of b_4x_8 and $\tau \neq q_8$ or b_{12}, then τ is a component of both $b_3\bar{c}_{15}$ and $b_3\varepsilon_{12}$. But we have either $\varepsilon_{12} \in B$ or any component of ε_{12} has degree at least 4 by (3.78). Hence, as a common component of $b_3\bar{c}_{15}$ and $b_3\varepsilon_{12}$, the degree of tau is at least 7. Furthermore, $b_4x_8 = b_{12} + q_8+$ the linear combination of some common components of $b_3\bar{c}_{15}$ and $b_3\varepsilon_{12}$. Thus, we must have that

$$b_3\bar{c}_{15} = t_9 + t_{12} + b_{12} + d_{12}, \quad b_4x_8 = b_{12} + d_{12} + q_8, \quad d_{12} \in B.$$

Hence from Lemmas 3.6 and 3.8, and $b_3^2\theta_{12} = b_3(b_3\theta_{12})$, we get that

$$b_5\theta_{12} = b_7 + b_8 + 2b_{12} + t_9 + d_{12}.$$

Note that $b_{12} \neq d_{12}$ or t_{12} by (3.74). Also $d_{12} \neq t_{12}$; otherwise $(\bar{b}_3t_{12}, \bar{c}_{15}) = 2$ and $(\bar{b}_3t_{12}, \bar{c}_{12}) = 1$, a contradiction. Thus, $(b_3\bar{c}_{15}, b_3\bar{c}_{15}) = 4$. $\quad\square$

From Lemmas 3.8 and 3.9, we have the following equalities:

$$\bar{b}_3t_9 = \bar{c}_{12} + \bar{c}_{15} \quad \text{and} \quad \bar{b}_3t_{12} = \bar{c}_{12} + \bar{c}_{15} + \Sigma_9, \quad \Sigma_9 \in \mathbb{N}(B). \tag{3.82}$$

Lemma 3.10 *There exist real elements* $s_4, s_9, s_{12} \in B$ *such that*

$$b_3c_{12} = \sigma_{11} + s_{12} + s_9 + s_4, \quad \bar{b}_3s_4 = c_{12}, \quad \bar{b}_3s_9 = c_{12} + c_{15}$$

and

$$b_4s_4 = b_7 + t_9, \quad b_5s_4 = p_8 + t_{12}.$$

Proof Recall that $(b_3c_{12}, \sigma_{11}) = 1$ and $(b_3c_{15}, \theta_{12}) = 1$ by Lemma 3.6. Since $(b_3c_{12}, b_3c_{12}) = (b_3\bar{c}_{12}, b_3\bar{c}_{12}) = 4$, $(b_3c_{15}, b_3c_{15}) = (b_3\bar{c}_{15}, b_3\bar{c}_{15}) = 4$, and $(b_3c_{12}, b_3c_{15}) = (b_3\bar{c}_{12}, b_3\bar{c}_{15}) = 2$, we see that b_3c_{12} and b_3c_{15} have two common components. Clearly these two components have degrees at least 9. Thus, besides σ_{11}, b_3c_{12} has two components of degrees at least 9, and its fourth component has degree at least 4. From the proof of Lemma 3.8, we have the following equations:

$$b_4b_7 + b_5b_7 = b_4 + \sigma_{11} + \theta_{12} + b_3c_{12},$$
$$b_4b_7 = b_4 + \sigma_{11} + \cdots,$$
$$b_5b_7 = \sigma_{11} + \theta_{12} + \cdots.$$

Thus, we must have $s_4, s_9, s_{12} \in B$ such that

$$b_3c_{12} = \sigma_{11} + s_{12} + s_9 + s_4,$$
$$b_4b_7 = b_4 + \sigma_{11} + s_9 + s_4, \tag{3.83}$$
$$b_5b_7 = \sigma_{11} + \theta_{12} + s_{12}.$$

Clearly $s_4, s_9,$ and s_{12} are all real, and

$$\bar{b}_3s_4 = c_{12}.$$

Furthermore, since we have already shown that $(b_3c_{15}, b_3c_{15}) = 4$, and that $(b_3c_{15}$ and $b_3\bar{c}_{12}$ have two common components, we must have that

$$b_3c_{15} = \theta_{12} + s_{12} + s_9 + e_{12}, \quad e_{12} \in B, \text{ and } \theta_{12}, s_{12}, e_{12} \text{ are distinct.} \quad (3.84)$$

Thus, $\bar{b}_3s_9 = c_{12} + c_{15}$.

Now from $b_3^2 s_4 = b_3(\bar{b}_3 s_4)$, $\bar{b}_3 s_4 = c_{12}$, and Lemma 3.8, we obtain that

$$b_4s_4 + b_5s_4 = b_7 + p_8 + t_9 + t_{12}.$$

Hence we must have that

$$b_4s_4 = b_7 + t_9 \quad \text{and} \quad b_5s_4 = p_8 + t_{12}.$$

So the lemma holds.   □

Lemma 3.11 *There exist $r_4, r_5 \in B$ such that*

$$\bar{b}_3 t_{12} = \bar{c}_{12} + \bar{c}_{15} + r_4 + r_5.$$

Proof It follows from $(b_3\bar{b}_3)s_4 = b_3(\bar{b}_3 s_4)$, (3.1), and Lemma 3.10 that

$$s_4b_8 = \sigma_{11} + s_{12} + s_9.$$

Thus, from (3.6), Lemma 3.10, (3.83), and $b_4(b_4s_4) = b_4^2 s_4$, we see that

$$s_4b_7 + s_{12} = b_4 + b_4t_9.$$

But $\bar{b}_3^2 t_9 = \bar{b}_3(\bar{b}_3 t_9)$, (3.82), and (3.84) yield that

$$b_4t_9 + b_5t_9 = \sigma_{11} + \theta_{12} + 2s_9 + 2s_{12} + s_4 + \bar{e}_{12}.$$

So e_{12} is real, and we must have that

$$b_4t_9 = \sigma_{11} + s_4 + s_{12} + s_9,$$
$$b_5t_9 = \theta_{12} + s_{12} + e_{12} + s_9,$$
$$s_4b_7 = b_4 + s_4 + s_9 + \sigma_{11}.$$

Hence from $\bar{b}_3(s_4b_7) = b_7(\bar{b}_3 s_4)$ we see that

$$b_7c_{12} = b_3 + 2x_9 + 3c_{12} + c_{15} + y_{12}.$$

Note that it follows from $(\bar{b}_3 b_4)s_4 = \bar{b}_3(b_4s_4)$ and (3.82) that

$$s_4x_9 = \bar{x}_9 + \bar{c}_{12} + \bar{c}_{15}.$$

Thus, $(b_7\bar{b}_3)s_4 = b_7(\bar{b}_3 s_4)$ yields that

$$s_4\bar{c}_{12} = b_3 + x_9 + 2c_{12} + y_{12}.$$

Therefore, $(\bar{b}_3 s_4)\bar{c}_{12} = \bar{b}_3(s_4 \bar{c}_{12})$ implies that

$$c_{12}\bar{c}_{12} = 1 + 3b_7 + 3b_8 + 2b_{12} + 3p_8 + q_8 + 2t_9 + 2t_{12}.$$

On the other hand, from $\bar{b}_3(b_5 s_4) = b_5(\bar{b}_3 s_4)$ and (3.82), we get that

$$b_5 \bar{c}_{12} = \bar{y}_{12} + 2\bar{c}_{12} + \bar{c}_{15} + \Sigma_9.$$

Since $(b_5 \bar{c}_{12}, b_5 \bar{c}_{12}) = (b_5^2, c_{12}\bar{c}_{12}) = 8$, the above equality forces that $(\Sigma_9, \Sigma_9) = 2$. But the degree of any component in Σ_9 has to be at least 4 by (3.82). Thus, $\Sigma_9 = r_4 + r_5$, $r_4, r_5 \in B$, and hence by (3.82),

$$\bar{b}_3 t_{12} = \bar{c}_{12} + \bar{c}_{15} + r_4 + r_5.$$

So the lemma holds.   □

Now we are ready to reach a contradiction for Subcase 1B. It follows from Lemma 3.11 that $b_3 r_4 = t_{12}$. Thus, $b_3^2 r_4 = b_3(b_3 r_4)$ implies that

$$b_4 r_4 + b_5 r_4 = c_{12} + c_{15} + \bar{r}_4 + \bar{r}_5.$$

Therefore,

$$b_4 r_4 = c_{12} + \bar{r}_4 \quad \text{and} \quad b_5 r_4 = c_{15} + \bar{r}_5.$$

So

$$(b_4 c_{12}, r_4) = 1. \tag{3.85}$$

But, on the other hand, $b_4(\bar{b}_3 s_4) = \bar{b}_3(b_4 s_4)$ forces that

$$b_4 c_{12} = \bar{x}_9 + 2\bar{c}_{12} + \bar{c}_{15},$$

a contradiction to (3.85).

3.4 Cases $z = z_4$, $z = z_5$, $z = z_6$, $z = z_7$, and $z = z_8$

In this section we derive contradictions for the cases $z = z_4$, $z = z_6$, $z = z_6$, $z = z_7$, and $z = z_8$.

Case 2 $z = z_4$.

So we have that

$$z = z_4, \qquad w = w_8, \qquad \bar{b}_3 z_4 = b_4 + b_5 + \sigma_3, \qquad \bar{b}_4 w_8 = b_4 + \theta_{20}.$$

Hence $(b_3 \bar{b}_3, z_4 \bar{z}_4) = (\bar{b}_3 z_4, \bar{b}_3 z_4) = 3$. So $(b_8, z_4 \bar{z}_4) = 2$, a contradiction.

Case 3 $z = z_5$.

In this case we have the following:

$$z = z_5, \qquad w = w_7, \qquad \bar{b}_3 z_5 = b_4 + b_5 + \sigma_6, \qquad \bar{b}_3 w_7 = b_4 + \theta_{17}. \quad (3.86)$$

Note that

$$z_5 \bar{b}_4 + w_7 \bar{b}_4 = 2b_3 + 2x_9 + y_{12} + \Sigma_{12}^*$$

by (3.40). Since $(z_5 \bar{b}_4, b_3) = 1$ and $(w_7 \bar{b}_4, b_3) = 1$ by (3.11), and any constituent of Σ_{12}^* has degree at least 4 by (3.38), we have either $\Sigma_{12}^* = \alpha_5 + \alpha_7$, $\alpha_5, \alpha_7 \in B$, and

$$z_5 \bar{b}_4 = b_3 + y_{12} + \alpha_5, \qquad w_7 \bar{b}_4 = b_3 + 2x_9 + \alpha_7,$$

or $\Sigma_{12}^* = \alpha_4 + \Sigma_8$, $\alpha_4 \in B$, $\Sigma_8 \in \mathbb{N}(B)$, and

$$z_5 \bar{b}_4 = b_3 + x_9 + \Sigma_8, \qquad w_7 \bar{b}_4 = b_3 + x_9 + y_{12} + \alpha_4.$$

If $w_7 \bar{b}_4 = b_3 + 2x_9 + \alpha_7$, then by (3.1) and (3.35),

$$(w_7 \bar{b}_4) \bar{b}_3 = 1 + 2b_7 + 3b_8 + 2b_{12} + \bar{b}_3 \alpha_7.$$

Hence $((w_7 \bar{b}_4) \bar{b}_3, b_8) \geq 3$. But if $w_7 \bar{b}_4 = b_3 + x_9 + y_{12} + \alpha_4$, then by (3.1) and (3.35),

$$(w_7 \bar{b}_4) \bar{b}_3 = 1 + b_7 + 2b_8 + b_{12} + \bar{b}_3 y_{12} + \bar{b}_3 \alpha_4.$$

Since $(\bar{b}_3 y_{12}, b_8) = 1$ by (3.5), we also have that $((w_7 \bar{b}_4) \bar{b}_3, b_8) \geq 3$. That is, we always have that

$$((w_7 \bar{b}_4) \bar{b}_3, b_8) \geq 3.$$

On the other hand, (3.86) and (3.6) imply that

$$(\bar{b}_3 w_7) \bar{b}_4 = b_4 \bar{b}_4 + \theta_{17} \bar{b}_4 = 1 + b_7 + b_8 + \theta_{17} \bar{b}_4.$$

Thus, $(\theta_{17} \bar{b}_4, b_8) \geq 2$. Hence $(b_4 b_8, \theta_{17}) \geq 2$, a contradiction.

Case 4 $z = z_6$.

So we have the following:

$$z = z_6, \qquad w = w_6, \qquad \bar{b}_3 z_6 = b_4 + b_5 + \sigma_9, \qquad \bar{b}_3 w_6 = b_4 + \theta_{14}. \quad (3.87)$$

Note that

$$z_6 \bar{b}_4 + w_6 \bar{b}_4 = 2b_3 + 2x_9 + y_{12} + \Sigma_{12}^* \qquad (3.88)$$

and

$$z_6 \bar{b}_4 + u_9 \bar{b}_4 = b_3 + x_9 + y_{12} + b_3 \bar{b}_{12} \qquad (3.89)$$

by (3.40) and (3.41), respectively. Since $(z_6\bar{b}_4, b_3) = 1$ and $(w_6\bar{b}_4, b_3) = 1$ by (3.11), and the degree of any constituent of Σ_{12}^* is at least 4 by (3.38), from (3.88) we have either

$$z_6\bar{b}_4 = b_3 + x_9 + y_{12}, \qquad w_6\bar{b}_4 = b_3 + x_9 + \Sigma_{12}^*$$

or

$$z_6\bar{b}_4 = b_3 + x_9 + \Sigma_{12}^*, \qquad w_6\bar{b}_4 = b_3 + x_9 + y_{12}.$$

Subcase 4A

$$z_6\bar{b}_4 = b_3 + x_9 + y_{12} \quad \text{and} \quad w_6\bar{b}_4 = b_3 + x_9 + \Sigma_{12}^*. \tag{3.90}$$

First we prove that $\Sigma_{12}^* \in B$. Toward a contradiction, assume that $\Sigma_{12}^* \notin B$. Then by (3.38),

$$(\bar{b}_3\Sigma_{12}^*, b_7) \geq 2. \tag{3.91}$$

Since $(\bar{b}_3 w_6)\bar{b}_4 = 1 + b_7 + b_8 + \bar{b}_4\theta_{14}$ by (3.87) and (3.6), and $(\bar{b}_4 w_6)\bar{b}_3 = 1 + b_7 + 2b_8 + b_{12} + \bar{b}_3\Sigma_{12}^*$ by (3.90), (3.1), and (3.35), we see that

$$\bar{b}_4\theta_{14} = b_8 + b_{12} + \bar{b}_3\Sigma_{12}^*.$$

Thus, $(\bar{b}_4\theta_{14}, b_7) \geq 2$ by (3.91), and hence $(b_4 b_7, \theta_{14}) \geq 2$. Therefore, we must have that $(b_4 b_7, \theta_{14}) = 2$, and $b_4 b_7 = 2\theta_{14}$. But, on the other hand, we also have $(b_4 b_7, b_4) = 1$ by (3.6), a contradiction. This proves that $\Sigma_{12}^* \in B$.

So we may assume that $\Sigma_{12}^* = c_{12} \in B$. Thus by (3.38), (3.39), and (3.90),

$$b_3 b_7 = x_9 + c_{12}, \qquad b_4\bar{x}_9 = b_3 + y_{12} + x_9 + c_{12}, \qquad w_6\bar{b}_4 = b_3 + x_9 + c_{12}. \tag{3.92}$$

From (3.4), (3.2), (3.12), (3.87), and $(b_3 b_5)\bar{b}_3 = (\bar{b}_3 b_5)b_3$, we get that

$$b_3 y_{12} = \sigma_9 + \bar{b}_3 u_9. \tag{3.93}$$

Hence $(\bar{b}_3\sigma_9, y_{12}) \geq 1$. But we also have $(\bar{b}_3\sigma_9, x_9) = 1$ by (3.42). So we may assume that $\bar{b}_3\sigma_9 = x_9 + y_{12} + v_6$, $v_6 \in \mathbb{N}(B)$. Hence $(b_3 x_9)\bar{b}_3 = b_3 + 2x_9 + y_{12} + v_6 + \bar{b}_3\theta_{14}$ by (3.3) and (3.95). But $(\bar{b}_3 x_9)b_3 = b_3 + 2x_9 + y_{12} + c_{12} + b_3 b_{12}$ by (3.35), (3.92), and (3.5). So we have that

$$v_6 + \bar{b}_3\theta_{14} = c_{12} + b_3 b_{12}. \tag{3.94}$$

Note that any constituent of $b_3 b_{12}$ has degree at least 4. So the above equality forces that $v_6 \in B$. Thus,

$$\bar{b}_3\sigma_9 = x_9 + y_{12} + v_6, \quad v_6 \in B. \tag{3.95}$$

Now (3.87), (3.3), (3.4), and (3.95) yield that $\bar{b}_3(\bar{b}_3 z_6) = 2b_3 + 2x_9 + 2y_{12} + v_6$, and (3.2) and (3.90) imply that $(\bar{b}_3\bar{b}_3)z_6 = b_3 + x_9 + y_{12} + \bar{b}_5 z_6$. Thus,

$$\bar{b}_5 z_6 = b_3 + x_9 + y_{12} + v_6. \tag{3.96}$$

Since $(b_3\bar{b}_3, z_6\bar{z}_6) = (\bar{b}_3 z_6, \bar{b}_3 z_6) = 3$ by (3.87), and $(b_4\bar{b}_5, z_6\bar{z}_6) = (\bar{b}_4 z_6, \bar{b}_5 z_6) = 3$ by (3.90) and (3.96), we see that

$$(z_6\bar{z}_6, b_8) = 2 \quad \text{and} \quad (z_6\bar{z}_6, b_{12}) = 1.$$

Hence we also have that $(z_6\bar{z}_6, \bar{b}_{12}) = 1$. Therefore,

$$b_{12} \quad \text{is a real element.}$$

Recall that $(b_3 b_{12}, x_9) = 1$ by (3.35), and b_{12} real implies that $(b_3 b_{12}, y_{12}) \geq 1$ by (3.37). Thus, since any constituent of $\bar{b}_3\theta_{14}$ has degree at least 6, from (3.94) we see that there is $v_9 \in B$ such that

$$b_3 b_{12} = x_9 + y_{12} + v_6 + v_9 \tag{3.97}$$

and

$$\bar{b}_3\theta_{14} = x_9 + y_{12} + c_{12} + v_9. \tag{3.98}$$

Since $(b_3 y_{12}, b_5) = 1$ by (3.4), $(b_3 y_{12}, \sigma_9) = 1$ by (3.95), and $(b_3 y_{12}\sigma_{14}) = 1$ by (3.98), we must have that $(b_3 y_{12}, \theta_{14}) = 1$ by (3.98). Thus,

$$b_3 y_{12} = b_5 + \sigma_9 + \theta_{14} + t_8, \quad t_8 \in \mathbb{N}(B). \tag{3.99}$$

So by (3.93),

$$\bar{b}_3 u_9 = b_5 + \theta_{14} + t_8. \tag{3.100}$$

If $t_8 \notin B$, since any constituent of $b_3 y_{12}$ has degree at least 4, we see that $t_8 = c_4 + d_4$, $c_4, d_4 \in B$. Hence (3.99) forces that $\bar{b}_3 c_4 = y_{12}$. Thus, $(b_3 c_4, b_3 c_4) = (\bar{b}_3 c_4, \bar{b}_3 c_4) = 1$. But, on the other hand, we also have $(b_3 c_4, u_9) = 1$ by (3.100), a contradiction. Therefore, we have proved that

$$t_8 \in B.$$

Since b_{12} is real, (3.89), (3.90), and (3.97) yield that $u_9\bar{b}_4 = b_3 b_{12} = x_9 + y_{12} + v_6 + v_9$. Hence by (3.2),

$$\bar{b}_3^2 u_9 = x_9 + y_{12} + v_6 + v_9 + \bar{b}_5 u_9.$$

On the other hand, from (3.100), (3.4), and (3.98), we get that

$$\bar{b}_3(\bar{b}_3 u_9) = b_3 + x_9 + 2y_{12} + c_{12} + v_9 + \bar{b}_3 t_8.$$

Thus, $(\bar{b}_3 t_8, v_6) = 1$. So $(b_3 v_6, t_8) = 1$. But we also have $(b_3 v_6, \sigma_9) = 1$ by (3.95). Hence

$$b_3 v_6 = t_8 + \sigma_9 + \cdots.$$

Since $L_1(B) = \{1\}$, the above equality cannot be true, a contradiction.

Subcase 4B

$$z_6\bar{b}_4 = b_3 + x_9 + \Sigma_{12}^* \quad \text{and} \quad w_6\bar{b}_4 = b_3 + x_9 + y_{12}. \tag{3.101}$$

So by (3.1), (3.35), and (3.101), $(w_6\bar{b}_4)\bar{b}_3 = 1 + b_7 + 2b_8 + b_{12} + \bar{b}_3 y_{12}$. But $(\bar{b}_3 w_6)\bar{b}_4 = 1 + b_7 + b_8 + \bar{b}_4\theta_{14}$ by (3.87) and (3.6). Hence

$$\bar{b}_4\theta_{14} = b_8 + b_{12} + \bar{b}_3 y_{12}.$$

Since $(\bar{b}_3 y_{12}, b_8) = 1$ by (3.5), we see that $(\bar{b}_4\theta_{14}, b_8) = 2$. Thus,

$$(b_4 b_8, \theta_{14}) = 2.$$

But, on the other hand, from $(b_3\bar{b}_3)b_4 = (\bar{b}_3 b_4)b_3$, (3.1), (3.3), (3.2), and (3.42), we see that

$$b_4 b_8 = b_5 + b_3 x_9 = b_4 + b_5 + \sigma_9 + \theta_{14},$$

a contradiction.

Case 5 $z = z_7$.

In this case we assume that

$$z = z_7, \qquad w = w_5, \qquad \bar{b}_3 z_7 = b_4 + b_5 + \sigma_{12}, \quad \text{and} \quad \bar{b}_3 w_5 = b_4 + \theta_{11}. \tag{3.102}$$

Recall that any constituent of Σ_{12}^* has degree at least 4 by (3.38). So from (3.40), we have either $\Sigma_{12}^* = \alpha_4 + \Sigma_8, \alpha_4 \in B, \Sigma_8 \in \mathbb{N}(B)$, and

$$z_7\bar{b}_4 = b_3 + x_9 + y_{12} + \alpha_4, \qquad w_5\bar{b}_4 = b_3 + x_9 + \Sigma_8 \tag{3.103}$$

or $\Sigma_{12}^* = \alpha_5 + \alpha_7, \alpha_5, \alpha_7 \in B$, and

$$z_7\bar{b}_4 = b_3 + 2x_9 + \alpha_7, \qquad w_5\bar{b}_4 = b_3 + y_{12} + \alpha_5. \tag{3.104}$$

If (3.103) holds, then $\bar{b}_3(z_7\bar{b}_4) = 1 + b_7 + 2b_8 + b_{12} + \bar{b}_3 y_{12} + \bar{b}_3\alpha_4$ by (3.1) and (3.35), and $(\bar{b}_3 z_7)\bar{b}_4 = 1 + b_7 + 2b_8 + \bar{b}_{12} + \bar{b}_4\sigma_{12}$ by (3.102), (3.6), and (3.36). Thus,

$$\bar{b}_{12} + \bar{b}_4\sigma_{12} = b_{12} + \bar{b}_3 y_{12} + \bar{b}_3\alpha_4.$$

Since $(\bar{b}_3\alpha_4, b_7) = 1$ by (3.38), the above equality implies that $(\bar{b}_4\sigma_{12}, b_7) = 1$. Hence

$$(b_4 b_7, \sigma_{12}) = 1.$$

On the other hand, we also have that $(\bar{b}_3 w_5)\bar{b}_4 = 1 + b_7 + b_8 + \bar{b}_4\theta_{11}$ by (3.102) and (3.6), and $(w_5\bar{b}_4)\bar{b}_3 = 1 + b_7 + 2b_8 + b_{12} + \bar{b}_3\Sigma_8$ by (3.103), (3.1), and (3.35). So

$$\bar{b}_4\theta_{11} = b_8 + b_{12} + \bar{b}_3\Sigma_8.$$

Since $(\bar{b}_3 \Sigma_8, b_7) \geq 1$ by (3.38), the above equality implies that $(\bar{b}_4 \theta_{11}, b_7) \geq 1$. Hence $(b_4 b_7, \theta_{11}) \geq 1$. But we already have $(b_4 b_7, \sigma_{12}) = 1$. So $(b_4 b_7, \theta_{11}) = 1$. Note that $(b_4 b_7, b_4) = 1$ by (3.6). Thus,

$$b_4 b_7 = b_4 + \theta_{11} + \sigma_{12} + \cdots .$$

Since $L_1(B) = \{1\}$, the above equality cannot be true, a contradiction.

   If (3.104) holds, then $(b_4 x_9, z_7) = 2$. Recall that $(b_4 x_9, x_9) = 1$ by (3.17). So $b_4 x_9 = 2 z_7 + x_9 + \cdots$, and hence $(b_4 x_9, b_4 x_9) \geq 6$. But, on the other hand, since $\bar{b}_4 x_9 = b_3 + x_9 + y_{12} + \alpha_5 + \alpha_7$ by (3.17), we see that $(b_4 x_9, b_4 x_9) = (\bar{b}_4 x_9, \bar{b}_4 x_9) = 5$, a contradiction.

Case 6 $z = z_8$.

   In this case we assume that

$$z = z_8, \qquad w = w_4, \qquad \bar{b}_3 z_8 = b_4 + b_5 + \sigma_{15}, \quad \text{and} \quad \bar{b}_3 w_4 = b_4 + \theta_8. \quad (3.105)$$

Recall that by (3.40),

$$z_8 \bar{b}_4 + w_4 \bar{b}_4 = 2 b_3 + 2 x_9 + y_{12} + \Sigma_{12}^*.$$

Since $(w_4 \bar{b}_4, b_3) = 1$ by (3.11), and $L_1(B) = \{1\}$, the above equality forces that

$$\Sigma_{12}^* = \alpha_4 + \Sigma_8, \qquad \alpha_4 \in B, \ \Sigma_8 \in \mathbb{N}(B), \qquad\qquad (3.106)$$

$$w_4 \bar{b}_4 = b_3 + x_9 + \alpha_4, \qquad\qquad (3.107)$$

and

$$z_8 \bar{b}_4 = b_3 + x_9 + y_{12} + \Sigma_8. \qquad\qquad (3.108)$$

Thus, from (3.105), (3.6), (3.8), (3.108), (3.1), (3.9), and $(\bar{b}_3 z_8) \bar{b}_4 = (z_8 \bar{b}_4) \bar{b}_3$, we get that

$$\bar{b}_{12} + \bar{b}_4 \sigma_{15} = b_{12} + \bar{b}_3 y_{12} + \bar{b}_3 \Sigma_8.$$

But $(\bar{b}_3 \Sigma_8, b_7) \geq 1$ by (3.38) and (3.106). So $(\bar{b}_4 \sigma_{15}, b_7) \geq 1$, and hence $(b_4 b_7, \sigma_{15}) \geq 1$. Therefore, we must have that

$$(b_4 b_7, \sigma_{15}) = 1. \qquad\qquad (3.109)$$

On the other hand, $(\bar{b}_3 w_4) \bar{b}_4 = (\bar{b}_4 w_4) \bar{b}_3$ yields that

$$\bar{b}_4 \theta_8 = b_8 + b_{12} + \bar{b}_3 \alpha_4,$$

by (3.105), (3.107), (3.6), (3.1), and (3.35). Since $(\bar{b}_3 \alpha_4, b_7) = 1$ by (3.38) and (3.106), the above equality implies that $(\bar{b}_4 \theta_8, b_7) = 1$. Hence

$$(b_4 b_7, \theta_8) = 1. \qquad\qquad (3.110)$$

But we also have that $(b_4 b_7, b_4) = 1$ by (3.6). So by (3.109) and (3.110),

$$b_4 b_7 = b_4 + \theta_8 + \sigma_{15} + \cdots .$$

However, $L_1(B) = \{1\}$ implies that the above equality cannot be true, a contradiction.

3.5 Case $z = z_9$

In this section we assume the following

$$z = z_9, \qquad w = w_3, \qquad u = u_6, \qquad \bar{b}_3 z_9 = b_4 + b_5 + \sigma_{18}, \qquad \bar{b}_3 w_3 = b_4 + \theta_5.$$
$$\text{(3.111)}$$

Since $(b_3 \bar{b}_3, w_3 \bar{w}_3) = (\bar{b}_3 w_3, \bar{b}_3 w_3) = 2$ by (3.111), we see that

$$w_3 \bar{w}_3 = 1 + b_8. \tag{3.112}$$

Thus, from $(b_3 \bar{w}_3) w_3 = b_3 (w_3 \bar{w}_3)$, (3.111), and (3.5), we see that $\bar{b}_4 w_3 + \bar{\theta}_5 w_3 = 2 b_3 + x_9 + y_{12}$. Note that $(\bar{b}_4 w_3, b_3) = 1$ and $(\bar{\theta}_5 w_3, b_3) = 1$ by (3.111). So we must have that

$$\bar{b}_4 w_3 = b_3 + x_9 \quad \text{and} \quad \bar{\theta}_5 w_3 = b_3 + y_{12}.$$

Thus, (3.112) yields that

$$(b_3 \theta_5, b_3 \theta_5) = (b_3 \bar{b}_3, \theta_5 \bar{\theta}_5) = (w_3 \bar{w}_3, \theta_5 \bar{\theta}_5) = (\bar{\theta}_5 w_3, \bar{\theta}_5 w_3) = 2.$$

So we may set

$$b_3 \theta_5 = w_3 + \gamma_{12}, \qquad \gamma_{12} \in B. \tag{3.113}$$

Furthermore, since $(b_3 w_3, b_3 w_3) = (\bar{b}_3 w_3, \bar{b}_3 w_3) = 2$ by (3.111), we may assume that $b_3 w_3 = \alpha + \beta$, $\alpha, \beta \in B$ and $|\alpha| < |\beta|$. Then from (3.11), (3.111), (3.113), and $\bar{b}_3 (b_3 w_3) = b_3 (\bar{b}_3 w_3)$, we get that

$$\bar{b}_3 \alpha + \bar{b}_3 \beta = 2 w_3 + z_9 + \gamma_{12}. \tag{3.114}$$

The above equality forces that either $\alpha = \alpha_4$ and $\beta = \beta_5$, or $\alpha = 1$ and $\beta = \beta_8$.

Subcase 7A $\alpha = \alpha_4$ and $\beta = \beta_5$.

Thus,

$$b_3 w_3 = \alpha_4 + \beta_5, \tag{3.115}$$

and by (3.114),

$$\bar{b}_3 \alpha_4 = w_3 + z_9, \tag{3.116}$$

and

$$\bar{b}_3\beta_5 = w_3 + \gamma_{12}. \tag{3.117}$$

Therefore, from (3.115), (3.112), (3.5), and $(b_3w_3)\bar{w}_3 = b_3(w_3\bar{w}_3)$, we get that

$$\alpha_4\bar{w}_3 + \beta_5\bar{w}_3 = 2b_3 + x_9 + y_{12}.$$

Since $(\alpha_4\bar{w}_3, b_3) = 1$ by (3.116), and $(\beta_5\bar{w}_3, b_3) = 1$ by (3.117), the above equality yields that

$$\alpha_4\bar{w}_3 = b_3 + x_9 \quad \text{and} \quad \beta_5\bar{w}_3 = b_3 + y_{12}.$$

Thus, (3.111), (3.2), (3.42), and $\alpha_4(\bar{w}_3 b_3) = (\alpha_4\bar{w}_3)b_3$ force that

$$\alpha_4\bar{b}_4 + \alpha_4\bar{\theta}_5 = 2b_4 + b_5 + \sigma_{18} + \theta_5.$$

However, the above equality cannot be true, a contradiction.

Subcase 7B $\alpha = 1$ and $\beta = \beta_8$.

Thus, we must have that $w_3 = \bar{b}_3$ and $\beta_8 = b_8$. So (3.114) yields that $\bar{b}_3 b_8 = \bar{b}_3 + z_9 + \gamma_{12}$. Hence (3.5) forces that

$$z_9 = \bar{x}_9 \quad \text{and} \quad \gamma_{12} = \bar{y}_{12}.$$

Therefore,

$$b_3 b_4 = \bar{x}_9 + \bar{b}_3 \quad \text{and} \quad b_3 b_5 = \bar{x}_9 + u_6 \tag{3.118}$$

by (3.11) and (3.12), respectively. Furthermore, from (3.111) we see that $\bar{b}_3\bar{x}_9 = b_4 + b_5 + \sigma_{18}$. But, on the other hand, $b_3 x_9 = b_4 + \sigma_{18} + \theta_5$ by (3.42). So we must have that

$$b_4 = \bar{b}_4, \qquad \theta_5 = \bar{b}_5, \qquad \sigma_{18} = \bar{\sigma}_{18}. \tag{3.119}$$

Thus, (3.42) implies that

$$b_3 x_9 = b_4 + \bar{b}_5 + \sigma_{18}. \tag{3.120}$$

Note that $\bar{b}_3(b_3 x_9) = b_3 + 2x_9 + \bar{u}_6 + \bar{b}_3\sigma_{18}$ by (3.120), (3.118), and (3.3). On the other hand, $b_3(\bar{b}_3 x_9) = b_3 + 2x_9 + y_{12} + \Sigma_{12}^* + b_3 b_{12}$ by (3.35), (3.38), and (3.5). Thus,

$$\bar{u}_6 + \bar{b}_3\sigma_{18} = y_{12} + \Sigma_{12}^* + b_3 b_{12}. \tag{3.121}$$

The above equality forces that $(b_3 b_{12}, \bar{u}_6) \geq 1$. Thus, $(b_3 u_6, \bar{b}_{12}) = 1$, and hence $b_3 u_6 = \bar{b}_{12} + t_6$, $t_6 \in \mathbb{N}(B)$. If $t_6 \notin B$, then $t_6 = c_3 + d_3$, $c_3, d_3 \in B$, and $b_3 u_6 = \bar{b}_{12} + c_3 + d_3$. Therefore, $(\bar{b}_3 c_3, u_6) = 1$ and $(\bar{b}_3 d_3, u_6) = 1$. So $(\bar{b}_3 c_3, \bar{b}_3 d_3) \geq 1$. Hence $(b_3\bar{b}_3, c_3\bar{d}_3) = (\bar{b}_3 c_3, \bar{b}_3 d_3)$ implies that

$$(b_3\bar{b}_3, c_3\bar{d}_3) \geq 1.$$

Recall that $L_1(B) = \{1\}$. So the above equality forces that $c_3 = d_3$. Thus, $\bar{b}_3 u_6 = \bar{b}_{12} + 2c_3$, and hence $(c_3\bar{b}_3, u_6) = (\bar{b}_3 u_6, c_3) = 2$, a contradiction. This proves that

$$\bar{b}_3 u_6 = \bar{b}_{12} + t_6, \quad t_6 \in B. \tag{3.122}$$

Note that $(\bar{b}_3 u_6, b_5) = 1$ by (3.118), and $(\bar{b}_3 u_6, \bar{b}_3 u_6) = (\bar{b}_3 u_6, b_3 u_6) = 2$ by (3.122). So we may set

$$\bar{b}_3 u_6 = b_5 + s_{13}, \quad s_{13} \in B. \tag{3.123}$$

From (3.118), (3.119), (3.120), and (3.123), we get that $\bar{b}_3(b_3 b_5) = b_4 + 2b_5 + \sigma_{18} + s_{13}$. But we also have $b_3(\bar{b}_3 b_5) = b_4 + b_5 + b_3 y_{12}$ by (3.4) and (3.2). Thus,

$$b_3 y_{12} = \sigma_{18} + b_5 + s_{13}. \tag{3.124}$$

Therefore, $(\bar{b}_3 y_{12}, \bar{b}_3 y_{12}) = (b_3 y_{12}, b_3 y_{12}) = 3$. Recall that $1 + \bar{b}_3 y_{12} = \bar{b}_{12} + b_5 \bar{b}_5$ by (3.37), and $(\bar{b}_3 y_{12}, b_8) = 1$ by (3.5). So we may assume that

$$\bar{b}_3 y_{12} = \bar{b}_{12} + b_8 + r_{16}, \quad r_{16} \in B. \tag{3.125}$$

Hence by (3.37),

$$b_5 \bar{b}_5 = 1 + b_8 + r_{16}. \tag{3.126}$$

Therefore, from $(b_3 \bar{b}_3) b_8 = \bar{b}_3 (b_3 b_8)$, (3.1), (3.5), (3.35), and (3.125), we get that

$$b_8^2 = 1 + b_7 + 2b_8 + b_{12} + \bar{b}_{12} + r_{16}. \tag{3.127}$$

Furthermore, $(b_3 \bar{b}_3) b_4 = (\bar{b}_3 b_4) b_3$ yields that

$$b_4 b_8 = b_4 + b_5 + \bar{b}_5 + \sigma_{18} \tag{3.128}$$

by (3.1), (3.3), (3.2), and (3.120). Also from (3.1), (3.4), (3.2), (3.124), and $(b_3 \bar{b}_3) b_5 = (\bar{b}_3 b_5) b_3$, we see that

$$b_5 b_8 = b_4 + b_5 + \sigma_{18} + s_{13}. \tag{3.129}$$

Note that $(b_4 \bar{b}_5, b_8^2) = (b_4 b_8, b_5 b_8) = 3$ by (3.128) and (3.129). So (3.36) and (3.127) force that

$$b_{12} \neq \bar{b}_{12}. \tag{3.130}$$

Moreover, it follows from $b_3^2 b_5 = b_3(b_3 b_5)$, (3.2), (3.119), (3.36), (3.118), (3.35), and (3.122) that

$$b_5^2 = b_7 + \bar{b}_{12} + t_6. \tag{3.131}$$

Now we claim that $\Sigma_{12}^* \in B$. If $\Sigma_{12}^* \notin B$, then $(\bar{b}_3 \Sigma_{12}^*, b_7) \geq 2$ by (3.38). Since $z_9 = \bar{x}_9$ and $b_4 = \bar{b}_4$, we have that

$$b_3 \bar{b}_{12} = \Sigma_{12}^* + u_6 b_4 \tag{3.132}$$

by (3.41) and (3.39). Thus, $\Sigma_{12}^* \notin B$ forces that $(\bar{b}_3 \Sigma_{12}^*, \bar{b}_{12}) \geq 2$. Hence $\bar{b}_3 \Sigma_{12}^* = 2b_7 + 2\bar{b}_{12} + \cdots$, a contradiction. Therefore, we must have that $\Sigma_{12}^* \in B$. So we may assume that $\Sigma_{12}^* = c_{12} \in B$. Thus,

$$b_3 b_7 = x_9 + c_{12} \tag{3.133}$$

by (3.38), and

$$b_4 \bar{x}_9 = b_3 + x_9 + y_{12} + c_{12} \tag{3.134}$$

by (3.39). From $\bar{b}_3(b_4 \bar{x}_9) = b_4(\bar{b}_3 \bar{x}_9)$, (3.134), (3.1), (3.35), (3.125), (3.120), (3.119), (3.6), and (3.36), we get that

$$b_4 \sigma_{18} = b_8 + b_{12} + r_{16} + \bar{b}_3 c_{12}. \tag{3.135}$$

Since $(\bar{b}_3 c_{12}, b_7) = 1$ by (3.133), the above equality implies that $(b_4 \sigma_{18}, b_7) = 1$. Thus, $(b_4 b_7, \sigma_{18}) = 1$ by (3.119). But we also have $(b_4 b_7, b_4) = 1$ by (3.6). Hence $b_4 b_7 = \sigma_{18} + b_4 + v_6$, $v_6 \in \mathbb{N}(B)$. By $b_3^2 b_7 = b_3(b_3 b_7)$, (3.2), (3.133), and (3.120), we see that

$$b_4 b_7 + b_5 b_7 = b_4 + \bar{b}_5 + \sigma_{18} + b_3 c_{12}. \tag{3.136}$$

Since any constituent of $b_3 c_{12}$ has degree at least 4, the above equality forces that $v_6 \in B$. Thus

$$b_4 b_7 = \sigma_{18} + b_4 + v_6, \quad v_6 \in B. \tag{3.137}$$

Note that $(b_3 \bar{b}_3, u_6 \bar{u}_6) = (b_3 u_6, b_3 u_6) = 2$ by (3.122). So $(u_6 \bar{u}_6, b_8) = 1$. Hence we must have that $(b_5 \bar{b}_5, u_6 \bar{u}_6) \leq 3$ by (3.126). Thus, $(b_5 u_6, b_5 u_6) = (b_5 \bar{b}_5, u_6 \bar{u}_6) \leq 3$. But from (3.122), (3.2), (3.132), and $b_3^2 u_6 = b_3(b_3 u_6)$, we get that

$$b_5 u_6 = c_{12} + b_3 t_6. \tag{3.138}$$

Since $(\bar{b}_3 t_6, u_6) = 1$ by (3.122), $(\bar{b}_3 t_6, \bar{b}_3 t_6) \geq 2$. Hence $(b_3 t_6, b_3 t_6) = (\bar{b}_3 t_6, \bar{b}_3 t_6) \geq 2$. So $(b_5 u_6, b_5 u_6) \geq 3$ by (3.138). Thus, we must have that $(b_5 u_6, b_5 u_6) = 3$, and $(b_3 t_6, b_3 t_6) = 2$. Hence $(\bar{b}_3 t_6, \bar{b}_3 t_6) = 2$, and by (3.122) we may set

$$\bar{b}_3 t_6 = u_6 + d_{12}, \quad d_{12} \in B. \tag{3.139}$$

From (3.2), (3.133), (3.120), (3.137), and $b_3^2 b_7 = b_3(b_3 b_7)$, we see that

$$v_6 + b_5 b_7 = \bar{b}_5 + b_3 c_{12}.$$

Hence $(\bar{b}_3 v_6, c_{12}) = (v_6, b_3 c_{12}) \geq 1$. Thus, we may assume that $\bar{b}_3 v_6 = c_{12} + p_6$, $p_6 \in \mathbb{N}(B)$. As in the proof of (3.122), we can prove that $p_6 \in B$. Therefore,

$$\bar{b}_3 v_6 = c_{12} + p_6, \quad p_6 \in B. \tag{3.140}$$

Note that $\bar{b}_3(b_4 b_7) = (\bar{b}_3 b_4) b_7$ implies that

$$b_7 x_9 = \bar{b}_3 \sigma_{18} + b_3 + p_6 \tag{3.141}$$

by (3.137), (3.3), (3.140), and (3.133). Since $(\bar{b}_3\sigma_{18}, x_9) = 1$ by (3.120), $(b_7 x_9, x_9) = 1$. So $(x_9\bar{x}_9, b_7) = 1$. Similarly, $\bar{b}_3(b_4 b_8) = (\bar{b}_3 b_4) b_8$ yields that

$$b_8 x_9 = \bar{b}_3\sigma_{18} + b_3 + x_9 + \bar{u}_6 \tag{3.142}$$

by (3.128), (3.3), (3.4), (3.118), and (3.5). Thus, $(\bar{b}_3\sigma_{18}, x_9) = 1$ implies that $(b_8 x_9, x_9) = 2$, and hence $(x_9\bar{x}_9, b_8) = 2$. Recall that $(b_3\bar{b}_{12}, y_{12}) = 1$ by (3.125). Note that $c_{12} \neq y_{12}$ by (3.132), (3.125), and (3.130). So $(b_4 u_6, y_{12}) = 1$ by (3.132). Hence we may assume that

$$b_4 u_6 = y_{12} + q_{12}, \tag{3.143}$$

where $q_{12} \in \mathbb{N}(B)$. So by (3.132),

$$b_3\bar{b}_{12} = c_{12} + y_{12} + q_{12}. \tag{3.144}$$

It follows from $b_3 b_5^2 = (b_3 b_5) b_5$, (3.131), (3.133), (3.144), (3.118), and (3.138) that

$$b_5\bar{x}_9 = x_9 + c_{12} + y_{12} + q_{12}. \tag{3.145}$$

Thus, $(b_4 b_5, x_9\bar{x}_9) = (b_4\bar{x}_9, b_5\bar{x}_9) = 3$ by (3.134) and (3.145). So $(x_9\bar{x}_9, b_8) = 2$ and (3.36) force that $(x_9\bar{x}_9, \bar{b}_{12}) = 1$. If $q_{12} \notin B$, then (3.145) implies that $(b_5\bar{x}_9, b_5\bar{x}_9) \geq 5$. Hence $(b_5\bar{b}_5, x_9\bar{x}_9) \geq 5$. But we have already proved that $(x_9\bar{x}_9, b_8) = 2$. So $(x_9\bar{x}_9, r_{16}) \geq 2$. Recall that b_{12} is not real by (3.130). Then the above discussions yield that

$$x_9\bar{x}_9 = 1 + b_7 + 2b_8 + b_{12} + \bar{b}_{12} + 2r_{16} + \cdots .$$

Since $L_1(B) = \{1\}$, the above equality cannot be true. This proves that

$$q_{12} \in B$$

and $(b_5\bar{b}_5, x_9\bar{x}_9) = 4$. So c_{12}, y_{12}, and q_{12} are distinct, and $(b_3\bar{b}_{12}, b_3\bar{b}_{12}) = 3$ by (3.144). Hence $(b_3 b_{12}, b_3 b_{12}) = (b_3\bar{b}_{12}, b_3\bar{b}_{12}) = 3$. Note that $(b_3 b_{12}, \bar{u}_6) = 1$ by (3.122), and $(b_3 b_{12}, x_9) = 1$ by (3.35). So we may set

$$b_3 b_{12} = \bar{u}_6 + x_9 + n_{21}, \quad n_{21} \in B. \tag{3.146}$$

Hence by (3.121),

$$\bar{b}_3\sigma_{18} = c_{12} + y_{12} + x_9 + n_{21}. \tag{3.147}$$

Now (3.141) and (3.147) imply that

$$b_7 x_9 = b_3 + p_6 + c_{12} + y_{12} + x_9 + n_{21}, \tag{3.148}$$

and (3.142) and (3.147) yield that

$$b_8 x_9 = b_3 + \bar{u}_6 + c_{12} + y_{12} + 2x_9 + n_{21}. \tag{3.149}$$

Recall that $(b_3\bar{b}_3, b_7^2) = (b_3b_7, b_3b_7) = 2$ by (3.133), and $(b_4\bar{b}_4, b_7^2) = (b_4b_7, b_4b_7) = 3$ by (3.137). So $(b_7^2, b_7) = 1$, and $(b_7^2, b_8) = 1$. Since $(\bar{b}_3c_{12}, b_7) = 1$ by (3.133), and $(\bar{b}_3c_{12}, \bar{b}_{12}) = 1$ by (3.144), we see that $(\bar{b}_3c_{12}, \bar{b}_3c_{12}) \geq 3$. Hence $(b_3c_{12}, b_3c_{12}) = (\bar{b}_3c_{12}, \bar{b}_3c_{12}) \geq 3$. Therefore (3.136), (3.137), and (3.140) force that $(b_5b_7, b_5b_7) = (b_3c_{12}, b_3c_{12}) \geq 3$. So $(b_5\bar{b}_5, b_7^2) = (b_5b_7, b_5b_7) \geq 3$. But we have just proved that $(b_7^2, b_8) = 1$. Thus, $(b_7^2, r_{16}) \geq 1$. Therefore, $L_1(B) = \{1\}$ forces that $(b_7^2, r_{16}) = 1$, and hence $(b_3c_{12}, b_3c_{12}) = 3$. So we may set

$$\bar{b}_3c_{12} = b_7 + \bar{b}_{12} + m_{17}, \quad m_{17} \in B. \tag{3.150}$$

Hence by (3.135),

$$b_4\sigma_{18} = b_7 + b_8 + b_{12} + \bar{b}_{12} + r_{16} + m_{17}. \tag{3.151}$$

Now it follows from $(b_3\bar{b}_3)b_7 = \bar{b}_3(b_3b_7)$, (3.1), (3.133), (3.35), and (3.150) that

$$b_7b_8 = b_7 + b_8 + b_{12} + \bar{b}_{12} + m_{17}. \tag{3.152}$$

Furthermore, from (3.134), (3.1), (3.35), (3.125), (3.150), (3.3), and $\bar{b}_3(b_4\bar{x}_9) = (\bar{b}_3b_4)\bar{x}_9$, we get that

$$x_9\bar{x}_9 = 1 + b_7 + 2b_8 + b_{12} + \bar{b}_{12} + r_{16} + m_{17}. \tag{3.153}$$

Since we have already known that $(b_7^2, b_7) = (b_7^2, b_8) = (b_7^2, r_{16}) = 1$, and $(b_7^2, x_9\bar{x}_9) = (b_7x_9, b_7x_9) = 6$, by (3.153) we must have that

$$b_7^2 = 1 + b_7 + b_8 + r_{16} + m_{17}. \tag{3.154}$$

Therefore, since b_4 is real, from (3.137), (3.6), (3.151), (3.154), (3.152), and $b_4(b_4b_7) = b_4^2b_7$, we get that

$$b_4v_6 = b_7 + m_{17}. \tag{3.155}$$

On the other hand, since $(b_3, v_6\bar{p}_6) = 1$ by (3.140), (3.3) yields that $(\bar{b}_3b_4, v_6\bar{p}_6) = (b_3 + x_9, v_6\bar{p}_6) \geq 1$. But b_4 real implies that $(\bar{b}_3p_6, b_4v_6) = (\bar{b}_3b_4, v_6\bar{p}_6)$. Thus,

$$(\bar{b}_3p_6, b_4v_6) \geq 1. \tag{3.156}$$

Note that $(\bar{b}_3p_6, b_7) = 0$ by (3.133). So (3.155) and (3.156) force that

$$(\bar{b}_3p_6, m_{17}) = 1.$$

Since $L_1(B) = \{1\}$, the above equality cannot be true, a contradiction.

Now we have derived a contradiction for each case in Table 3.1 before the end of Sect. 3.2. Hence, Theorem 3.1 holds.

Chapter 4
Preliminary Classification of Sub-case (3)

4.1 Introduction

In Chap. 4 we shall freely use the definitions and notation used in Chap. 2. The Main Theorem 2 of Chap. 3 left two unsolved problems for the complete classification of the Normalized Integral Table Algebra (NITA) (A, B) generated by a faithful non-real element of degree 3 with $L_1(B) = 1$ and $L_2(B) = \emptyset$. In this chapter we solve the second problem where $(b_3b_{10}, b_3b_{10}) \neq 2$, and prove that the NITA that satisfies case (2) of the Main Theorem 1 of Chap. 2 where $(b_3b_{10}, b_3b_{10}) \neq 2$ does not exist. Consequently, we can state the Main Theorem 3 of this chapter as follows.

Main Theorem 3 *Let (A, B) be a NITA generated by a non-real element $b_3 \in B$ of degree 3 and without non-identity basis element of degree 1 and 2. Then $b_3\bar{b}_3 = 1 + b_8$, $b_8 \in B$, and one of the following holds:*

(1) *There exists a real element $b_6 \in B$ such that $b_3^2 = \bar{b}_3 + b_6$ and $(A, B) \cong_x$ $(CH(PSL(2, 7)), Irr(PSL(2, 7)))$.*

(2) *There exist $b_6, b_{10}, b_{15} \in B$, where b_6 is non-real, such that $b_3^2 = \bar{b}_3 + b_6$, $\bar{b}_3b_6 = b_3 + b_{15}$, $b_3b_6 = b_8 + b_{10}$, and $(b_3b_8, b_3b_8) = 3$. Moreover, if b_{10} is real then $(A, B) \cong_x (CH(3 \cdot A_6), Irr(3 \cdot A_6))$ of dimension 17, and if b_{10} is non-real then $(b_3b_{10}, b_3b_{10}) = 2$ and b_{15} is a non-real element.*

(3) *There exist $c_3, b_6 \in B$, $c_3 \neq b_3$ or $\bar{b}_3$, such that $b_3^2 = c_3 + b_6$ and either $(b_3b_8, b_3b_8) = 3$ or 4. If $(b_3b_8, b_3b_8) = 3$ and c_3 is non-real, then $(A, B) \cong_x$ $(A(3 \cdot A_6 \cdot 2), B_{32})$ of dimension 32. (See Theorem 2.9 of Chap. 2 for the definition of this specific NITA.) If $(b_3b_8, b_3b_8) = 3$ and c_3 is real, then $(A, B) \cong_x$ $(A(7 \cdot 5 \cdot 10), B_{22})$ of dimension 22. (See Theorem 2.10 of Chap. 2 for the definition of this specific NITA.)*

In the above Main Theorem 3 we still have 2 open problems in the cases (2) and (3). In case (2) we must classify the NITA such that b_{10} is non-real and $(b_3b_{10}, b_3b_{10}) = 2$, and in case (3) we must classify the NITA such that $(b_3b_8, b_3b_8) = 4$. In Chap. 5, the open case (2) will be solved.

Z. Arad et al., *On Normalized Integral Table Algebras (Fusion Rings)*,
Algebra and Applications 16, DOI 10.1007/978-0-85729-850-8_4,
© Springer-Verlag London Limited 2011

Let us emphasize that the NITA's of dimension 7 and 17 in the cases (1) and (2) of the Main Theorem 2 are strictly isomorphic to the NITA's induced from finite groups G via the basis of the irreducible characters of G. However, the NITA's of dimensions 22 and 32 are not induced from finite groups as described in Chap. 2.

The Main Theorem 3 follows from the Main Theorem 2 in Chap. 3 and the next theorem.

Theorem 4.1   *Let (A, B) be a NITA generated by a non-real element $b_3 \in B$ of degree 3 and without non-identity basis element of degree 1 or 2. Then $b_3\bar{b}_3 = 1 + b_8$, $b_8 \in B$. Assume that*

$$b_3^2 = \bar{b}_3 + b_6, \quad \bar{b}_3 b_6 = b_3 + b_{15}, \quad b_3 b_6 = b_8 + b_{10}, \quad and \quad b_3 b_8 = b_3 + \bar{b}_6 + b_{15}$$

where b_6, b_{10} are non-real elements in B and $b_{15} \in B$. Then b_{15} is a non-real element and $(b_3 b_{10}, b_3 b_{10}) = 2$.

The rest of this chapter is devoted to proving the above theorem.

4.2 Preliminary Results

For the rest of this chapter, we shall always assume that (A, B) is a NITA generated by a non-real element $b_3 \in B$ of degree 3 and without non-identity basis element of degree 1 and 2 such that

$$b_3^2 = \bar{b}_3 + b_6, \tag{4.1}$$

$$b_3\bar{b}_3 = 1 + b_8, \quad b_8 \in B, \tag{4.2}$$

$$\bar{b}_3 b_6 = b_3 + b_{15}, \quad b_{15} \in B, \tag{4.3}$$

$$b_3 b_6 = b_8 + b_{10}, \quad b_{10} \in B, \tag{4.4}$$

and

$$b_3 b_8 = b_3 + \bar{b}_6 + b_{15}. \tag{4.5}$$

By (4.4), (4.5) and $\bar{b}_3(b_3 b_6) = (b_3\bar{b}_3)b_6$, we obtain that

$$b_6 b_8 = \bar{b}_3 + \bar{b}_{15} + \bar{b}_3 b_{10}. \tag{4.6}$$

By (4.1), (4.3), (4.4), (4.5) and $b_3(b_3 b_6) = b_3^2 b_6$ we obtain that

$$b_6^2 = \bar{b}_6 + b_3 b_{10}. \tag{4.7}$$

By (4.2) and (4.3), we obtain that

$$(\bar{b}_3 b_6, b_3 b_{10}) = (b_3, b_3 b_{10}) + (b_{15}, b_3 b_{10}) = (b_{15}, b_3 b_{10});$$

on the other hand, (4.1) and (4.4) imply that

$$(\bar{b}_3 b_6, b_3 b_{10}) = (b_3^2, b_6 \bar{b}_{10}) = (\bar{b}_3, b_6 \bar{b}_{10}) + (b_6, b_6 \bar{b}_{10}) = 1 + (b_6, b_6 \bar{b}_{10}).$$

Hence $1 + (b_6, b_6 \bar{b}_{10}) = (b_{15}, b_3 b_{10})$ which implies that

$$b_3 b_{10} = b_{15} + R_{15} \quad \text{where } R_{15} \in \mathbb{N}B. \tag{4.8}$$

The degrees appearing in R_{15} must all be ≥ 5.
   By (4.1), (4.3), (4.5) and $b_3(b_3 b_8) = b_3^2 b_8$, we obtain that

$$b_6 b_8 = \bar{b}_3 + b_3 b_{15}. \tag{4.9}$$

Now (4.6) implies that

$$b_3 b_{15} = \bar{b}_{15} + \bar{b}_3 b_{10}. \tag{4.10}$$

By (4.1), (4.2), (4.3), (4.4) and $(b_3 \bar{b}_3)(b_3 \bar{b}_3) = b_3^2 \bar{b}_3^2$, we obtain that

$$b_8^2 = b_8 + b_{10} + \bar{b}_{10} + b_6 \bar{b}_6. \tag{4.11}$$

By (4.1), (4.2), (4.3), (4.4) and $\bar{b}_6 b_3^2 = (b_3 \bar{b}_6)b_3$, we obtain that

$$\bar{b}_{10} + b_6 \bar{b}_6 = 1 + b_3 \bar{b}_{15}. \tag{4.12}$$

By (4.1), (4.8), (4.10) and $b_3(b_3 b_{10}) = b_3^2 b_{10}$, we obtain that

$$b_6 b_{10} = \bar{b}_{15} + b_3 R_{15}. \tag{4.13}$$

The proof that b_{15} is non-real is simple.
   Assume henceforth that b_{15} is a real element and we shall derive a contradiction. By (4.3) we have that $(b_6, b_3 b_{15}) = 1$ and together with (4.12), we obtain that $(\bar{b}_6, b_3 b_{15}) = 1$ which implies that $(b_{15}, b_3 b_6) = 1$, and we have a contradiction to (4.4). Therefore

$$b_{15} \quad \text{is a non-real element.} \tag{4.14}$$

The remainder of this chapter will consist of the proof that $(b_3 b_{10}, b_3 b_{10}) = 2$.
   By (4.4) we obtain that $(b_6, \bar{b}_3 b_{10}) = 1$ which implies that $(b_3 b_{10}, b_3 b_{10}) \geq 2$. By (4.4) we obtain that $(\bar{b}_6, b_3 b_{10}) = 0$. Now by (4.3), (4.4) and (4.7) we obtain that $(b_3 b_{10}, b_3 b_{10}) = (b_6^2, b_3 b_{10}) = (\bar{b}_3 b_6, \bar{b}_6 b_{10}) = (b_3, \bar{b}_6 b_{10}) + (b_{15}, \bar{b}_6 b_{10}) = 1 + (b_{15}, \bar{b}_6 b_{10})$. Since $(b_3, \bar{b}_6 b_{10}) = 1$, we have that $(b_{15}, \bar{b}_6 b_{10}) \leq 3$ which implies that $(b_3 b_{10}, b_3 b_{10}) \leq 4$. Therefore

$$2 \leq (b_3 b_{10}, b_3 b_{10}) \leq 4. \tag{4.15}$$

Assume henceforth that $b_{15} = R_{15}$ and we shall derive a contradiction. Then by (4.8) we obtain that $b_3 b_{10} = 2b_{15}$ which implies that $(\bar{b}_{10}, b_3 \bar{b}_{15}) = 2$, and by (4.12) we obtain that $(\bar{b}_{10}, b_6 \bar{b}_6) = 1$ which implies that $(b_{10}, b_6 \bar{b}_6) = 1$. Now (4.12)

Table 4.1 Splitting to five
cases

Case 1	$R_{15} = x_5 + x_{10}, x_5, x_{10} \in B$
Case 2	$R_{15} = x_6 + x_9, x_6, x_9 \in B$
Case 3	$R_{15} = x_7 + x_8, x_7, x_8 \in B$
Case 4	$R_{15} = x_5 + y_5 + z_5, x_5, y_5, z_5 \in B$
Case 5	$R_{15} \in B, b_{15} \neq R_{15}$

implies that $(\bar{b}_{15}, b_3\bar{b}_{10}) = (b_{10}, b_3\bar{b}_{15}) = 1$. By (4.4) we obtain that $(b_6, \bar{b}_3 b_{10}) = 1$ and since $(b_3 b_{10}, b_3 b_{10}) = 4$, we obtain that $\bar{b}_3 b_{10} = b_6 + \bar{b}_{15} + \alpha + \beta$ where $\alpha, \beta \in B$, $|\alpha| + |\beta| = 9$ and $|\alpha|, |\beta| \geq 5$, which is a contradiction to our assumption that $b_{15} = R_{15}$.

Now by (4.8) and (4.15), we have five cases (see Table 4.1).

In the rest of this chapter, we derive a contradiction for Cases 1–4 in the above table. In Case 5, by (4.8) we then have that $(b_3 b_{10}, b_3 b_{10}) = (b_{15}, b_{15}) + (R_{15}, R_{15}) = 2$. Then we will have proven Theorem 1.1.

4.3 Case $R_{15} = x_5 + x_{10}$

In this section we assume that $R_{15} = x_5 + x_{10}$. Then by (4.7), (4.8) and (4.13), we have that

$$b_3 b_{10} = b_{15} + x_5 + x_{10}, \tag{4.16}$$

$$b_6^2 = \bar{b}_6 + b_{15} + x_5 + x_{10} \tag{4.17}$$

and

$$b_6 b_{10} = \bar{b}_{15} + b_3 x_5 + b_3 x_{10}. \tag{4.18}$$

Now by (4.16) we obtain that $(b_{10}, \bar{b}_3 x_5) = 1$ which implies that

$$\bar{b}_3 x_5 = b_{10} + y_5, \quad \text{where } y_5 \in B. \tag{4.19}$$

Now (4.1) and $\bar{b}_3^2 x_5 = \bar{b}_3 (\bar{b}_3 x_5)$ imply that

$$b_3 x_5 + \bar{b}_6 x_5 = \bar{b}_3 b_{10} + \bar{b}_3 y_5. \tag{4.20}$$

By (4.2), (4.16), (4.19) and $b_3(\bar{b}_3 x_5) = (b_3 \bar{b}_3) x_5$, we obtain that

$$b_8 x_5 = x_{10} + b_{15} + b_3 y_5. \tag{4.21}$$

Now by (4.4), (4.19) and $(\bar{b}_3 \bar{b}_6) x_5 = \bar{b}_6 (\bar{b}_3 x_5)$, we obtain that

$$\bar{b}_6 b_{10} + \bar{b}_6 y_5 = x_{10} + b_{15} + b_3 y_5 + \bar{b}_{10} x_5. \tag{4.22}$$

By (4.1) and (4.19) we obtain that $(\bar{b}_3 y_5, b_3 x_5) = (b_3^2, y_5 \bar{x}_5) = (\bar{b}_3, y_5 \bar{x}_5) + (b_6, y_5 \bar{x}_5) = 1 + (b_6, \bar{x}_5 y_5)$. Hence

$$(\bar{b}_3 y_5, b_3 x_5) = 1 + (b_6, \bar{x}_5 y_5). \tag{4.23}$$

By (4.2) and (4.19) we obtain that $2 = (b_3 x_5, b_3 x_5) = (b_3 \bar{b}_3, x_5 \bar{x}_5) = 1 + (b_8, x_5 \bar{x}_5)$. Hence

$$(b_8, x_5 \bar{x}_5) = 1 \tag{4.24}$$

and together with $b_{10} \neq \bar{b}_{10}$, we obtain that

$$(b_{10}, x_5 \bar{x}_5) = 0. \tag{4.25}$$

Now (4.4) implies that $(\bar{b}_3 \bar{x}_5, b_6 \bar{x}_5) = (b_3 b_6, x_5 \bar{x}_5) = (b_8, x_5 \bar{x}_5) + (b_{10}, x_5 \bar{x}_5) = 1$. Hence

$$(\bar{b}_3 \bar{x}_5, b_6 \bar{x}_5) = 1. \tag{4.26}$$

By (4.3) and (4.19) we obtain that $(\bar{b}_3 y_5, \bar{b}_6 x_5) = (b_3 b_6, y_5 \bar{x}_5) = (\bar{b}_3, y_5 \bar{x}_5) + (\bar{b}_{15}, y_5 \bar{x}_5) = 1 + (\bar{b}_{15}, y_5 \bar{x}_5)$. Therefore

$$(\bar{b}_3 y_5, \bar{b}_6 x_5) = 1 + (\bar{b}_{15}, y_5 \bar{x}_5) \tag{4.27}$$

which implies that

$$(\bar{b}_3 y_5, \bar{b}_6 x_5) \leq 2. \tag{4.28}$$

By (4.2) we obtain that $(b_3 y_5, b_3 y_5) = (b_3 \bar{b}_3, y_5 \bar{y}_5) = 1 + (b_8, y_5 \bar{y}_5)$ which implies that

$$(b_3 y_5, b_3 y_5) \leq 4. \tag{4.29}$$

Assume henceforth that $(\bar{b}_3 y_5, \bar{b}_3 y_5) = 4$ and we shall derive a contradiction. Then

$$\bar{b}_3 y_5 = x + y + z + v \quad \text{where } x, y, z, v \in B.$$

Now, since $(b_3 x_5, b_3 x_5) = 2$, by (4.23) we have that $(b_3 x_5, \bar{b}_3 y_5) = 1$. Hence

$$b_3 x_5 = x + w \quad \text{where } w \in B,$$

and together with (4.20), we get that

$$x + w + \bar{b}_6 x_5 = \bar{b}_3 b_{10} + x + y + z + v.$$

Thus $(\bar{b}_6 x_5, \bar{b}_3 y_5) \geq 3$, and we have a contradiction to (4.28).

Assume henceforth that $(\bar{b}_3 y_5, \bar{b}_3 y_5) = 3$ and we shall derive a contradiction. Then

$$\bar{b}_3 y_5 = x + y + z \quad \text{where } x, y, z \in B$$

and

$$b_3 x_5 = x + w \quad \text{where } w \in B.$$

By (4.20) we have that

$$x + w + \bar{b}_6 x_5 = \bar{b}_3 b_{10} + x + y + z.$$

Now (4.28) implies that $(x, \bar{b}_6 x_5) = 0$ and by (4.26), we obtain that $(b_3 x_5, \bar{b}_6 x_5) = 1$ which implies that $(w, \bar{b}_6 x_5) = 1$ and $(w, \bar{b}_3 b_{10}) = 2$. Since $(b_3 b_{10}, b_3 b_{10}) = 3$ we have a contradiction.

Assume henceforth that $(\bar{b}_3 y_5, \bar{b}_3 y_5) = 1$ and we shall derive a contradiction. Then $\bar{b}_3 y_5 = c_{15}$ and by (4.23) we obtain that $b_3 x_5 = c_{15}$, which is a contradiction to (4.19). Now (4.29) implies that

$$(b_3 y_5, b_3 y_5) = 2. \tag{4.30}$$

Assume henceforth that $(\bar{b}_{15}, y_5 \bar{x}_5) = 1$ and we shall derive a contradiction. Then (4.27) implies that $(\bar{b}_3 y_5, \bar{b}_6 x_5) = 2$. Now (4.17) and (4.20) imply that

$$\bar{b}_6 x_5 = b_6 + 2 x_{12} \quad \text{where } x_{12} \in B$$

and

$$\bar{b}_3 y_5 = x_{12} + c_3 \text{ where } c_3 \in B.$$

Now by (4.4) we obtain that $(b_6, \bar{b}_3 \bar{x}_5) = 0$, which is a contradiction to (4.26). Hence

$$(\bar{b}_{15}, y_5 \bar{x}_5) = 0. \tag{4.31}$$

By (4.21) and (4.30) we obtain that $(b_8 x_5, b_3 y_5) \geq 2$. In addition, (4.5), (4.19) and (4.31) imply that $(b_8 x_5, b_3 y_5) = (\bar{b}_3 b_8, y_5 \bar{x}_5) = (\bar{b}_3, y_5 \bar{x}_5) + (b_6, y_5 \bar{x}_5) + (\bar{b}_{15}, y_5 \bar{x}_5) = 1 + (b_6, y_5 \bar{x}_5)$. Hence

$$(b_6, y_5 \bar{x}_5) \geq 1.$$

By (4.19), (4.23) and (4.30) we have that

$$(b_6, y_5 \bar{x}_5) \leq 1.$$

Therefore

$$(b_6, y_5 \bar{x}_5) = 1 \quad \text{and} \quad (\bar{b}_3 y_5, b_3 x_5) = 2 \tag{4.32}$$

which implies that

$$\bar{b}_3 y_5 = b_3 x_5. \tag{4.33}$$

In addition, (4.20) implies that

$$\bar{b}_6 x_5 = \bar{b}_3 b_{10}. \tag{4.34}$$

By (4.19) we obtain that $(x_5, b_3 y_5) = 1$. Now (4.30) implies that

$$b_3 y_5 = x_5 + y_{10}, \quad \text{where } y_{10} \in B. \tag{4.35}$$

Now (4.33) implies that

$$\bar{b}_3 y_5 = b_3 x_5 = x + y \quad \text{where } x, y \in B \tag{4.36}$$

and (4.21) implies that

$$b_8 x_5 = x_5 + x_{10} + y_{10} + b_{15}. \tag{4.37}$$

Now by (4.2) and $\bar{b}_3(b_3 x_5) = (\bar{b}_3 b_3) x_5$, we obtain that

$$\bar{b}_3 x + \bar{b}_3 y = 2 x_5 + x_{10} + y_{10} + b_{15}. \tag{4.38}$$

Thus we can assume that $x = z_5$ and $y = z_{10}$, which imply that

$$\bar{b}_3 y_5 = b_3 x_5 = z_5 + z_{10}, \tag{4.39}$$

and one of the two following cases hold:

Case 1

$$\bar{b}_3 z_5 = x_5 + x_{10} \quad \text{and} \quad \bar{b}_3 z_{10} = x_5 + y_{10} + b_{15}. \tag{4.40}$$

Case 2

$$\bar{b}_3 z_5 = x_5 + y_{10} \quad \text{and} \quad \bar{b}_3 z_{10} = x_5 + x_{10} + b_{15}. \tag{4.41}$$

Thus $(z_{10}, b_3 b_{15}) = 1$, by (4.3) we have that $(b_6, b_3 b_{15}) = 1$ and together with (4.10), we obtain that

$$b_3 b_{15} = b_6 + z_{10} + x_{14} + \bar{b}_{15} \quad \text{where } x_{14} \in B, \tag{4.42}$$

$$\bar{b}_3 b_{10} = b_6 + z_{10} + x_{14}, \tag{4.43}$$

and by (4.34) we obtain that

$$\bar{b}_6 x_5 = b_6 + z_{10} + x_{14}. \tag{4.44}$$

By (4.43) we obtain that $(b_{10}, b_3 z_{10}) = 1$, by (4.39) we obtain that $(y_5, b_3 z_{10}) = 1$, and by (4.40) and (4.41) we obtain that $(b_3 z_{10}, b_3 z_{10}) = 3$. Thus

$$b_3 z_{10} = y_5 + b_{10} + x_{15} \quad \text{where } x_{15} \in B. \tag{4.45}$$

By (4.9) and (4.42), we obtain that

$$b_6 b_8 = \bar{b}_3 + \bar{b}_{15} + b_6 + z_{10} + x_{14}. \tag{4.46}$$

Now by (4.4) we obtain that $(b_6\bar{b}_6, b_3 z_{10}) = (\bar{b}_3 b_6, \bar{b}_6 z_{10}) = (b_8, \bar{b}_6 z_{10}) + (\bar{b}_{10}, \bar{b}_6 z_{10}) = 1 + (\bar{b}_{10}, \bar{b}_6 z_{10})$ which implies that $(b_6\bar{b}_6, b_3 z_{10}) \geq 1$. By (4.17) we have that $(b_6\bar{b}_6, b_6\bar{b}_6) = 4$. Thus either

$$b_6\bar{b}_6 = 1 + b_8 + x_{12} + x_{15} \quad \text{where } x_{12}, x_{15} \in B$$

or

$$b_6\bar{b}_6 = 1 + b_8 + y_5 + x_{22} \quad \text{where } y_5, x_{22} \in B.$$

Assume henceforth that $b_6\bar{b}_6 = 1 + b_8 + y_5 + x_{22}$ and we shall derive a contradiction. Then by (4.12) we obtain that $\bar{b}_3 y_5 = \bar{b}_{15}$, and we have a contradiction to (4.30). Hence

$$b_6\bar{b}_6 = 1 + b_8 + x_{12} + x_{15} \tag{4.47}$$

and

$$b_3\bar{b}_{15} = \bar{b}_{10} + b_8 + x_{12} + x_{15}. \tag{4.48}$$

By (4.1), (4.35), (4.39) and $b_3(b_3 y_5) = b_3^2 y_5$, we obtain that

$$b_3 y_{10} = b_6 y_5. \tag{4.49}$$

By (4.39) we obtain that $(y_5, b_3 z_5) = 1$, and by (4.40) and (4.41) we obtain that $(b_3 z_5, b_3 z_5) = 2$ which implies that

$$b_3 z_5 = y_5 + t_{10} \quad \text{where } t_{10} \in B. \tag{4.50}$$

By (4.19), (4.35), (4.39), (4.45), (4.50) and $b_3(\bar{b}_3 y_5) = \bar{b}_3(b_3 y_5)$, we obtain that

$$\bar{b}_3 y_{10} = y_5 + t_{10} + x_{15}. \tag{4.51}$$

By (4.4), (4.8), (4.12), (4.16), (4.19), (4.43), (4.45), (4.48) and $\bar{b}_3(b_3 b_{10}) = b_3(\bar{b}_3 b_{10})$, we obtain that

$$x_{12} + \bar{b}_3 x_{10} = b_3 x_{14} \tag{4.52}$$

which implies that $(x_{14}, \bar{b}_3 x_{12}) = 1$. Now by (4.48) we obtain that

$$\bar{b}_3 x_{12} = t_7 + x_{14} + \bar{b}_{15} \quad \text{where } t_7 \in B. \tag{4.53}$$

By (4.34) we obtain that $3 = (b_6 x_5, b_6 x_5) = (b_6\bar{b}_6, x_5\bar{x}_5)$. Now by (4.47) we obtain that

$$x_5\bar{x}_5 = 1 + b_8 + x_{12} + t_4 \quad \text{where } t_4 \in B. \tag{4.54}$$

By (4.5), (4.19) ,(4.53), (4.54) and $b_3(x_5\bar{x}_5) = (b_3\bar{x}_5)x_5$, we obtain that

$$\bar{b}_{10}x_5 + \bar{y}_5 x_5 = 2b_3 + b_6 + 2b_{15} + \bar{t}_7 + \bar{x}_{14} + b_3 t_4. \tag{4.55}$$

By (4.35) we obtain that $(b_3, \bar{y}_5 x_5) = 1$, and by (4.32) we obtain that $(b_{15}, \bar{y}_5 x_5) = (\bar{x}_{14}, \bar{y}_5 x_5) = 0$ which implies that

$$\bar{b}_{10} x_5 = b_3 + c_3 + 2b_{15} + \bar{x}_{14} \quad \text{where } c_3 \in B$$

and

$$(c_3, b_3 t_4) = 1$$

which implies that $(\bar{b}_3 c_3, \bar{b}_3 c_3) = 2$. By (4.2) we obtain that $(\bar{b}_3 c_3, \bar{b}_3 c_3) = (b_3 \bar{b}_3, c_3 \bar{c}_3) = 1 + (b_8, c_3 \bar{c}_3)$ which implies that

$$c_3 \bar{c}_3 = 1 + b_8$$

and

$$\bar{b}_3 c_3 = t_4 + u_5 \quad \text{where } u_5 \in B. \tag{4.56}$$

Now by (4.5) and $b_3 (c_3 \bar{c}_3) = (b_3 \bar{c}_3) c_3$ we obtain that

$$c_3 t_4 + c_3 \bar{u}_5 = 2b_3 + \bar{b}_6 + b_{15}$$

which implies that

$$c_3 \bar{u}_5 = b_{15}.$$

By (4.56) we obtain that $(b_3, c_3 \bar{u}_5) = 1$, which is a contradiction.

4.4 Case $R_{15} = x_6 + x_9$

Throughout this section we assume that $R_{15} = x_6 + x_9$. Then by (4.7), (4.8) and (4.13), we have that

$$b_3 b_{10} = b_{15} + x_6 + x_9, \tag{4.57}$$

$$b_6^2 = \bar{b}_6 + b_{15} + x_6 + x_9, \tag{4.58}$$

$$b_6 b_{10} = \bar{b}_{15} + b_3 x_6 + b_3 x_9. \tag{4.59}$$

By (4.57) we obtain that

$$(b_{10}, \bar{b}_3 x_6) = 1 \tag{4.60}$$

which implies that either $2 \le (\bar{b}_3 x_6, \bar{b}_3 x_6) \le 3$ or $\bar{b}_3 x_6 = b_{10} + 2b_4$.

Assume henceforth that $\bar{b}_3 x_6 = b_{10} + 2b_4$ and we shall derive a contradiction. Then $b_3 b_4 = 2x_6$ which implies by (4.2) that $4 = (b_3 b_4, b_3 b_4) = (b_3 \bar{b}_3, b_4 \bar{b}_4) = 1 + (b_8, b_4 \bar{b}_4)$. Thus $(b_8, b_4 \bar{b}_4) = 3$, and we have a contradiction.

Now we have that $2 \le (\bar{b}_3 x_6, \bar{b}_3 x_6) \le 3$. Assume that $(b_3 x_6, b_3 x_6) = 2$, so we can write that

$$b_3 x_6 = c + d \quad \text{where } c, d \in B. \tag{4.61}$$

By (4.60) we obtain that

$$\bar{b}_3 x_6 = b_{10} + y_8 \quad \text{where } y_8 \in B. \tag{4.62}$$

Now by (4.2), (4.57) and $b_3(\bar{b}_3 x_6) = (b_3 \bar{b}_3) x_6$ we obtain that

$$b_8 x_6 = x_9 + b_{15} + b_3 y_8. \tag{4.63}$$

By (4.1), (4.61), (4.62) and $\bar{b}_3(\bar{b}_3 x_6) = \bar{b}_3^2 x_6$ we obtain that

$$\bar{b}_3 b_{10} + \bar{b}_3 y_8 = c + d + \bar{b}_6 x_6. \tag{4.64}$$

By (4.58) we obtain that $(b_6 \bar{b}_6, b_6 \bar{b}_6) = 4$, by (4.4) we obtain that $2 = (b_3 b_6, b_3 b_6) = (b_3 \bar{b}_3, b_6 \bar{b}_6)$ which implies that

$$b_6 \bar{b}_6 = 1 + b_8 + s + t \quad \text{where } s, t \text{ are real elements in } B. \tag{4.65}$$

By (4.61) we obtain that $2 = (b_3 x_6, b_3 x_6) = (b_3 \bar{b}_3, x_6 \bar{x}_6)$. Now (4.2) implies that

$$(b_8, x_6 \bar{x}_6) = 1 \tag{4.66}$$

and together with (4.4), we obtain that $(b_3 x_6, \bar{b}_6 x_6) = (b_3 b_6, x_6 \bar{x}_6) = (b_8, x_6 \bar{x}_6) + (b_{10}, x_6 \bar{x}_6) = 1 + (b_{10}, x_6 \bar{x}_6)$. Thus

$$(b_3 x_6, \bar{b}_6 x_6) = 1 + (b_{10}, x_6 \bar{x}_6). \tag{4.67}$$

By (4.12) and (4.65) we obtain that

$$b_3 \bar{b}_{15} = b_8 + \bar{b}_{10} + s + t, \tag{4.68}$$

which implies that $|s|, |t| \geq 5$.

Assume henceforth that $s = s_5$ and we shall derive a contradiction. Then by (4.68) we obtain that $b_3 \bar{b}_{15} = b_8 + \bar{b}_{10} + s_5 + t_{22}$ and $b_3 s_5 = b_{15}$. Hence (4.2) and $b_3(b_3 s_5) = (b_3 \bar{b}_3) s_5$ imply that $b_8 s_5 = b_8 + b_{10} + t_{22}$. Since b_8 and s_5 are real elements and b_{10} is non-real, we have a contradiction.

Assume henceforth that $s = s_6$ and we shall derive a contradiction. Then by (4.68) we obtain that $b_3 \bar{b}_{15} = b_8 + \bar{b}_{10} + s_6 + t_{21}$ and $b_3 s_6 = b_{15} + c_3$ where $c_3 \in B$. Now $\bar{b}_3(b_3 s_6) = (b_3 \bar{b}_3) s_6$ implies that $b_8 s_6 = b_8 + b_{10} + t_{21} + \bar{b}_3 c_3$. Since b_8 and s_6 are reals and b_{10} is non-real, we obtain that $(\bar{b}_{10}, \bar{b}_3 c_3) = 1$ and we have a contradiction, which implies that

$$|s| \geq 7. \tag{4.69}$$

In the same way we can show that

$$|t| \geq 7. \tag{4.70}$$

Now (4.65) implies that $(b_6 \bar{b}_6, x_6 \bar{x}_6) \leq 5$. By (4.58) we obtain that $(b_6, \bar{b}_6 x_6) = 1$ and by (4.64) we obtain that $(b_6 \bar{b}_6, x_6 \bar{x}_6) = (\bar{b}_6 x_6, \bar{b}_6 x_6) \geq 3$. Therefore

$$3 \leq (b_6 \bar{b}_6, x_6 \bar{x}_6) \leq 5. \tag{4.71}$$

Assume henceforth that $(b_{10}, x_6\bar{x}_6) = 1$ and we shall derive a contradiction. Then $x_6\bar{x}_6 = 1 + b_8 + b_{10} + \bar{b}_{10} + z_7$, where $z_7 \in B$. Now (4.65) implies that $(\bar{b}_6x_6, \bar{b}_6x_6) \leq 3$ and together with (4.71) we obtain that $(\bar{b}_6x_6, \bar{b}_6x_6) = 3$, and by (4.61), (4.67) and $(b_{10}, x_6\bar{x}_6) = 1$ we obtain that $\bar{b}_6x_6 = b_6 + c + d$, which is a contradiction. Thus (4.67) implies that

$$(b_{10}, x_6\bar{x}_6) = 0 \tag{4.72}$$

and

$$(b_3x_6, \bar{b}_6x_6) = 1. \tag{4.73}$$

Now we can assume that

$$(c, \bar{b}_6x_6) = 1. \tag{4.74}$$

In addition, (4.1) and (4.62) imply that $(\bar{b}_3y_8, b_3x_6) = (b_3^2, y_8\bar{x}_6) = (\bar{b}_3, y_8\bar{x}_6) + (b_6, y_8\bar{x}_6) = (y_8, \bar{b}_3x_6) + (y_8, b_6x_6) = 1 + (y_8, b_6x_6)$. By (4.62) and (4.73), we obtain that $1 = (b_3x_6, \bar{b}_6x_6) = (\bar{b}_3x_6, b_6x_6) = (b_{10}, b_6x_6) + (y_8, b_6x_6)$ which implies that either $(\bar{b}_3y_8, b_3x_6) = 1$ or $(\bar{b}_3y_8, b_3x_6) = 2$. Assume that $(\bar{b}_3y_8, b_3x_6) = 1$. Then (4.4), (4.64), (4.74) and $(b_3b_{10}, b_3b_{10}) = 3$ imply that $\bar{b}_3b_{10} = b_6 + c + d$ which is a contradiction, and we obtain that $(\bar{b}_3y_8, b_3x_6) = 2$. Now we have that either $(c, \bar{b}_3y_8) = 2$ or $(c, \bar{b}_3y_8) = (d, \bar{b}_3y_8) = 1$, which implies that $(\bar{b}_3y_8, b_3y_8) \geq 3$. By (4.11), (4.66) and (4.72), we obtain that $(b_8x_6, b_8x_6) = (b_8^2, x_6\bar{x}_6) = 1 + (b_6\bar{b}_6, x_6\bar{x}_6)$. Now (4.71) implies that $(b_8x_6, b_8x_6) \leq 6$ together with (4.63) and since $(\bar{b}_3y_8, b_3y_8) \geq 3$, we have that

$$(x_9, b_3y_8) = (b_{15}, b_3y_8) = 0, \tag{4.75}$$

which implies that

$$(b_8x_6, b_8x_6) = 2 + (b_3y_8, b_3y_8). \tag{4.76}$$

Therefore $(b_3y_8, b_3y_8) \leq 4$. If $(c, \bar{b}_3y_8) = 2$ then $\bar{b}_3y_8 = 2c_{12}$ and $(\bar{b}_6x_6, \bar{b}_6x_6) = 5$. Now (4.64) implies that $\bar{b}_3b_{10} = b_6 + d_6 + f_{18}$ where $f_{18} \in B$ and $\bar{b}_6x_6 = b_6 + c_{12} + f_{18}$, which is a contradiction.

Now we have that $(c, \bar{b}_3y_8) = (d, \bar{b}_3y_8) = 1$ which implies that

$$\bar{b}_3y_8 = c + d + e_6 \quad \text{where } e_6 \in B. \tag{4.77}$$

Now (4.64) and (4.74) imply that

$$\bar{b}_3b_{10} = b_6 + c + f \quad \text{where } f \in B, \tag{4.78}$$

$$\bar{b}_6x_6 = b_6 + e_6 + c + f, \tag{4.79}$$

and by (4.10) we obtain that

$$b_3b_{15} = \bar{b}_{15} + b_6 + c + f. \tag{4.80}$$

By (4.2), (4.61), (4.63) and $\bar{b}_3(b_3x_6) = (b_3\bar{b}_3)x_6$, we obtain that

$$\bar{b}_3c + \bar{b}_3d = x_6 + x_9 + b_{15} + b_3y_8. \tag{4.81}$$

Now we can assume that $(b_{15}, \bar{b}_3c) = 1$. By (4.61) we obtain that $(x_6, \bar{b}_3c) = (x_6, \bar{b}_3d) = 1$ and by (4.77) we obtain that $(b_3y_8, b_3y_8) = 3$ which implies by (4.75) and (4.81) that $2 \leq (b_3c, b_3c) \leq 4$. By (4.77) and (4.78), we obtain that $(b_{10}, b_3c) = (y_8, b_3c) = 1$ which implies that $(b_3c, b_3c) \neq 2$.
    Therefore

$$3 \leq (b_3c, b_3c) \leq 4. \tag{4.82}$$

By (4.77) we obtain that $(b_3y_8, b_3y_8) = 3$ and by (4.62), we obtain that $(y_8, \bar{b}_3x_6) = 1$ which implies that

$$b_3y_8 = x_6 + \alpha + \beta \quad \text{where } \alpha, \beta \in B. \tag{4.83}$$

Now by (4.81) we obtain that

$$\bar{b}_3c + \bar{b}_3d = 2x_6 + x_9 + b_{15} + \alpha + \beta. \tag{4.84}$$

Assume that $(b_3c, b_3c) = 3$. Then by (4.61) and (4.80) we obtain that

$$\bar{b}_3c = x_6 + b_{15} + g, \quad \text{where } g \in B \tag{4.85}$$

which implies that $|c| \geq 9$, and by (4.84) we obtain that $g + \bar{b}_3d = x_6 + x_9 + \alpha + \beta$. If $g = x_9$ then by (4.85) we obtain that $|c| = 10$, and if $g \neq x_9$ then we obtain that either $g = \alpha$ or $g = \beta$. If $\alpha = 3 = \alpha_3$, then (4.5) and (4.83) imply that $\alpha = b_3$, $y_8 = b_8$ and $x_6 = \bar{b}_6$ which implies by (4.57) that $b_3b_{10} = b_{15} + \bar{b}_6 + x_9$. Thus $(\bar{b}_{10}, b_3b_6) = 1$, and we have a contradiction to (4.4). In the same way, we can show that $|\beta| \neq 3$ which implies by (4.83) that

$$4 \leq |\alpha|, |\beta| \leq 14. \tag{4.86}$$

Now (4.85) implies that $|g| \leq 12$ and $|c| \leq 11$. Therefore

$$9 \leq |c| \leq 11. \tag{4.87}$$

Assume that $(b_3c, b_3c) = 4$. Then by (4.62) and by (4.68), we obtain that

$$\bar{b}_3c = x_6 + b_{15} + g + h \quad \text{where } g, h \in B. \tag{4.88}$$

Now by (4.81) and (4.83) we obtain that $\bar{b}_3c + \bar{b}_3d = 2x_6 + x_9 + b_{15} + \alpha + \beta$, which implies that

$$g + h + \bar{b}_3d = x_6 + x_9 + \alpha + \beta. \tag{4.89}$$

If $(x_9, \bar{b}_3d) = 1$, then $g + h = \alpha + \beta$ which implies by (4.83) and (4.88) that $|c| = 13$. If $(x_9, \bar{b}_3d) = 0$ then we can assume that $g + h = x_9 + \alpha$. Now from $4 \leq |\alpha| \leq 14$ and by (4.88) we obtain that

$$12 \leq |c| \leq 14. \tag{4.90}$$

Table 4.2 Splitting to seven cases

Case 1	$c = c_9$	$d = d_9$	$f = f_{15}$	$(b_3c_9, b_3c_9) = 3$
Case 2	$c = c_{10}$	$d = d_8$	$f = f_{14}$	$(b_3c_{10}, b_3c_{10}) = 3$
Case 3	$c = c_{11}$	$d = d_7$	$f = f_{13}$	$(b_3c_{11}, b_3c_{11}) = 3$
Case 4	$c = c_{12}$	$d = d_6$	$f = f_{12}$	$(b_3c_{12}, b_3c_{12}) = 4$
Case 5	$c = c_{13}$	$d = d_5$	$f = f_{11}$	$(b_3c_{13}, b_3c_{13}) = 4$
Case 6	$c = c_{14}$	$d = d_4$	$f = f_{10}$	$(b_3c_{14}, b_3c_{14}) = 4$
Case 7	$(b_3x_6, b_3x_6) = 3$			

From (4.61), (4.78), (4.87) and (4.90) we have seven cases (see Table 4.2).

In the following seven sections we derive a contradiction for each case in the above table, and in the following three sections we assume that

$$(b_3c, b_3c) = 3, \tag{4.91}$$

so (4.87) implies that $9 \le |c| \le 11$.

4.4.1 Case $c = c_9$

In this section we assume that $c = c_9$. Then (4.77), (4.78) and (4.91) imply that $b_3c_9 = y_8 + b_{10} + q_9$ where $q_9 \in B$ which implies that

$$(c_9, \bar{b}_3q_9) = 1. \tag{4.92}$$

By (4.85) we obtain that $\bar{b}_3c_9 = x_6 + b_{15} + g_6$. Now by (4.61), (4.77), (4.78), (4.80) and $\bar{b}_3(b_3c_9) = b_3(\bar{b}_3c_9)$, we obtain that $b_{15} + b_3g_6 = e_6 + \bar{b}_3q_9$ which implies that $(\bar{b}_{15}, \bar{b}_3h_9) = 1$, and together with (4.92) we obtain that $\bar{b}_3h_9 = c_9 + \bar{b}_{15} + c_3$ where $c_3 \in B$, $b_3c_3 = h_9$ and $(c_3, b_3g_6) = 1$. Since $b_3c_3 = h_9$ we obtain that $(b_3c_3, b_3c_3) = 1$, which is a contradiction to $(c_3, b_3g_6) = 1$.

4.4.2 Case $c = c_{10}$

In this section we assume that $c = c_{10}$. Then (4.77), (4.78) and (4.91) imply that

$$b_3c_{10} = y_8 + b_{10} + q_{12} \quad \text{where } q_{12} \in B. \tag{4.93}$$

By (4.85) we obtain that

$$\bar{b}_3c_{10} = x_6 + b_{15} + g_9. \tag{4.94}$$

Now by (4.61), (4.77), (4.78), (4.80) and $\bar{b}_3(b_3c_{10}) = b_3(\bar{b}_3c_{10})$, we obtain that $b_{15} + b_3g_9 = e_6 + \bar{b}_3q_{12}$ which implies that

$$(q_{12}, b_3\bar{b}_{15}) = 1, \tag{4.95}$$

and (4.12) implies that q_{12} is a real element. By (4.62), (4.77), (4.83), (4.93) and $b_3(\bar{b}_3 y_8) = \bar{b}_3(b_3 y_8)$ we obtain that $q_{12} + b_3 d_8 + b_3 e_6 = \bar{b}_3 \alpha + \bar{b}_3 \beta$. Now we can assume that $(q_{12}, \bar{b}_3 \alpha) = 1$ and by (4.83) we obtain that $(y_8, \bar{b}_3 \alpha) = 1$ which implies that $|\alpha| \geq 8$, and since q_{12} is a real element we obtain that

$$(\bar{\alpha}, \bar{b}_3 q_{12}) = 1. \tag{4.96}$$

Assume henceforth that $\alpha = \bar{c}_{10}$ and we shall derive a contradiction. Then by (4.81), (4.83) and (4.94) we obtain that $(\bar{c}_{10}, \bar{b}_3 d_8) = 1$, and we have a contradiction to (4.94). By (4.86) we obtain that $\alpha \neq b_{15}$. Now by (4.93), (4.95) and (4.96) we obtain that $(c_{10}, \bar{b}_3 q_{12}) = 1$, $(\bar{\alpha}, \bar{b}_3 q_{12}) = 1$ and $(\bar{b}_{15}, \bar{b}_3 q_{12}) = 1$ which imply that either $|\alpha| \leq 7$ or $|\alpha| = 11$. Now $|\alpha| \geq 8$ implies that $\alpha = \alpha_{11}$ and by (4.83) we obtain that $\beta = \beta_7$.

Now we have that

$$b_3 q_{12} = \bar{c}_{10} + b_{15} + \alpha_{11}. \tag{4.97}$$

By (4.5), (4.8) and (4.95) we obtain that $\bar{b}_3 b_{15} = b_8 + b_{10} + q_{12} + q_{15}$ where $q_{15} \in B$. Hence (4.2), (4.93) and $\bar{b}_3(b_3 q_{12}) = (b_3 \bar{b}_3) q_{12}$ imply that

$$b_8 q_{12} = \bar{y}_8 + \bar{b}_{10} + q_{12} + b_8 + b_{10} + q_{15} + \bar{b}_3 \alpha_{11}. \tag{4.98}$$

By (4.62), (4.77), (4.83), (4.93) and $b_3(\bar{b}_3 y_8) = \bar{b}_3(b_3 y_8)$ we obtain that

$$\bar{b}_3 \alpha_{11} + \bar{b}_3 \beta_7 = q_{12} + b_3 d_8 + b_3 e_6 \tag{4.99}$$

and

$$(y_8, \bar{b}_3 \alpha_{11}) = (y_8, b_3 e_6) = 1. \tag{4.100}$$

If $(\bar{b}_3 \alpha_{11}, b_3 e_6) = 1$, then by (4.97) we obtain that $\bar{b}_3 \alpha_{11} = q_{12} + y_8 + \Sigma_{13}$ where $\Sigma_{13} \in \mathbb{N}B$ and $b_3 d_8 = y_8 + \Sigma_{13} + v_3$ where $v_3 \in B$ which implies that $(v_3, \bar{b}_3 \beta_7) = 1$ and we have a contradiction. Now we have that $(\bar{b}_3 \alpha_{11}, b_3 e_6) \geq 2$.

If $b_3 e_6 = y_8 + r_{10}$ where $r_{10} \in B$, then $\bar{b}_3 \alpha_{11} = c_3 + y_8 + r_{10} + q_{12}$ and we have a contradiction.

If $b_3 e_6 = y_8 + r_4 + r_6$ where $r_4, r_6 \in B$, then $(r_6, \bar{b}_3 \alpha_{11}) = 1$ and (4.98) implies that r_6 is a real element, and we have that $(\bar{e}_6, b_3 r_6) = (\alpha_{11}, b_3 r_6) = 1$, which is a contradiction.

If $b_3 e_6 = y_8 + r_5 + \gamma_5$ where $r_5, \gamma_5 \in B$ then we have that $(\bar{e}_6, b_3 r_5) = (\alpha_{11}, b_3 r_5) = 1$ and we have a contradiction.

4.4.3 Case $c = c_{11}$

In this section we assume that $c = c_{11}$. Then (4.77), (4.78) and (4.91) imply that

$$b_3 c_{11} = y_8 + b_{10} + q_{15} \quad \text{where } q_{15} \in B. \tag{4.101}$$

By (4.85) we obtain that

$$\bar{b}_3 c_{11} = x_6 + b_{15} + g_{12}. \tag{4.102}$$

Now by (4.61), (4.77), (4.78), (4.80) and $\bar{b}_3(b_3 c_{11}) = b_3(\bar{b}_3 c_{11})$, we obtain that

$$\bar{b}_{15} + b_3 g_{12} = e_6 + \bar{b}_3 q_{15} \tag{4.103}$$

which implies that

$$(q_{15}, b_3 \bar{b}_{15}) = 1. \tag{4.104}$$

Now (4.12) implies that q_{15} is a real element, and we obtain that

$$(b_{15}, b_3 q_{15}) = 1. \tag{4.105}$$

By (4.81), (4.83) and (4.102) we obtain that

$$b_3 y_8 = x_6 + \alpha_6 + g_{12}. \tag{4.106}$$

Thus $\beta = g_{12}$.

By (4.62), (4.77), (4.101) and $b_3(\bar{b}_3 y_8) = \bar{b}_3(b_3 y_8)$, we obtain that

$$\bar{b}_3 \alpha_6 + \bar{b}_3 g_{12} = q_{15} + b_3 d_7 + b_3 e_6. \tag{4.107}$$

Now by (4.106) we obtain that $(y_8, \bar{b}_3 \alpha_6) = 1$. Hence $(q_{15}, \bar{b}_3 \alpha_6) = 0$, which implies that

$$1 = (q_{15}, \bar{b}_3 g_{12}) = (b_3 q_{15}, g_{12}). \tag{4.108}$$

Since q_{15} is a real element, (4.101) implies that

$$(\bar{c}_{11}, b_3 q_{15}) = 1.$$

Now by (4.105) and (4.108) we obtain that

$$b_3 q_{15} = \bar{c}_{11} + g_{12} + b_{15} + z_7 \quad \text{where } z_7 \in B \tag{4.109}$$

which implies that

$$\bar{b}_3 z_7 = q_{15} + z_6 \quad \text{where } z_6 \in B. \tag{4.110}$$

By (4.103), (4.109) and since q_{15} is a real element, we have that

$$b_3 g_{12} = e_6 + \bar{z}_7 + c_{11} + \bar{g}_{12},$$

which implies that

$$b_3 z_7 = \bar{g}_{12} + r_9, \quad \text{where } r_9 \in B.$$

Now $b_3(\bar{b}_3 z_7) = \bar{b}_3(b_3 z_7)$ implies that

$$b_{15} + b_3 z_6 = \bar{e}_6 + \bar{b}_3 r_9.$$

Hence $1 = (b_{15}, \bar{b}_3 r_9) = (r_9, b_3 b_{15})$ and we have a contradiction to (4.80).

In the following three sections, we assume that

$$(b_3c, b_3c) = 4, \tag{4.111}$$

so (4.90) implies that $12 \leq |c| \leq 14$.

4.4.4 Case $c = c_{12}$

In this section we assume that $c = c_{12}$. Then by (4.61), (4.80), (4.81) and (4.111), we obtain that

$$\bar{b}_3c_{12} = x_6 + x_9 + b_{15} + g_6 \quad \text{where } g_6 \in B \tag{4.112}$$

and

$$b_3y_8 = x_6 + g_6 + \beta_{12} \quad \text{where } \beta_{12} \in B. \tag{4.113}$$

Assume henceforth that $(b_3g_6, b_3g_6) = 3$ and we shall derive a contradiction. Then (4.112) implies that $b_3g_6 = c_{12} + v_3 + w_3$ and $\bar{b}_3g_6 = y_8 + v + \mu$ where $v_3, w_3, v, \mu \in B$. Now by (4.112), (4.113) and $\bar{b}_3(b_3g_6) = b_3(\bar{b}_3g_6)$, we obtain that $x_9 + b_{15} + \bar{b}_3v_3 + \bar{b}_3w_3 = \beta_{12} + b_3v + b_3\mu$, which is a contradiction. Now we have that $(b_3g_6, b_3g_6) = 2$, by (4.112) and (4.113) we obtain that

$$b_3g_6 = c_{12} + v_6 \quad \text{where } v_6 \in B \tag{4.114}$$

and

$$\bar{b}_3g_6 = y_8 + v_{10} \quad \text{where } v_{10} \in B. \tag{4.115}$$

Now by (4.112), (4.113) and $\bar{b}_3(b_3g_6) = b_3(\bar{b}_3g_6)$ we obtain that

$$x_9 + b_{15} + \bar{b}_3v_6 = \beta_{12} + b_3v_{10},$$

which implies that

$$(v_{10}, \bar{b}_3b_{15}) = 1. \tag{4.116}$$

In addition, we have that $(b_{10}, \bar{b}_3b_{15}) = (b_8, \bar{b}_3b_{15}) = 1$ which implies that

$$\bar{b}_3b_{15} = b_8 + b_{10} + v_{10} + v_{17} \quad \text{where } v_{17} \in B. \tag{4.117}$$

Now by (4.12) we obtain that v_{10} and v_{17} are real elements. By (4.77), (4.78) and (4.111) we obtain that

$$b_3c_{12} = y_8 + b_{10} + q + u \quad \text{where } q, u \in B. \tag{4.118}$$

Now by (4.61), (4.77), (4.78), (4.80), (4.112), (4.114) and $\bar{b}_3(b_3c_{12}) = b_3(\bar{b}_3c_{12})$, we obtain that

$$e_6 + \bar{b}_3q + \bar{b}_3u = \bar{b}_{15} + c_{12} + v_6 + b_3x_9.$$

Now we can assume that $(q, b_3\bar{b}_{15}) = 1$; in addition $q \neq y_8, b_{10}$, and by (4.118) we obtain that $|q| \neq 17$. Thus $q \neq v_{17}$ which implies that $q = v_{10}$. Now since v_{10} is a real element, (4.118) implies that $(\bar{c}_{12}, b_3v_{10}) = 1$ and by (4.115) we obtain that $(g_6, b_3v_{10}) = 1$, which is a contradiction to (4.116).

4.4.5   Case $c = c_{13}$

In this section we assume that $c = c_{13}$. Then by (4.77) we obtain that

$$\bar{b}_3y_8 = d_5 + e_6 + c_{13}. \tag{4.119}$$

By (4.61) and (4.81) we obtain that either

$$\bar{b}_3d_5 = x_6 + x_9 \tag{4.120}$$

or

$$\bar{b}_3d_5 = x_6 + z_9 \quad \text{where } z_9 \in B. \tag{4.121}$$

Now we have that $(b_3d_5, b_3d_5) = 2$ and together with (4.119), we obtain that

$$b_3d_5 = y_8 + z_7 \quad \text{where } z_7 \in B. \tag{4.122}$$

Assume that $\bar{b}_3d_5 = x_6 + z_9$. Then by (4.81) we obtain that

$$\bar{b}_3c_{13} = x_6 + x_9 + b_{15} + r_9 \quad \text{where } r_9 \in B \tag{4.123}$$

and

$$b_3y_8 = x_6 + z_9 + r_9. \tag{4.124}$$

Now by (4.1), (4.121), (4.122) and $b_3(b_3d_5) = b_3^2d_5$, we obtain that $r_9 + b_3z_7 = b_6d_5$. By (4.65), (4.69) and (4.70), we obtain that $(b_6d_5, b_6d_5) \leq 3$ which implies that

$$(r_9, b_3z_7) = 0. \tag{4.125}$$

By (4.77), (4.78) and (4.123), we obtain that

$$b_3c_{13} = y_8 + b_{10} + q + u \quad \text{where } q, u \in B. \tag{4.126}$$

Now by (4.62), (4.119), (4.122), (4.124), and $b_3(\bar{b}_3y_8) = \bar{b}_3(b_3y_8)$ we obtain that

$$q + u + y_8 + z_7 + b_3e_6 = \bar{b}_3z_9 + \bar{b}_3r_9, \tag{4.127}$$

and together with (4.125), we obtain that

$$(z_9, b_3z_7) = 1. \tag{4.128}$$

By (4.126) we obtain that

$$|q| + |u| = 21,$$ (4.129)

and by (4.124) we obtain that $(y_8, \bar{b}_3 z_9) = (y_8, \bar{b}_3 r_9) = 1$. Now we can assume by (4.127) that

$$(q, \bar{b}_3 r_9) = (u, \bar{b}_3 z_9) = 1,$$ (4.130)

and together with (4.128) we obtain that $|u| \le 12$.

By (4.80) we obtain that $b_3 b_{15} = \bar{b}_{15} + b_6 + c_{13} + f_{11}$, and by (4.78) we obtain that $\bar{b}_3 b_{10} = b_6 + c_{13} + f_{11}$. Now by (4.61), (4.119), (4.123), (4.126) and $b_3(\bar{b}_3 c_{13}) = \bar{b}_3(b_3 c_{13})$, we obtain that

$$\bar{b}_{15} + b_3 x_9 + b_3 r_9 = e_6 + \bar{b}_3 q + \bar{b}_3 u$$

which implies that either $(\bar{b}_{15}, \bar{b}_3 q) = 1$ or $(\bar{b}_{15}, \bar{b}_3 u) = 1$.

Assume henceforth that $(u, b_3 \bar{b}_{15}) = 1$ and we shall derive a contradiction. Then by (4.12) we obtain that u is a real element. By (4.130) we obtain that $(z_9, b_3 u) = 1$, by (4.126) we obtain that $(\bar{c}_{13}, b_3 u) = 1$, and by the assumption we obtain that $(b_{15}, b_3 u) = 1$ which implies that $|u| \ge 14$, and we have a contradiction to $|u| \le 12$.

Now we obtain that

$$(q, b_3 \bar{b}_{15}) = 1$$ (4.131)

which implies by (4.12) that q is a real element. By (4.130) we obtain that $(r_9, b_3 q) = 1$, by (4.126) we obtain that $(\bar{c}_{13}, b_3 q) = 1$, and together with $(b_{15}, b_3 q) = 1$ we obtain that $|q| \ge 14$.

Since $(c_{13}, \bar{b}_3 u) = 1$ we obtain that $|u| \ge 6$. Now (4.129) implies that $|q| \le 15$, and we obtain that

$$14 \le |q| \le 15$$ (4.132)

and

$$6 \le |u| \le 7.$$ (4.133)

By (4.123), (4.124) and (4.130) we obtain that

$$\bar{b}_3 r_9 = y_8 + q + \lambda \quad \text{where } \lambda \in B \text{ and } 4 \le \lambda \le 5$$

and

$$b_3 r_9 = c_{13} + m + n \quad \text{where } m, n \in B.$$

Now by (4.123), (4.124) and $b_3(\bar{b}_3 r_9) = \bar{b}_3(b_3 r_9)$, we obtain that

$$z_9 + b_3 q + b_3 \lambda = x_9 + b_{15} + \bar{b}_3 m + \bar{b}_3 n.$$

Since $(r_9, b_3 \lambda) = 1$ and $4 \le \lambda \le 5$, we obtain that $(x_9, b_3 \lambda) = 0$ which implies that either $x_9 = z_9$ or $(x_9, b_3 q) = 1$. If $x_9 = z_9$ then by (4.124), (4.128) and (4.57), we obtain that $(y_8, \bar{b}_3 x_9) = (z_7, \bar{b}_3 x_9) = (b_{10}, \bar{b}_3 x_9) = 1$ and we have a contradiction.

Now we have that $(x_9, b_3q) = 1$. By (4.126), (4.130) and (4.131) we obtain that $(b_{15}, b_3q) = (\bar{c}_{13}, b_3q) = (r_9, b_3q) = 1$, and we have a contradiction to $(x_9, b_3q) = 1$.

Now we have that

$$\bar{b}_3 d_5 = x_6 + x_9. \tag{4.134}$$

By (4.81) and (4.83) we obtain that

$$\bar{b}_3 c_{13} = x_6 + b_{15} + \alpha + \beta \tag{4.135}$$

and

$$b_3 y_8 = x_6 + \alpha + \beta. \tag{4.136}$$

Now by (4.62), (4.119), (4.122), (4.126) and $\bar{b}_3(b_3 y_8) = b_3(\bar{b}_3 y_8)$ we obtain that

$$y_8 + z_7 + q + u + b_3 e_6 = \bar{b}_3 \alpha + \bar{b}_3 \beta.$$

Now we can assume that

$$(\alpha, b_3 z_7) = 1. \tag{4.137}$$

By (4.1), (4.122), (4.134), (4.136) and $b_3(\bar{b}_3 d_5) = b_3^2 d_5$, we obtain that

$$x_9 + b_6 d_5 = \alpha + \beta + b_3 z_7.$$

By (4.65), (4.69) and (4.70) we obtain that $(b_6 d_5, b_6 d_5) \le 3$ which implies that $x_9 = \alpha$. By (4.57), (4.136) and (4.137) we obtain that $(z_7, \bar{b}_3 x_9) = (b_{10}, \bar{b}_3 x_9) = (y_8, \bar{b}_3 x_9) = 1$, which is a contradiction.

4.4.6 Case $c = c_{14}$

In this section we assume that $c = c_{14}$. Then by (4.77) we obtain that

$$\bar{b}_3 y_8 = d_4 + e_6 + c_{14} \tag{4.138}$$

which implies

$$b_3 d_4 = y_8 + y_4 \quad \text{where } y_4 \in B, \tag{4.139}$$

and by (4.78) we obtain that

$$b_3 c_{14} = b_{10} + y_8 + q + u \quad \text{where } q, u \in B. \tag{4.140}$$

Now by (4.61) we obtain that

$$\bar{b}_3 d_4 = x_6 + y_6 \quad \text{where } y_6 \in B \tag{4.141}$$

and by (4.81) we obtain that

$$b_3 y_8 = x_6 + y_6 + y_{12} \quad \text{where } y_{12} \in B. \tag{4.142}$$

Now by (4.62) and $\bar{b}_3(b_3 y_8) = b_3(\bar{b}_3 y_8)$ we obtain that

$$\bar{b}_3 y_6 + \bar{b}_3 y_{12} = q + u + y_8 + y_4 + b_3 e_6. \tag{4.143}$$

By (4.139) we obtain that

$$(b_3 y_4, b_3 y_4) \neq 1 \tag{4.144}$$

which implies that

$$(y_4, \bar{b}_3 y_{12}) = 0, \tag{4.145}$$

and by (4.143) we obtain that

$$(y_4, \bar{b}_3 y_6) = 1. \tag{4.146}$$

By (4.140) we obtain that $|q| + |u| = 24$, and by (4.142) we obtain that $(y_8, \bar{b}_3 y_{12}) = 1$.

Assume henceforth that $(u, \bar{b}_3 y_{12}) = (q, \bar{b}_3 y_{12}) = 1$ and we shall derive a contradiction. Then $\bar{b}_3 y_{12} = q + u + y_8 + z_4$ where $z_4 \in B$ which implies that $b_3 z_4 = y_{12}$ and by (4.143) and (4.145) we obtain that $(z_4, b_3 e_6) = 1$, which is a contradiction to $b_3 z_4 = y_{12}$.

Now we can assume that $(u, \bar{b}_3 y_{12}) = 1$ and $(q, \bar{b}_3 y_6) = 1$. By (4.142) and (4.146) we obtain that

$$\bar{b}_3 y_6 = y_4 + y_8 + q_6,$$

and by (4.140) we obtain that

$$b_3 c_{14} = b_{10} + y_8 + q_6 + u_{18}$$

which implies that

$$\bar{b}_3 q_6 = c_{14} + z_4 \quad \text{where } z_4 \in B$$

and

$$b_3 q_6 = y_6 + z_{12} \quad \text{where } z_{12} \in B.$$

Now by $b_3(\bar{b}_3 q_6) = \bar{b}_3(b_3 q_6)$ we obtain that

$$b_{10} + u_{18} + b_3 z_4 = e_4 + \bar{b}_3 z_{12},$$

and we have a contradiction to (4.57).

4.4.7 Case $(b_3x_6, b_3x_6) = 3$

In this section we assume that $(b_3x_6, b_3x_6) = 3$ Then

$$b_3x_6 = c + d + e \quad \text{where } c, d, e \in B \tag{4.147}$$

and by (4.57) we obtain that

$$\bar{b}_3x_6 = b_{10} + v + w \quad \text{where } v, w \in B. \tag{4.148}$$

Now by (4.57) and $b_3(\bar{b}_3x_6) = \bar{b}_3(b_3x_6)$ we obtain that

$$\bar{b}_3c + \bar{b}_3d + \bar{b}_3e = b_{15} + x_6 + x_9 + b_3v + b_3w. \tag{4.149}$$

Now we can assume

$$(b_{15}, \bar{b}_3c) = 1 \tag{4.150}$$

and by (4.147) we obtain that $(x_6, \bar{b}_3c) = 1$ which implies that $|c| \geq 7$, and by (4.147) we obtain that $|c| \leq 12$. Now we have that

$$7 \leq |c| \leq 12. \tag{4.151}$$

By (4.4), (4.10) and (4.150) we obtain that

$$\bar{b}_3b_{10} = b_6 + c + f \quad \text{where } f \in B \tag{4.152}$$

and

$$\bar{b}_3b_{15} = \bar{b}_{15} + b_6 + c + f. \tag{4.153}$$

Assume henceforth that $v = v_3$ and we shall derive a contradiction. Then by (4.148) we obtain that

$$\bar{b}_3x_6 = b_{10} + v_3 + w_5,$$

$$b_3v_3 = x_6 + z_3 \quad \text{where } z_3 \in B$$

and

$$\bar{b}_3v_3 = \alpha + \beta \quad \text{where } \alpha, \beta \in B. \tag{4.154}$$

Now by $b_3(\bar{b}_3v_3) = \bar{b}_3(b_3v_3)$ we obtain that

$$b_3\alpha + b_3\beta = b_{10} + v_3 + w_5 + \bar{b}_3z_3$$

and we can assume that

$$(\alpha, \bar{b}_3b_{10}) = 1. \tag{4.155}$$

By (4.4) and (4.154) we obtain that $\alpha \neq b_6$ and $|\alpha| \leq 6$, so (4.151) and (4.152) imply that $\alpha \neq c$ and $\alpha \neq f$. Now we have a contradiction to (4.152) and (4.155).

Hence

$$\bar{b}_3 x_6 = b_{10} + v_4 + w_4 \tag{4.156}$$

which implies that

$$(x_6, b_3 b_{10}) = (x_6, b_3 v_4) = 1, \tag{4.157}$$

and we obtain

$$(b_3 v_4, b_3 v_4) = 2 \tag{4.158}$$

and $1 = (b_3 b_{10}, b_3 v_4) = (\bar{b}_3 b_{10}, \bar{b}_3 v_4)$. By (4.4) we obtain that $(b_6, \bar{b}_3 v_4) = 0$. By (4.151) and (4.152) we obtain that $12 \le f \le 17$, and together with $(b_3 v_4, b_3 v_4) = 2$ we obtain that $(f, \bar{b}_3 v_4) = 0$. Now we have that

$$(c, \bar{b}_3 v_4) = 1 \tag{4.159}$$

which implies that $|c| \le 9$, and together with (4.151) we obtain that

$$7 \le c \le 9.$$

If $c = c_7$ then by (4.147) and (4.153) we obtain that $\bar{b}_3 c_7 = b_{15} + x_6$, and by (4.152) and (4.159) we obtain that $b_3 c_7 = v_4 + b_{10}$, which is a contradiction.

If $c = c_8$ then by (4.147) and (4.153), we obtain that $\bar{b}_3 c_8 = b_{15} + x_6 + r_3$ where $r_3 \in B$. Now we have that $r_3 = \bar{b}_3$ and $b_8 = c_8$, since b_{15} is a non-real element we have a contradiction to (4.5).

Now we have that $c = c_9$ then by (4.147) and (4.153), we obtain that

$$\bar{b}_3 c_9 = b_{15} + x_6 + \Sigma_6 \quad \text{where } \Sigma_6 \in \mathbb{N}B, \tag{4.160}$$

and by (4.152) and (4.159) we obtain that

$$b_3 c_9 = v_4 + b_{10} + \Sigma_{13} \quad \text{where } \Sigma_{13} \in \mathbb{N}B. \tag{4.161}$$

By (4.157), (4.158) and (4.159) we obtain that

$$b_3 v_4 = x_6 + m_6 \quad \text{where } m_6 \in B$$

and

$$\bar{b}_3 v_4 = c_9 + m_3 \quad \text{where } m_3 \in B.$$

Now by (4.156) and $\bar{b}_3 (b_3 v_4) = b_3 (\bar{b}_3 v_4)$ we obtain that

$$w_4 + \bar{b}_3 m_6 = \Sigma_{13} + b_3 m_3.$$

Now $(v_4, b_3 m_3) = 1$ implies that $(w_4, b_3 m_3) = 0$, so $(w_4, \Sigma_{13}) = 1$. By (4.161) we obtain that

$$b_3 c_9 = v_4 + b_{10} + w_4 + w_9 \quad \text{where } w_9 \in B,$$

and by (4.160) we obtain that

$$\bar{b}_3 c_9 = b_{15} + x_6 + l_3 + n_3 \quad \text{where } l_3, n_3 \in B.$$

Now by (4.149) we obtain that $b_3 w_4 = x_6 + l_3 + n_3$, which is a contradiction to $(b_3 w_4, b_3 w_4) \leq 2$.

4.5 Case $R_{15} = x_7 + x_8$

Throughout this section we assume that $R_{15} = x_7 + x_8$. Then by (4.7) and (4.8) we have that

$$b_3 b_{10} = b_{15} + x_7 + x_8, \tag{4.162}$$

$$b_6^2 = \bar{b}_6 + b_{15} + x_7 + x_8. \tag{4.163}$$

Now $(b_6 \bar{b}_6, b_6 \bar{b}_6) = 4$ and by (4.4) and (4.2), we obtain that

$$b_6 \bar{b}_6 = 1 + b_8 + s + t \quad \text{where } s, t \text{ are real elements in } B \tag{4.164}$$

and by (4.12), we obtain that

$$\bar{b}_3 b_{15} = b_8 + b_{10} + s + t. \tag{4.165}$$

Assume henceforth that $s = s_{21}$ and we shall derive a contradiction. Then by (4.165) $t = t_6$ and $b_3 t_6 = b_{15} + c_3$ where $c_3 \in B$. Now $\bar{b}_3(b_3 t_6) = (b_3 \bar{b}_3)t_6$ implies that $b_8 t_6 = b_8 + b_{10} + s_{21} + \bar{b}_3 c_3$. Since b_8 and t_6 are real elements and b_{10} is a non-real element, we obtain that $(\bar{b}_{10}, \bar{b}_3 c_3) = 1$ and we have a contradiction. Now we have that

$$|s| \neq 21. \tag{4.166}$$

By (4.13) we obtain that

$$b_6 b_{10} = \bar{b}_{15} + b_3 x_7 + b_3 x_8. \tag{4.167}$$

By (4.162) we obtain that

$$(b_{10}, \bar{b}_3 x_7) = 1 \tag{4.168}$$

which implies that $2 \leq (\bar{b}_3 x_7, \bar{b}_3 x_7) \leq 3$. Assume that $(b_3 x_7, b_3 x_7) = 2$. Then we obtain that

$$b_3 x_7 = c + d \quad \text{where } c, d \in B, \tag{4.169}$$

and by (4.162) we obtain that

$$\bar{b}_3 x_7 = b_{10} + y_{11} \quad \text{where } y_{11} \in B. \tag{4.170}$$

By (4.1) and (4.162) we obtain that $(b_3 x_7, \bar{b}_3 b_{10}) = (b_3^2, \bar{x}_7 b_{10}) = (\bar{b}_3, \bar{x}_7 b_{10}) + (b_6, \bar{x}_7 b_{10}) = (x_7, b_3 b_{10}) + (b_6, \bar{x}_7 b_{10}) = 1 + (b_6, \bar{x}_7 b_{10})$ which implies that $(b_3 x_7, \bar{b}_3 b_{10}) \geq 1$. By (4.4) we obtain that $(b_6, \bar{b}_3 b_{10}) = 1$ which implies by (4.169) that $(b_3 x_7, \bar{b}_3 b_{10}) \neq 2$. Therefore

$$(b_3 x_7, \bar{b}_3 b_{10}) = 1.$$

Now we can assume that

$$\bar{b}_3 b_{10} = b_6 + c + e \quad \text{where } e \in B, e \neq c \text{ and } e \neq d, \tag{4.171}$$

and by (4.10) we obtain that

$$b_3 b_{15} = \bar{b}_{15} + b_6 + c + e. \tag{4.172}$$

Now we obtain by (4.169) that $(b_{15}, \bar{b}_3 c) = (x_7, \bar{b}_3 c) = 1$ which implies that

$$9 \leq |c| \leq 17. \tag{4.173}$$

By (4.162), (4.169, (4.170) and $b_3(\bar{b}_3 x_7) = \bar{b}_3(b_3 x_7)$ we obtain that

$$\bar{b}_3 c + \bar{b}_3 d = b_{15} + x_7 + x_8 + b_3 y_{11}. \tag{4.174}$$

Assume henceforth that $(b_{15}, b_3 y_{11}) = 1$ and we shall derive a contradiction. Then we have that $(b_{15}, \bar{b}_3 d) = 1$. By (4.171) we obtain that $e \neq d$, by (4.169) and (4.3) we obtain that $b_6 \neq d$ and by (4.173) and (4.169) we obtain that $d \neq \bar{b}_{15}$, which is a contradiction to (4.172). Now we have that

$$(b_{15}, b_3 y_{11}) = 0. \tag{4.175}$$

Assume henceforth that $(\bar{b}_{15}, \bar{b}_3 y_{11}) = 1$ and we shall derive a contradiction. Then by (4.12) we obtain that y_{11} is a real element which implies that $(b_{15}, b_3 y_{11}) = 1$, and we have a contradiction to (4.175). Now we have that

$$(\bar{b}_{15}, \bar{b}_3 y_{11}) = 0. \tag{4.176}$$

By (4.1), (4.169), (4.170), (4.171) and $\bar{b}_3(\bar{b}_3 x_7) = \bar{b}_3^2 x_7$ we obtain that

$$b_6 + e + \bar{b}_3 y_{11} = d + \bar{b}_6 x_7. \tag{4.177}$$

Assume that $(u, \bar{b}_3 c) = 1$ where $u \in B$. Then by (4.174) we obtain that $|u| \geq 5$.

Assume henceforth that $u = u_5$ and we shall derive a contradiction. Then by (4.174) we obtain that $\bar{b}_3 u_5 = y_{11} + y_4$ where $y_4 \in B$ and $b_3 u_5 = c + q$ where $q \in B$. Now by $b_3(\bar{b}_3 u_5) = \bar{b}_3(b_3 u_5)$ we obtain that $\bar{b}_3 c + \bar{b}_3 q = b_3 y_{11} + b_3 y_4$. Now by (4.172) we obtain that $(b_{15}, \bar{b}_3 c) = 1$ and by (4.175) we obtain that $(b_{15}, b_3 y_{11}) = 0$ which implies that $(b_{15}, b_3 y_4) = 1$, and we have a contradiction. Hence

$$\text{If } (u, \bar{b}_3 c) = 1 \quad \text{where } u \in B \quad \text{then } |u| \geq 6. \tag{4.178}$$

Assume henceforth that $c = c_9$ and we shall derive a contradiction. Then by (4.169) and (4.172) we obtain that $\bar{b}_3 c_9 = x_7 + b_{15} + u_5$ where $u_5 \in B$, which is a contradiction to (4.178). Now (4.173) implies that

$$10 \le |c| \le 17. \tag{4.179}$$

Assume henceforth that

$$(y_{11}, b_3 e) = 1 \tag{4.180}$$

and we shall derive a contradiction. Then by (4.171) and (4.177) we obtain that $(b_6 \bar{b}_6, x_7 \bar{x}_7) = (\bar{b}_6 x_7, \bar{b}_6 x_7) \ge 6$. By assumption $(b_3 \bar{b}_3, x_7 \bar{x}_7) = 2$ which implies by (4.2) that $(b_8, x_7 \bar{x}_7) = 1$. Now by (4.164) we obtain that $6 \le (b_6 \bar{b}_6, x_7 \bar{x}_7) = 2 + (s, x_7 \bar{x}_7) + (t, x_7 \bar{x}_7)$. Hence $(s, x_7 \bar{x}_7) + (t, x_7 \bar{x}_7) \ge 4$ which implies that

$$\text{either} \quad |s| \le 10 \quad \text{or} \quad |t| \le 10. \tag{4.181}$$

By (4.3) and (4.169), we obtain that $d \ne b_6$ and by (4.171) we obtain that $d \ne e$. Now (4.177) implies that $(d, \bar{b}_3 y_{11}) = 1$ and by (4.180) we obtain that

$$\bar{b}_3 y_{11} = d + e + \Sigma \quad \text{where } \Sigma \in \mathbb{N}B. \tag{4.182}$$

Now (4.171) implies that

$$b_3 e = b_{10} + y_{11} + C \quad \text{where } C \in \mathbb{N}B \tag{4.183}$$

and by (4.172) we obtain that

$$\bar{b}_3 e = b_{15} + D \quad \text{where } D \in \mathbb{N}B. \tag{4.184}$$

Now by (4.171), (4.172) and $b_3(\bar{b}_3 e) = \bar{b}_3(b_3 e)$ we obtain that

$$b_{15} + b_3 D = \bar{b}_3 y_{11} + \bar{b}_3 C. \tag{4.185}$$

Now (4.176) implies that

$$(\bar{b}_{15}, \bar{b}_3 C) = (b_3 \bar{b}_{15}, C) = 1. \tag{4.186}$$

By (4.165), (4.170), (4.183) and since b_{10} is a non-real element we obtain that $e \ne \bar{x}_7$, $e \ne \bar{x}_8$ and $e \ne \bar{b}_{15}$. Thus $(\bar{b}_{10}, b_3 e) = (\bar{b}_3 b_{10}, e) = 0$ and $(b_8, b_3 e) = (\bar{b}_3 b_8, e) = 0$. Now by (4.165), (4.183) and (4.186) we can assume that $(s, C) = 1$ which implies that

$$(s, b_3 e) = 1 \quad \text{and} \quad |s| \le |C|. \tag{4.187}$$

By (4.183) we obtain that $|C| = 3|e| - 21$ which implies that

$$|s| \le 3|e| - 21. \tag{4.188}$$

By (4.165) and (4.187) we obtain that

$$(\bar{b}_3 s, e) = (\bar{b}_3 s, \bar{b}_{15}) = 1, \qquad (4.189)$$

and together with (4.188) we obtain that $11 \leq |e|$.

By (4.171) and (4.179) we obtain that $|e| \leq 14$. Therefore

$$11 \leq |e| \leq 14. \qquad (4.190)$$

Assume henceforth that $e = e_{11}$ and we shall derive a contradiction. Then by (4.189) we obtain that $11 \leq |s|$ and by (4.183) we obtain that $|C| = 12$ which implies by (4.188) that $|s| \leq 12$. Hence $11 \leq |s| \leq 12$ which implies by (4.164) that $15 \leq |t| \leq 16$, and we have a contradiction to (4.181).

Assume henceforth that $e = e_{14}$ and we shall derive a contradiction. Then by (4.189) we obtain that $12 \leq |s|$ and by (4.183) we obtain that $|C| = 21$ which implies by (4.188) that $|s| \leq 21$. Therefore

$$12 \leq |s| \leq 21.$$

By (4.166) and (4.183) we obtain that $12 \leq |s| \leq 15$, and by (4.164) we obtain that $12 \leq |t| \leq 15$, which is a contradiction to (4.181).

Assume henceforth that $e = e_{12}$ and we shall derive a contradiction. Then by (4.189) we obtain that $9 \leq |s|$ and by (4.188) we obtain that $|s| \leq 15$. Hence $9 \leq |s| \leq 15$. Now (4.164) and (4.181) imply that $9 \leq |s| \leq 10$. By (4.189) it is impossible that $s = s_{10}$. Therefore $s = s_9$ and $\bar{b}_3 s_9 = e_{12} + \bar{b}_{15}$ and by (4.183) we obtain that

$$b_3 e_{12} = s_9 + y_{11} + b_{10} + w_6 \quad \text{where } w_6 \in B.$$

Now by (4.2), (4.165) and $b_3(\bar{b}_3 s_9) = (b_3 \bar{b}_3) s_9$ we obtain that

$$b_8 s_9 = b_8 + s_9 + b_{10} + w_6 + \bar{b}_{10} + t_{18} + y_{11}. \qquad \cdot$$

Since b_8 and s_9 are real elements we obtain that y_{11} is a real element. By (4.172) we obtain that $c = c_{12}$ and by (4.169) $d = d_9$. Now (4.170) and (4.182) imply that

$$\bar{b}_3 y_{11} = e_{12} + d_9 + \bar{x}_7 + w_5 \quad \text{where } w_5 \in B. \qquad (4.191)$$

Now we have that

$$b_3 w_5 = y_{11} + v_4 \quad \text{where } v_4 \in B \qquad (4.192)$$

and by (4.174) we obtain that either $(\bar{w}_5, \bar{b}_3 c_{12}) = 1$ or $(\bar{w}_5, \bar{b}_3 d_9) = 1$. Now by (4.178) we obtain that $(\bar{w}_5, \bar{b}_3 d_9) = 1$ which implies that

$$\bar{b}_3 w_5 = \bar{d}_9 + w_6 \quad \text{where } w_6 \in B.$$

Now by (4.191), (4.192) and $\bar{b}_3(b_3 w_5) = b_3(\bar{b}_3 w_5)$ we obtain that $e_{12} + d_9 + \bar{x}_7 + w_5 + \bar{b}_3 v_4 = b_3 \bar{d}_9 + b_3 w_6$. Since $(w_5, b_3 w_6) = 1$ we obtain that $(e_{12}, b_3 w_6) = 0$

which implies that $(e_{12}, b_3\bar{d}_9) = 1$. By (4.169) we obtain that $(\bar{x}_7, b_3\bar{d}_9) = 1$ and together with $(w_5, b_3\bar{d}_9) = 1$, we obtain that

$$b_3\bar{d}_9 = w_5 + \bar{x}_7 + e_{12} + r_3 \quad \text{where } r_3 \in B$$

which implies that $(r_3, \bar{b}_3 v_4) = 1$ and $b_3\bar{r}_3 = d_9$, which is a contradiction.

Now we have that $e = e_{13}$ which implies by (4.188) that $|s| \le 18$, and by (4.189) we obtain that $11 \le |s|$. Therefore $11 \le |s| \le 18$. Now (4.181) and (4.164) imply that $17 \le |s| \le 18$. By (4.183) and (4.187) we obtain that $|s| \ne 17$. Thus $s = s_{18}$ and by (4.165) we obtain that $t = t_9$ and

$$\bar{b}_3 t_9 = \bar{b}_{15} + E_{12} \quad \text{where } E_{12} \in \mathbb{N}B.$$

Now by (4.2), (4.165) and $b_3(\bar{b}_3 t_9) = (b_3\bar{b}_3)t_9$ we obtain that

$$b_8 t_9 = b_8 + \bar{b}_{10} + s_{18} + b_3 E_{12}.$$

Now since b_8 and t_9 are real elements and b_{10} is a non-real element, we obtain that $(\bar{b}_{10}, b_3 E_{12}) = 1$, which is a contradiction to (4.171).

Now we have that

$$(y_{11}, b_3 e) = 0. \tag{4.193}$$

By (4.4), (4.162), (4.170), (4.171) and $b_3(\bar{b}_3 b_{10}) = \bar{b}_3(b_3 b_{10})$ we obtain that

$$b_8 + b_3 c + b_3 e = \bar{b}_3 b_{15} + \bar{b}_3 x_8 + y_{11}. \tag{4.194}$$

Now (4.193) implies that $(y_{11}, b_3 c) = 1$ and together with (4.171) we obtain that

$$b_3 c = b_{10} + y_{11} + F \quad \text{where } F \in \mathbb{N}B. \tag{4.195}$$

Now by (4.177) we obtain that

$$\bar{b}_3 y_{11} = c + d + G_{12} \quad \text{where } G_{12} \in \mathbb{N}B. \tag{4.196}$$

By (4.169) and (4.172) we obtain that

$$\bar{b}_3 c = x_7 + b_{15} + H \quad \text{where } H \in \mathbb{N}B. \tag{4.197}$$

Now by (4.169), (4.171), (4.172) and $b_3(\bar{b}_3 c) = \bar{b}_3(b_3 c)$ we obtain that

$$\bar{b}_{15} + b_3 H = G_{12} + \bar{b}_3 F. \tag{4.198}$$

If $c = c_{10}$ then by (4.197) we obtain that $H = H_8 \in B$ and $(c_{10}, b_3 H_8) = 1$, which is a contradiction to (4.198), and if $c = c_{17}$ then (4.169) implies that $d = d_4$, which is a contradiction to (4.196). Now (4.179) implies that

$$11 \le |c| \le 16.$$

From (4.169) and (4.171) we have seven cases (see Table 4.3). In the following seven sections, we derive a contradiction for each case in Table 4.3.

Table 4.3 Splitting to seven cases

Case 1	$c = c_{11}$	$d = d_{10}$	$e = e_{13}$
Case 2	$c = c_{12}$	$d = d_9$	$e = e_{12}$
Case 3	$c = c_{13}$	$d = d_8$	$e = e_{11}$
Case 4	$c = c_{14}$	$d = d_7$	$e = e_{10}$
Case 5	$c = c_{15}$	$d = d_6$	$e = e_9$
Case 6	$c = c_{16}$	$d = d_5$	$e = e_8$
Case 7	$(b_3 x_7, b_3 x_7) = 3$		

4.5.1 Case $c = c_{11}$

In this section we assume that $c = c_{11}$. Then (4.178) and (4.197) imply that

$$\bar{b}_3 c_{11} = x_7 + b_{15} + f_{11} \quad \text{where } f_{11} \in B. \tag{4.199}$$

Now (4.195) implies that

$$b_3 c_{11} = b_{10} + y_{11} + g_{12} \quad \text{where } g_{12} \in B. \tag{4.200}$$

By (4.198) we obtain that

$$\bar{b}_{15} + b_3 f_{11} = G_{12} + \bar{b}_3 g_{12} \tag{4.201}$$

which implies that $(g_{12}, b_3 \bar{b}_{15}) = 1$, and by (4.165) we obtain that

$$b_3 \bar{b}_{15} = b_8 + \bar{b}_{10} + g_{12} + g_{15} \quad \text{where } g_{15} \in B. \tag{4.202}$$

By (4.170), (4.174) and (4.199) we obtain that

$$b_3 y_{11} = x_7 + f_{11} + \Sigma_{15} \quad \text{where } \Sigma_{15} \in \mathbb{N}B. \tag{4.203}$$

By (4.196) we obtain that

$$\bar{b}_3 y_{11} = c_{11} + d_{10} + G_{12}. \tag{4.204}$$

Assume henceforth that y_{11} is a real element and we shall derive a contradiction. Then by (4.203) and (4.204) we obtain that $c_{11} = \bar{f}_{11}$ and

$$b_3 y_{11} = x_7 + \bar{d}_{10} + \bar{c}_{11} + h_5 \quad \text{where } h_5 \in B \tag{4.205}$$

which implies that

$$\bar{b}_3 h_5 = y_{11} + h_4 \quad \text{where } h_4 \in B. \tag{4.206}$$

By (4.174), (4.199) and (4.205) we obtain that

$$\bar{b}_3 d_{10} = x_7 + x_8 + \bar{d}_{10} + h_5 \tag{4.207}$$

which implies that

$$b_3 h_5 = d_{10} + v_5 \quad \text{where } v_5 \in B. \tag{4.208}$$

Now $b_3(\bar{b}_3 h_5) = \bar{b}_3(b_3 h_5)$ implies that $\bar{c}_{11} + b_3 h_4 = x_8 + \bar{b}_3 v_5$. Hence $(v_5, b_3 \bar{c}_{11}) = 1$, which is a contradiction to (4.199).

Now we have that y_{11} is a non-real element. By (4.200) and (4.201) we obtain that

$$\bar{b}_3 g_{12} = \bar{b}_{15} + c_{11} + \Sigma_{10} \quad \text{where } \Sigma_{10} \in \mathbb{N}B. \tag{4.209}$$

Now by (4.2), (4.200), (4.202) and $b_3(\bar{b}_3 g_{12}) = (b_3 \bar{b}_3) g_{12}$, we obtain that

$$b_8 g_{12} = b_8 + b_{10} + \bar{b}_{10} + y_{11} + g_{12} + g_{15} + b_3 \Sigma_{10}. \tag{4.210}$$

Since b_8 and g_{12} are real elements and y_{11} is a non-real element, we obtain that $(\bar{y}_{11}, b_3 \Sigma_{10}) = 1$. Now (4.209) implies that $\Sigma_{10} \in B$ and by (4.203) we obtain that

$$b_3 y_{11} = \bar{\Sigma}_{10} + x_7 + f_{11} + r_5 \quad \text{where } r_5 \in B. \tag{4.211}$$

Now by (4.174) and (4.199) we obtain that

$$\bar{b}_3 d_{10} = r_5 + x_7 + x_8 + \bar{\Sigma}_{10}. \tag{4.212}$$

Now

$$\bar{b}_3 r_5 = y_{11} + h_4 \quad \text{where } h_4 \in B$$

and

$$b_3 r_5 = d_{10} + v_5 \quad \text{where } v_5 \in B. \tag{4.213}$$

By $b_3(\bar{b}_3 r_5) = \bar{b}_3(b_3 r_5)$ we obtain that

$$f_{11} + b_3 h_4 = x_8 + \bar{b}_3 v_5$$

which implies that $(f_{11}, \bar{b}_3 v_5) = 1$, and we have a contradiction to (4.213).

4.5.2 Case $c = c_{12}$

In this section we assume that $c = c_{12}$. Then (4.195) and (4.197) imply that

$$\bar{b}_3 c_{12} = x_7 + b_{15} + H_{14} \tag{4.214}$$

and

$$b_3 c_{12} = b_{10} + y_{11} + F_{15}. \tag{4.215}$$

Assume henceforth that y_{11} is a real element and we shall derive a contradiction. Then by (4.170) and (4.196) we obtain that

$$b_3 y_{11} = x_7 + \bar{d}_9 + \bar{c}_{12} + h_5 \quad \text{where } h_5 \in B. \tag{4.216}$$

Now by (4.174) and (4.214) we obtain that

$$H_{14} + \bar{b}_3 d_9 = x_8 + \bar{c}_{12} + \bar{d}_9 + x_7 + h_5$$

which implies that

$$\bar{b}_3 c_{12} = x_7 + b_{15} + \bar{d}_9 + h_5,$$

and we have a contradiction to (4.178).

Now we have that y_{11} is a non-real element. By (4.198) we obtain that

$$\bar{b}_{15} + b_3 H_{14} = G_{12} + \bar{b}_3 F_{15}, \tag{4.217}$$

which implies that $\bar{b}_{15} \in Irr(\bar{b}_3 F_{15})$. By (4.5) we obtain that $(b_8, F_{15}) = (b_8, b_3 c_{12}) = 0$, and by (4.162) we obtain that $(\bar{b}_{10}, F_{15}) = (\bar{b}_{10}, b_3 c_{12}) = 0$. Hence by (4.165) we can assume that

$$s \in Irr(F_{15}) \tag{4.218}$$

which implies that $|s| \leq 15$, and by (4.164) s is a real element. Now we have that

$$(s, b_3 c_{12}) = 1 \tag{4.219}$$

and by (4.165) we obtain that

$$(s, b_3 \bar{b}_{15}) = 1 \tag{4.220}$$

which implies that $9 \leq |s|$. Hence

$$9 \leq |s| \leq 15. \tag{4.221}$$

Assume henceforth that $s = s_{15}$ and we shall derive a contradiction. Then by (4.215) and (4.218) we obtain that

$$b_3 c_{12} = b_{10} + y_{11} + s_{15}. \tag{4.222}$$

Now (4.165) implies that

$$\bar{b}_3 s_{15} = c_{12} + \bar{b}_{15} + \Sigma_{18} \quad \text{where } \Sigma_{18} \in \mathbb{N}B \tag{4.223}$$

and (4.214) implies that

$$\bar{b}_3 c_{12} = x_7 + b_{15} + f_{14} \quad \text{where } f_{14} \in B. \tag{4.224}$$

Now by (4.2), (4.165) and $b_3(\bar{b}_3 s_{15}) = (b_3 \bar{b}_3) s_{15}$ we obtain that

$$b_8 s_{15} = b_8 + b_{10} + \bar{b}_{10} + y_{11} + t_{12} + s_{15} + b_3 \Sigma_{18}.$$

Since b_8 and s_{15} are real elements and y_{11} is a non-real element, we obtain that

$$\bar{y}_{11} \in Irr(b_3 \Sigma_{18}).$$

By (4.174) we obtain that $(f_{14}, b_3 y_{11}) = 1$, and together with (4.170) we obtain that

$$b_3 y_{11} = x_7 + f_{14} + S_{12} \quad \text{where } S_{12} \in \mathbb{N}B. \tag{4.225}$$

Since $\bar{y}_{11} \in Irr(b_3 \Sigma_{18})$ we obtain that there exists an element α in B such that $\alpha \in Irr(\bar{\Sigma}_{18}) \cap Irr(S_{12})$ which implies that $10 \le |\alpha|$. Hence $S_{12} = \alpha_{12}$. Now we have that

$$\bar{b}_3 s_{15} = c_{12} + \bar{b}_{15} + \bar{\alpha}_{12} + \alpha_6 \quad \text{where } \alpha_6 \in B \tag{4.226}$$

which implies that

$$b_3 \alpha_6 = s_{15} + \alpha_3 \quad \text{where } \alpha_3 \in B, \tag{4.227}$$

and by (4.225) we obtain that

$$b_3 y_{11} = x_7 + f_{14} + \alpha_{12}.$$

By (4.196) we obtain that

$$\bar{b}_3 y_{11} = c_{12} + d_9 + \beta_{12} \quad \text{where } \beta_{12} \in B.$$

By (4.169), (4.171) (4.172), (4.224), (4.226) and $b_3(\bar{b}_3 c_{12}) = \bar{b}_3(b_3 c_{12})$, we obtain that

$$b_3 f_{14} = \beta_{12} + c_{12} + \bar{\alpha}_{12} + \alpha_6.$$

Now we have that

$$\bar{b}_3 \alpha_6 = f_{14} + \alpha_4 \quad \text{where } \alpha_4 \in B.$$

Now by (4.226), (4.227) and $b_3(\bar{b}_3 \alpha_6) = \bar{b}_3(b_3 \alpha_6)$ we obtain that $\bar{b}_{15} + \bar{b}_3 \alpha_3 = \beta_{12} + b_3 \alpha_4$, and we have a contradiction.

Now we have that $|s| \ne 15$, so (4.215), (4.219) and (4.221) imply that

$$b_3 c_{12} = b_{10} + y_{11} + s + k \quad \text{where } k \in B \text{ and } 9 \le |s| \le 11.$$

Now (4.219) and (4.220) imply that

$$\bar{b}_3 s = c_{12} + \bar{b}_{15} + T \quad \text{where } T \in \mathbb{N}B \text{ and } |T| \le 6, \tag{4.228}$$

and together with (4.165) and $b_3(\bar{b}_3 s) = (b_3 \bar{b}_3)s$ we obtain that

$$b_8 s = b_8 + b_{10} + \bar{b}_{10} + y_{11} + s + t + k + b_3 T.$$

Since b_8 and s are real elements and y_{11} is a non-real element, we obtain that $\bar{y}_{11} \in Irr(b_3 T)$, which is a contradiction to (4.228).

4.5.3 Case $c = c_{13}$

In this section we assume that $c = c_{13}$. Then (4.195) and (4.197) imply that

$$\bar{b}_3 c_{13} = x_7 + b_{15} + H_{17} \tag{4.229}$$

and

$$b_3 c_{13} = b_{10} + y_{11} + F_{18}. \tag{4.230}$$

Assume henceforth that y_{11} is a real element and we shall derive a contradiction. Then by (4.170) and (4.196) we obtain that

$$b_3 y_{11} = x_7 + \bar{d}_8 + \bar{c}_{13} + h_5 \quad \text{where } h_5 \in B. \tag{4.231}$$

Now by (4.174) and (4.229) we obtain that

$$H_{17} + \bar{b}_3 d_8 = x_8 + \bar{c}_{13} + \bar{d}_8 + x_7 + h_5,$$

and we have a contradiction.

Now we have that y_{11} is a non-real element. By (4.198) we obtain that

$$\bar{b}_{15} + b_3 H_{17} = G_{12} + \bar{b}_3 F_{18}.$$

Now we have that $\bar{b}_{15} \in Irr(\bar{b}_3 F_{18})$. By (4.5), (4.162) and (4.230) we can assume that

$$s \in Irr(F_{18}) \tag{4.232}$$

and together with (4.165) we obtain that

$$11 \le |s| \le 18. \tag{4.233}$$

Assume henceforth that $s = s_{18}$ and we shall derive a contradiction. Then (4.230) and (4.232) imply that

$$b_3 c_{13} = b_{10} + y_{11} + s_{18}. \tag{4.234}$$

By (4.229) we obtain that

$$\bar{b}_3 c_{13} = x_7 + b_{15} + f_{17} \tag{4.235}$$

and by (4.165) we obtain that

$$b_3 \bar{b}_{15} = b_8 + \bar{b}_{10} + s_{18} + t_9. \tag{4.236}$$

Now we have that

$$\bar{b}_3 s_{18} = c_{13} + \bar{b}_{15} + \Sigma_{26} \quad \text{where } \Sigma_{26} \in \mathbb{N}B. \tag{4.237}$$

By (4.2) and $b_3(\bar{b}_3 s_{18}) = (b_3\bar{b}_3)s_{18}$ we obtain that

$$b_8 s_{18} = b_8 + t_9 + b_{10} + \bar{b}_{10} + y_{11} + s_{18} + b_3 \Sigma_{26}.$$

Since b_8 and s_{18} are real elements and y_{11} is a non-real element, we obtain that

$$\bar{y}_{11} \in Irr(b_3 \Sigma_{26}). \tag{4.238}$$

By (4.174) and (4.235) we obtain that

$$\bar{b}_3 d_8 = x_7 + x_8 + g_9 \quad \text{where } g_9 \in B \tag{4.239}$$

and

$$b_3 y_{11} = x_7 + f_{17} + g_9. \tag{4.240}$$

Now (4.196) implies that

$$\bar{b}_3 y_{11} = c_{13} + d_8 + h_{12} \quad \text{where } h_{12} \in B. \tag{4.241}$$

By (4.237) and (4.240) we obtain that $\bar{g}_9 \notin \Sigma_{26}$, and by (4.170) and (4.237) we obtain that $\bar{x}_7 \notin \Sigma_{26}$. Now (4.238) and (4.240) imply that $\bar{f}_{17} \in \Sigma_{26}$. By (4.237) we obtain that

$$\bar{b}_3 s_{18} = c_{13} + \bar{b}_{15} + \bar{f}_{17} + f_9 \quad \text{where } f_9 \in B. \tag{4.242}$$

By (4.198) we obtain that

$$b_3 f_{17} = h_{12} + c_{13} + \bar{f}_{17} + f_9 \tag{4.243}$$

which implies that

$$\bar{b}_3 f_9 = f_{17} + \Sigma_{10} \quad \text{where } \Sigma_{10} \in \mathbb{N}B \tag{4.244}$$

and

$$b_3 f_9 = s_{18} + \Sigma_9 \quad \text{where } \Sigma_9 \in \mathbb{N}B.$$

Now by (4.242), (4.243) and $b_3(\bar{b}_3 f_9) = \bar{b}_3(b_3 f_9)$ we obtain that

$$h_{12} + b_3 \Sigma_{10} = \bar{b}_{15} + \bar{b}_3 \Sigma_9$$

which implies that $\bar{b}_{15} \in Irr(b_3 \Sigma_{10})$, and by (4.172) we obtain that $\bar{b}_6 \in \Sigma_{10}$. Now (4.244) implies that $(f_9, b_3\bar{b}_6) = 1$, which is a contradiction to (4.3).

Now we have that $|s| \neq 18$ which implies by (4.230) and (4.233) that $11 \leq |s| \leq 12$. By (4.165), (4.230) and (4.232) we obtain that

$$\bar{b}_3 s = c_{13} + \bar{b}_{15} + T \quad \text{where } T \in \mathbb{N}B \text{ and } |T| \leq 8. \tag{4.245}$$

Now by (4.2), (4.230), (4.165) and $b_3(\bar{b}_3 s) = (b_3\bar{b}_3)s$ we obtain that

$$b_8 s = b_8 + b_{10} + \bar{b}_{10} + y_{11} + t + F_{18} + b_3 T.$$

Since b_8 and s are real elements and y_{11} is a non-real element, we obtain that $s \neq \bar{y}_{11}$; and by (4.232) and $11 \leq |s| \leq 12$, we have that $\bar{y}_{11} \notin F_{18}$. Hence $\bar{y}_{11} \in b_3 T$, which is a contradiction to (4.245).

4.5.4 Case $c = c_{14}$

In this section we assume that $c = c_{14}$. Then (4.195) and (4.197) imply that

$$\bar{b}_3 c_{14} = x_7 + b_{15} + H_{20} \tag{4.246}$$

and

$$\bar{b}_3 c_{14} = b_{10} + y_{11} + F_{21}. \tag{4.247}$$

By (4.169) we obtain that

$$\bar{b}_3 d_7 = x_7 + \Sigma_{14} \quad \text{where } \Sigma_{14} \in \mathbb{N}B. \tag{4.248}$$

By (4.198) we obtain that

$$\bar{b}_{15} + b_3 H_{20} = G_{12} + \bar{b}_3 F_{21}$$

which implies that $\bar{b}_{15} \in Irr(\bar{b}_3 F_{21})$. By (4.5) and (4.162) we obtain that $b_8 \notin Irr(b_3 c_{14})$ and $\bar{b}_{10} \notin Irr(b_3 c_{14})$, and by (4.165) and (4.247) we can assume that

$$s \in Irr(F_{21}). \tag{4.249}$$

By (4.165) we obtain that $(s, b_3 \bar{b}_{15}) = 1$ and together with (4.166) we obtain that

$$12 \leq |s| \leq 15 \tag{4.250}$$

and

$$\bar{b}_3 s = \bar{b}_{15} + c_{14} + T \quad \text{where } T \in \mathbb{N}B. \tag{4.251}$$

Assume henceforth that y_{11} is a real element and we shall derive a contradiction. Then by (4.170) and (4.196) we obtain that either

$$b_3 y_{11} = x_7 + \bar{d}_7 + \bar{c}_{14} + h_5 \quad \text{where } h_5 \in B \tag{4.252}$$

or $x_7 = \bar{d}_7$ and then

$$b_3 y_{11} = x_7 + \bar{c}_{14} + \Sigma_{12} \quad \text{where } \Sigma_{12} \in \mathbb{N}B. \tag{4.253}$$

If (4.252) holds then by (4.174) and (4.246), we obtain that

$$H_{20} + \bar{b}_3 d_7 = x_8 + \bar{c}_{14} + \bar{d}_7 + x_7 + h_5.$$

By (4.169) we obtain that $(x_7, \bar{b}_3 d_7) = 1$ which implies that $H_{20} = x_8 + \bar{d}_7 + h_5$, and we have a contradiction to (4.246).

Now we have that $x_7 = \bar{d}_7$. By (4.247), (4.249) and (4.250) we obtain that

$$b_3 c_{14} = b_{10} + y_{11} + s + g \quad \text{where } g \in B \text{ and } 6 \leq |g| \leq 9. \tag{4.254}$$

By (4.162), (4.169), (4.170), (4.246), (4.252) and $b_3(\bar{b}_3 x_7) = \bar{b}_3(b_3 x_7)$, we obtain that

$$x_8 + \Sigma_{12} = H_{20}.$$

By (4.254) we obtain that $(b_3 c_{14}, b_3 c_{14}) = 4$. Therefore $\Sigma_{12} = h_{12}$ where $h_{12} \in B$ and $H_{20} = x_8 + h_{12}$. Therefore

$$\bar{b}_3 c_{14} = x_7 + b_{15} + x_8 + h_{12} \tag{4.255}$$

and

$$b_3 y_{11} = x_7 + \bar{c}_{14} + h_{12}. \tag{4.256}$$

By (4.194), (4.165) and (4.254) we obtain that

$$g + b_3 e_{10} = t + \bar{b}_3 x_8.$$

It is obvious that $g \neq t$. By (4.165) and (4.250) we obtain that $12 \leq |t| \leq 15$. Now (4.171) implies that

$$b_3 e_{10} = b_{10} + t + k \quad \text{where } k \in B \tag{4.257}$$

and

$$\bar{b}_3 x_8 = b_{10} + g + k. \tag{4.258}$$

By (4.255) we obtain that

$$b_3 x_8 = c_{14} + \alpha + \beta \quad \text{where } \alpha, \beta \in B. \tag{4.259}$$

Assume henceforth that g is a real element and we shall derive a contradiction. By (4.254) and (4.258) we obtain that $(\bar{c}_{14}, b_3 g) = (x_8, b_3 g)$ which implies that $9 \leq |g|$. By (4.254) we obtain that $g = g_9$. Now we have that

$$b_3 g_9 = x_8 + \bar{c}_{14} + w_5 \quad \text{where } w_5 \in B.$$

By (4.162), (4.255), (4.258), (4.258), (4.259) and $b_3(\bar{b}_3 x_8) = \bar{b}_3(b_3 x_8)$, we obtain that $x_8 + \bar{c}_{14} + w_5 + b_3 k_5 = h_{12} + \bar{b}_3 \alpha + \bar{b}_3 \beta$. Hence $(h_{12}, b_3 k_5) = 1$, which is a contradiction to (4.258).

Now we have that g is a non-real element. By (4.2), (4.256), (4.170), (4.254) and $(b_3 \bar{b}_3) y_{11} = b_3(\bar{b}_3 y_{11})$, we obtain that

$$b_8 y_{11} = b_{10} + \bar{b}_{10} + y_{11} + s + g + b_3 \bar{h}_{12}.$$

Since b_8 and y_{11} are real elements and g is a non-real element, we obtain that $(h_{12}, b_3 g) = (\bar{g}, b_3 \bar{h}_{12}) = 1$ and by (4.258) we obtain that $(x_8, b_3 g) = 1$ which implies that $8 \le |g|$. By (4.254) we obtain that either $g = g_8$ or $g = g_9$.

Assume henceforth that $g = g_9$ and we shall derive a contradiction. Then by (4.258) and (4.257) we obtain that

$$b_3 k_5 = x_8 + h_7 \quad \text{where } h_7 \in B$$

and

$$\bar{b}_3 k_5 = e_{10} + h_5 \quad \text{where } h_5 \in B. \qquad (4.260)$$

Now by (4.258), (4.257) and $b_3(\bar{b}_3 k_5) = \bar{b}_3(b_3 k_5)$, we obtain that $t_{15} + b_3 h_5 = g_9 + \bar{b}_3 h_7$. Hence $(g_9, b_3 h_5) = 1$, which is a contradiction to (4.260).

Now we have that $g = g_8$. By (4.258) we obtain that $(x_8, b_3 g_8) = 1$, and since $(h_{12}, b_3 g_8) = 1$ we obtain that

$$b_3 g_8 = x_8 + h_{12} + h_4 \quad \text{where } h_4 \in B.$$

By (4.254) we obtain that

$$\bar{b}_3 g_8 = c_{14} + m + n \quad \text{where } m, n \in B.$$

Now by (4.254), (4.258) and $b_3(\bar{b}_3 g_8) = \bar{b}_3(b_3 g_8)$, we obtain that $y_{11} + s_{13} + b_3 m + b_3 n = k_6 + \bar{b}_3 h_4 + \bar{b}_3 h_{12}$ which implies that $\bar{b}_3 h_{12} = g_8 + y_{11} + s_{13} + \gamma_4$ where $\gamma_4 \in B$. Hence $b_3 \gamma_4 = h_{12}$ which implies that $(b_3 \gamma_4, b_3 \gamma_4) = 1$. On the other hand, we have that either $(m, \bar{b}_3 \gamma_4) = 1$ or $(n, \bar{b}_3 \gamma_4) = 1$ which implies that $(b_3 \gamma_4, b_4 \gamma_4) \ne 1$, and we have a contradiction.

Now we have that y_{11} is a non-real element. By (4.2), (4.165), (4.247), (4.251) and $b_3(\bar{b}_3 s) = (b_3 \bar{b}_3) s$ we obtain that

$$b_8 s = b_8 + b_{10} + \bar{b}_{10} + s + t + F_{21} + b_3 T.$$

By (4.249) and (4.250) we obtain that $\bar{y}_{11} \notin Irr(F_{21})$. Now since b_8 and s are real elements and y_{11} is a non-real element we obtain that

$$\bar{y}_{11} \in Irr(b_3 T). \qquad (4.261)$$

If $s = s_{12}$ then by (4.251) we obtain that

$$\bar{b}_3 s_{12} = \bar{b}_{15} + c_{14} + f_7 \quad \text{where } f_7 \in B.$$

Now we have that $(\bar{y}_{11}, b_3 f_7) = (s_{12}, b_3 f_7) = 1$, which is a contradiction.

If $s = s_{13}$ then by (4.251), we obtain that

$$\bar{b}_3 s_{13} = \bar{b}_{15} + c_{14} + f_{10} \quad \text{where } f_{10} \in B. \qquad (4.262)$$

Now (4.261) implies that

$$b_3 f_{10} = s_{13} + \bar{y}_{11} + h_6 \quad \text{where } h_6 \in B. \qquad (4.263)$$

By (4.170) we obtain that

$$b_3 y_{11} = x_7 + \bar{f}_{10} + \Sigma_{16} \quad \text{where } \Sigma_{16} \in NB.$$

Now by (4.174), (4.248) and (4.246), we obtain that $H_{20} + \Sigma_{14} = x_8 + \bar{f}_{10} + \Sigma_{16}$ which implies that

$$\bar{b}_3 c_{14} = x_7 + \bar{f}_{10} + g_{10} + b_{15} \quad \text{where } g_{10} \in B, \tag{4.264}$$

$$b_3 y_{11} = x_7 + \bar{f}_{10} + g_{10} + g_6 \quad \text{where } g_6 \in B \tag{4.265}$$

and

$$\bar{b}_3 d_7 = x_7 + x_8 + g_6. \tag{4.266}$$

By (4.247) and (4.249) we obtain that

$$b_3 c_{14} = b_{10} + y_{11} + s_{13} + g_8 \quad \text{where } g_8 \in B.$$

By (4.198) and (4.262) we obtain that

$$b_3 \bar{f}_{10} + b_3 g_{10} = G_{12} + c_{14} + f_{10} + \bar{b}_3 g_8.$$

By (4.264) we obtain that

$$\bar{b}_3 f_{10} = \bar{c}_{14} + Q_{16} \quad \text{where } Q_{16} \in NB. \tag{4.267}$$

Now by (4.264), (4.263), (4.262), (4.265) and $b_3(\bar{b}_3 f_{10}) = \bar{b}_3(b_3 f_{10})$ we obtain that

$$c_{14} + f_{10} + \bar{g}_6 + \bar{b}_3 h_6 = b_3 Q_{16} \tag{4.268}$$

which implies that $c_{14} \in Irr(b_3 Q_{16})$.

By (4.264) one of the following cases hold:

Case 1: $x_7 \in Irr(Q_{16})$.
Case 2: $g_{10} \in Irr(Q_{16})$.
Case 3: $\bar{f}_{10} \in Irr(Q_{16})$.

Case 1 By (4.169) and (4.268) we obtain that $(d_7, \bar{b}_3 h_6) = 1$ and by (4.263) we obtain that $(f_{10}, \bar{b}_3 h_6) = 1$, which is a contradiction.

Case 2 By (4.267) we obtain that

$$\bar{b}_3 f_{10} = \bar{c}_{14} + g_{10} + \alpha_6 \quad \text{where } \alpha_6 \in B.$$

Now (4.264) implies that

$$b_3 g_{10} = \beta_6 + f_{10} + c_{14} \quad \text{where } \beta_6 \in B. \tag{4.269}$$

By (4.263) we obtain that $(f_{10}, \bar{b}_3 h_6) = 1$ which implies that $(\beta_6, \bar{b}_3 h_6) = 0$. Since $g_{10} \in Irr(Q_{16})$ we obtain by (4.268) and (4.269) that $\beta_6 = \bar{g}_6$. Now by (4.269) we obtain that $(\bar{g}_{10}, b_3 g_6) = 1$, and by (4.266) we obtain that $(d_7, b_3 g_6) = 1$, which is a contradiction.

Case 3 By (4.267) we obtain that

$$\bar{b}_3 f_{10} = \bar{c}_{14} + \bar{f}_{10} + \gamma_6 \quad \text{where } \gamma_6 \in B.$$

By (4.263) we obtain that $(f_{10}, \bar{b}_3 h_6) = 1$ which implies that $(\gamma_6, \bar{b}_3 h_6) = 1$. Since $\bar{f}_{10} \in Q_{16}$ we obtain by (4.268) that $\gamma_6 = \bar{g}_6$ which implies that $(f_{10}, b_3 \bar{g}_6) = 1$, and by (4.265) we obtain that $(\bar{y}_{11}, b_3 \bar{g}_6) = 1$, which is a contradiction.
   If $s = s_{14}$ then by (4.251) we obtain that

$$\bar{b}_3 s_{14} = c_{14} + \bar{b}_{15} + T_{13}.$$

Now by (4.261) we obtain that there exists an element $f \in B$ in $Irr(T_{13})$ such that $10 \le |f|$ which implies that $f = T_{13}$. Hence

$$\bar{b}_3 s_{14} = c_{14} + \bar{b}_{15} + f_{13} \quad \text{where } f_{13} \in B. \tag{4.270}$$

By (4.170) and (4.261) we obtain that

$$b_3 y_{11} = x_7 + \bar{f}_{13} + \Sigma_{13} \quad \text{where } \Sigma_{13} \in \mathbb{N}B. \tag{4.271}$$

Now by (4.174), (4.246) and (4.248) we obtain that $H_{20} + \Sigma_{14} = x_8 + \bar{f}_{13} + \Sigma_{13}$ which implies that

$$\bar{b}_3 c_{14} = x_7 + g_7 + \bar{f}_{13} + b_{15} \quad \text{where } g_7 \in B,$$
$$\bar{b}_3 d_7 = x_7 + x_8 + g_6 \quad \text{where } g_6 \in B \tag{4.272}$$

and

$$b_3 y_{11} = \bar{f}_{13} + x_7 + g_6 + g_7. \tag{4.273}$$

By (4.247) and (4.249) we obtain that

$$b_3 c_{14} = b_{10} + y_{11} + s_{14} + h_7 \quad \text{where } h_7 \in B.$$

Now we have that

$$b_3 g_7 = c_{14} + \alpha_7 \quad \text{where } \alpha_7 \in B$$

and

$$\bar{b}_3 h_7 = c_{14} + \beta_7 \quad \text{where } \beta_7 \in B.$$

By (4.198) we obtain that

$$b_3 \bar{f}_{13} + \alpha_7 = G_{12} + c_{14} + f_{13} + \beta_7. \tag{4.274}$$

If $\alpha_7 \in Irr(G_{12})$ then there exists an element $\alpha_5 \in B$ such that $\alpha_5 \in Irr(G_{12})$ which implies that $(\alpha_5, b_3 \bar{f}_{13}) = 1$, and we have a contradiction. Hence $\alpha_7 = \beta_7$. By (4.273) we obtain that $3 \leq (b_3 y_{11}, b_3 y_{11})$. Now (4.196) and (4.274) imply that

$$b_3 y_{11} = c_{14} + d_7 + \alpha_6 + \beta_6 \quad \text{where } \alpha_6, \beta_6 \in B \tag{4.275}$$

and

$$b_3 \bar{f}_{13} = c_{14} + f_{13} + \alpha_6 + \beta_6. \tag{4.276}$$

By (4.270) and (4.273) we obtain that

$$b_3 f_{13} = \bar{y}_{11} + s_{14} + \Sigma_{14} \quad \text{where } \Sigma_{14} \in \mathbb{N}B.$$

Now by (4.270), (4.272), (4.273) and $b_3(\bar{b}_3 f_{13}) = \bar{b}_3(b_3 f_{13})$ we obtain that

$$\bar{g}_6 + \bar{b}_3 \Sigma_{14} = \alpha_6 + \beta_6 + b_3 \bar{\alpha}_6 + b_3 \bar{\beta}_6.$$

By (4.276) we obtain that $(f_{13}, b_3 \bar{\alpha}_6) = (f_{13}, b_3 \bar{\beta}_6) = 1$ which implies that $(\bar{g}_6, b_3 \bar{\alpha}_6) = (\bar{g}_6, b_3 \bar{\beta}_6) = 0$. Now we have that either $g_6 = \bar{\alpha}_6$ or $g_6 = \bar{\beta}_6$. By (4.273) we obtain that $(\bar{y}_{11}, b_3 \bar{g}_6) = 1$, and by (4.275) we obtain that $(y_{11}, b_3 \bar{g}_6) = 1$, which is a contradiction.

If $s = s_{15}$ then by (4.251) we obtain that

$$\bar{b}_3 s_{15} = \bar{b}_{15} + c_{14} + T_{16}.$$

Now by (4.261) we obtain that there exists an element $f \in B$ in $Irr(T_{16})$ such that $10 \leq |f|$ which implies that $f = T_{16}$. Hence

$$\bar{b}_3 s_{15} = c_{14} + \bar{b}_{15} + f_{16} \quad \text{where } f_{16} \in B. \tag{4.277}$$

By (4.261) and (4.170) we obtain that

$$b_3 y_{11} = x_7 + \bar{f}_{16} + \Sigma_{10} \quad \text{where } \Sigma_{10} \in \mathbb{N}B. \tag{4.278}$$

Now by (4.174), (4.246) and (4.248) we obtain that $H_{20} + \Sigma_{14} = x_8 + \bar{f}_{16} + \Sigma_{10}$ which implies that $\bar{f}_{16} \in Irr(H_{20})$ which implies by (4.246) that $\bar{b}_3 c_{14} = x_7 + b_{15} + \bar{f}_{16} + f_4$ where $f_4 \in B$, and we have a contradiction since $(f_4, \bar{b}_3 c_{14}) = 1$.

4.5.5 Case $c = c_{15}$

In this section we assume that $c = c_{15}$. Then (4.195) and (4.197) imply that

$$\bar{b}_3 c_{15} = x_7 + b_{15} + H_{23} \tag{4.279}$$

and

$$b_3 c_{15} = b_{10} + y_{11} + F_{24}. \tag{4.280}$$

By (4.169) we obtain that

$$\bar{b}_3 d_6 = x_7 + \Sigma_{11} \quad \text{where } \Sigma_{11} \in \mathbb{N}B. \tag{4.281}$$

Assume henceforth that y_{11} is a real element and we shall derive a contradiction. Then by (4.170) and (4.196) we obtain that

$$b_3 y_{11} = x_7 + \bar{d}_6 + \bar{c}_{15} + h_5 \quad \text{where } h_5 \in B. \tag{4.282}$$

Now by (4.174) and (4.279) we obtain that

$$H_{23} + \Sigma_{11} = x_8 + \bar{c}_{15} + \bar{d}_6 + h_5.$$

Now we have that $H_{23} = x_8 + \bar{c}_{15}$ and $\Sigma_{11} = \bar{d}_6 + h_5$ which imply by (4.279) and (4.281) that

$$\bar{b}_3 c_{15} = x_7 + x_8 + b_{15} + \bar{c}_{15}$$

and

$$\bar{b}_3 d_6 = x_7 + \bar{d}_6 + h_5. \tag{4.283}$$

Now we have by (4.198) that $\bar{b}_{15} + b_3 x_8 + \bar{x}_7 + \bar{x}_8 + \bar{b}_{15} + c_{15} = \bar{x}_7 + \bar{h}_5 + \bar{b}_3 F_{24}$ which implies that $(\bar{x}_8, b_3 h_5) = (\bar{h}_5, b_3 x_8) = 1$, and by (4.283) we obtain that $(d_6, b_3 h_5) = 1$, which is a contradiction.

Now we have that y_{11} is a non-real element. By (4.198) we obtain that

$$\bar{b}_{15} + b_3 H_{23} = G_{12} + \bar{b}_3 F_{24}$$

which implies that $\bar{b}_{15} \in Irr(\bar{b}_3 F_{24})$. By (4.280) we obtain that $(y_{11}, b_3 c_{15}) = 1$, and by (4.165) and since y_{11} is a non-real element we obtain that $(y_{11}, b_3 \bar{b}_{15}) = 0$ which implies that $c_{15} \neq \bar{b}_{15}$. Now by (4.5) and (4.162) we obtain that $b_8 \notin Irr(b_3 c_{15})$ and $\bar{b}_{10} \notin Irr(b_3 c_{15})$ and by (4.165) (4.280) we can assume that

$$s \in Irr(F_{24}). \tag{4.284}$$

By (4.165) we obtain that $(s, b_3 \bar{b}_{15}) = 1$ which implies that

$$\bar{b}_3 s = \bar{b}_{15} + c_{15} + T \quad \text{where } T \in \mathbb{N}B \tag{4.285}$$

which implies that

$$10 \leq |s|; \tag{4.286}$$

and by (4.280) we obtain that

$$b_3 c_{15} = b_{10} + y_{11} + s + S \quad \text{where } S \in \mathbb{N}B. \tag{4.287}$$

Now by (4.2), (4.165) and $b_3(\bar{b}_3 s) = (b_3 \bar{b}_3)s$ we obtain that

$$b_8 s = b_8 + b_{10} + \bar{b}_{10} + y_{11} + s + t + S + b_3 T.$$

Since b_8 and s are real elements and y_{11} is a non-real element we obtain that either $\bar{y}_{11} \in Irr(S)$ or $\bar{y}_{11} \in Irr(b_3 T)$. By (4.174) we obtain that

$$\bar{b}_3 c_{15} + \bar{b}_3 d_6 = b_{15} + x_7 + x_8 + b_3 y_{11}.$$

If $(x_8, \bar{b}_3 d_6) = 1$ then by (4.169), we obtain that

$$\bar{b}_3 d_6 = x_7 + x_8 + \alpha_3 \quad \text{where } \alpha_3 \in B$$

which implies that $(\alpha_3, b_3 y_{11}) = 1$, and we have a contradiction. Hence $(x_8, \bar{b}_3 c_{15}) = 1$ which implies that

$$b_3 x_8 = c_{15} + \Sigma_9 \quad \text{where } \Sigma_9 \in \mathbb{N}B. \tag{4.288}$$

In addition, we have that

$$2 \le (b_3 x_8, b_3 x_8) \le 3.$$

If $\bar{y}_{11} \in S$ then we obtain by (4.286) and by (4.287) that

$$b_3 c_{15} = b_{10} + y_{11} + \bar{y}_{11} + s_{13}.$$

Now by (4.285) we obtain that

$$\bar{b}_3 s_{13} = \bar{b}_{15} + c_{15} + \alpha_9 \quad \text{where } \alpha_9 \in B.$$

By (4.170) and since $\bar{y}_{11} \in S$ we obtain that

$$b_3 y_{11} = x_7 + \bar{c}_{15} + \Sigma_{11}^*.$$

By (4.174), (4.279), (4.281) and (4.288) we obtain that

$$\bar{b}_3 c_{15} = x_7 + x_8 + b_{15} + \bar{c}_{15}.$$

Now by (4.169), (4.171), (4.172), (4.196), (4.288) and $b_3(\bar{b}_3 c_{15}) = \bar{b}_3(b_3 c_{15})$ implies that $\Sigma_9 + \bar{x}_8 + \bar{b}_{15} = \Sigma_{12} + \alpha_9 + \Sigma_{11}^*$, which is a contradiction.

Now we have that $\bar{y}_{11} \in b_3 T$ which implies that $5 \le |T|$ and by (4.285) we obtain that $12 \le |s|$.

Assume henceforth that $(b_3 x_8, b_3 x_8) = 2$ and we shall derive a contradiction. Then by (4.162) we obtain that

$$\bar{b}_3 x_8 = b_{10} + \alpha_{14} \quad \text{where } \alpha_{14} \in B.$$

Now by (4.165), (4.194) and (4.287) we obtain that

$$S + b_3 e_9 = t + b_{10} + \alpha_{14}.$$

Since $12 \le |s|$ we obtain by (4.287) that $|S| \le 12$ which implies that $\alpha_{14} \in Irr(b_3 e_9)$ and by (4.171) we obtain that $(b_{10}, b_3 e_9) = 1$, which is a contradiction.

Now we have that $(b_3x_8, b_3x_8) = 3$. Then by (4.162) we obtain that

$$\bar{b}_3x_8 = b_{10} + \alpha + \beta \quad \text{where } \alpha, \beta \in B. \tag{4.289}$$

Now by (4.5) we obtain that $b_8 \neq x_8$, and by (4.288) we obtain that

$$b_3x_8 = c_{15} + v_4 + v_5 \quad \text{where } v_4, v_5 \in B. \tag{4.290}$$

Now by (4.165), (4.194) and (4.287) we obtain that

$$S + b_3e_9 = b_{10} + \alpha + \beta + t.$$

Now we can assume that

$$b_3e_9 = b_{10} + t + \beta \tag{4.291}$$

and

$$b_3c_{15} = b_{10} + y_{11} + s + \alpha \tag{4.292}$$

which implies that $5 \leq |\alpha|$, and by (4.289) we obtain that $|\alpha| \leq 10$. Therefore

$$5 \leq |\alpha| \leq 10.$$

Now (4.289) implies that

$$4 \leq |\beta| \leq 9.$$

By (4.165), (4.291) and (4.292) we obtain that

$$14 \leq |s| \leq 19$$

which implies that $\alpha \neq s$, $\alpha \neq y_{11}$ and $s \neq y_{11}$. Since $(b_3x_8, b_3x_8) = 3$ we obtain by (4.289) that $\alpha \neq b_{10}$. Now by (4.292) we obtain that $(b_3c_{15}, b_3c_{15}) = 4$. By (4.279) and (4.288) we obtain that

$$\bar{b}_3c_{15} = b_{15} + x_7 + x_8 + f_{15} \quad \text{where } f_{15} \in B. \tag{4.293}$$

By (4.198), (4.290) and (4.292) we obtain that

$$\bar{b}_{15} + c_{15} + v_4 + v_5 + b_3f_{15} = G_{12} + \bar{b}_3s + \bar{b}_3\alpha. \tag{4.294}$$

By (4.196) we obtain that $v_4 \notin Irr(G_{12})$ and since $14 \leq |s|$ we have that $v_4 \notin Irr(\bar{b}_3s)$. Therefore $(v_4, \bar{b}_3\alpha) = 1$ which implies that $|\alpha| \leq 9$ and by (4.292) we obtain that $(c_{15}, \bar{b}_3\alpha) = 1$ which implies that $8 \leq |\alpha|$. Thus

$$8 \leq |\alpha| \leq 9 \tag{4.295}$$

which implies by (4.289) that

$$5 \leq |\beta| \leq 6, \tag{4.296}$$

and by (4.292) we obtain that $15 \leq |s| \leq 16$. By (4.162), (4.289), (4.290), (4.293) and $b_3(\bar{b}_3 x_8) = \bar{b}_3(b_3 x_8)$ we obtain that

$$f_{15} + \bar{b}_3 v_4 + \bar{b}_3 v_5 = b_3 \alpha + b_3 \beta.$$

By (4.289) we obtain that $(x_8, b_3\alpha) = (x_8, b_3\beta) = 1$ which implies by (4.296) that $(f_{15}, b_3\beta) = 0$ and $(f_{15}, b_3\alpha) = 1$. So $9 \leq |\alpha|$ and by (4.295), we obtain that $\alpha = \alpha_9$ which implies by (4.292) that $s = s_{15}$. By (4.285) we obtain that

$$\bar{b}_3 s_{15} = c_{15} + \bar{b}_{15} + T_{15},$$

and by (4.294) we obtain that $v_4 + v_5 + b_3 f_{15} = G_{12} + \dot{T}_{15} + \bar{b}_3 \alpha_9$. By (4.292) we obtain that $(c_{15}, \bar{b}_3 \alpha_9) = 1$ which implies that either $b_3 f_{15} = G_{12} + T_{15} + c_{15} + h_3$ where $h_3 \in B$, which is impossible, or $v_5 \in Irr(G_{12})$.

Now we have that $v_5 \in Irr(G_{12})$. By (4.196) we obtain that

$$\bar{b}_3 y_{11} = c_{15} + d_6 + v_5 + v_7 \quad \text{where } v_7 \in B$$

which implies that

$$b_3 v_5 = y_{11} + m_4 \quad \text{where } m_4 \in B,$$

and by (4.290) we obtain that

$$\bar{b}_3 v_5 = x_8 + m_7 \quad \text{where } m_7 \in B.$$

Now by (4.290) and $b_3(\bar{b}_3 v_5) = \bar{b}_3(b_3 v_5)$ we obtain that $v_4 + b_3 m_7 = d_6 + v_7 + \bar{b}_3 m_4$. Thus

$$\bar{b}_3 m_4 = v_4 + v_5 + m_3 \quad \text{where } m_3 \in B$$

which implies that $(m_3, b_3 m_7) = 1$, and we have a contradiction.

4.5.6 Case $c = c_{16}$

In this section we assume that $c = c_{16}$. Then (4.195) and (4.197) imply that

$$\bar{b}_3 c_{16} = x_7 + b_{15} + H_{26} \tag{4.297}$$

and

$$b_3 c_{16} = b_{10} + y_{11} + F_{27}. \tag{4.298}$$

By (4.169) and (4.174) we obtain that

$$\bar{b}_3 d_5 = x_7 + g_8 \quad \text{where } g_8 \in B. \tag{4.299}$$

Assume henceforth that y_{11} is a real element and we shall derive a contradiction. Then by (4.170) and (4.196) we obtain that

$$b_3 y_{11} = x_7 + \bar{d}_5 + \bar{c}_{16} + h_5 \quad \text{where } h_5 \in B. \tag{4.300}$$

Now by (4.174) and (4.297) we obtain that

$$H_{26} + g_8 = x_8 + \bar{c}_{16} + \bar{d}_5 + h_5.$$

Now we have that $H_{26} = \bar{c}_{16} + \bar{d}_5 + h_5$ which implies by (4.297) that $h_5 \in Irr(\bar{b}_3 c_{16})$, which is a contradiction.

Now we have that y_{11} is a non-real element. By (4.198) we obtain that

$$\bar{b}_{15} + b_3 H_{26} = G_{12} + \bar{b}_3 F_{27}$$

which implies that $\bar{b}_{15} \in Irr(\bar{b}_3 F_{27})$. By (4.5) and (4.162) we obtain that $b_8 \notin Irr(b_3 c_{16})$ and $\bar{b}_{10} \notin Irr(b_3 c_{16})$ and by (4.298) we can assume that

$$s \in Irr(F_{27}). \tag{4.301}$$

By (4.165) we obtain that $(s, b_3 \bar{b}_{15}) = 1$ which implies that

$$\bar{b}_3 s = \bar{b}_{15} + c_{16} + T \quad \text{where } T \in \mathbb{N}B \tag{4.302}$$

which implies that

$$12 \leq |s|, \tag{4.303}$$

and by (4.298) we obtain that

$$b_3 c_{16} = b_{10} + y_{11} + s + S \quad \text{where } S \in \mathbb{N}B. \tag{4.304}$$

Now by (4.2), (4.165) and $b_3(\bar{b}_3 s) = (b_3 \bar{b}_3)s$ we obtain that

$$b_8 s = b_8 + b_{10} + \bar{b}_{10} + y_{11} + s + t + S + b_3 T. \tag{4.305}$$

By (4.171) we obtain that

$$b_3 e_8 = b_{10} + \Sigma_{14} \quad \text{where } \Sigma_{14} \in \mathbb{N}B \tag{4.306}$$

and by (4.162) we obtain that

$$\bar{b}_3 x_8 = b_{10} + \Sigma_{14}^*. \tag{4.307}$$

Now by (4.165) and (4.304) we obtain that

$$S + \Sigma_{14} = t + \Sigma_{14}^*.$$

Now we have the following three cases:

Case 1: $S = t$ and $\Sigma_{14} = \Sigma_{14}^*$.
Case 2: $S = \Sigma_{14}^*$ and $\Sigma_{14} = t_{14}$.
Case 3: $\Sigma_{14}^* = \alpha + \beta$ where $\alpha, \beta \in B$, $S = \alpha$ and $\Sigma_{14} = \beta + t$.

Case 1 By (4.304) we obtain that

$$b_3 c_{16} = b_{10} + y_{11} + s + t, \qquad (4.308)$$

and by (4.165) we obtain that

$$\bar{b}_3 t = \bar{b}_{15} + c_{16} + \Gamma \quad \text{where } \Gamma \in \mathbb{N}B \qquad (4.309)$$

which implies that $12 \leq |t|$. By (4.305) we obtain that

$$\bar{y}_{11} \in Irr(b_3 T), \qquad (4.310)$$

and by (4.302) we obtain that $s \in Irr(b_3 T)$ which implies that $8 \leq |T|$ and $13 \leq |s|$. In the same way we can show that

$$\bar{y}_{11} \in Irr(b_3 \Gamma) \qquad (4.311)$$

and $13 \leq |t|$. Now by (4.165) we can assume that $s = s_{13}$ and $t = t_{14}$. Now (4.302) and (4.309) imply that

$$\bar{b}_3 s_{13} = \bar{b}_{15} + c_{16} + v_8 \quad \text{where } v_8 \in B$$

and

$$\bar{b}_3 t_{14} = \bar{b}_{15} + c_{16} + v_{11} \quad \text{where } v_{11} \in B.$$

By (4.170), (4.310) and (4.311) we obtain that

$$b_3 y_{11} = x_7 + \bar{v}_8 + \bar{v}_{11} + f_7 \quad \text{where } f_7 \in B.$$

By (4.174), (4.297) and (4.299) we obtain that $H_{26} + g_8 = x_8 + \bar{v}_8 + \bar{v}_{11} + f_7$, and together with (4.297) we obtain that $(\bar{b}_3 c_{16}, \bar{b}_3 c_{16}) = 5$, which is a contradiction to (4.308).

Case 2 By (4.304) we obtain that

$$b_3 c_{16} = b_{10} + y_{11} + s_{13} + \Sigma_{14}^*,$$

by (4.306) we obtain that

$$b_3 e_8 = b_{10} + t_{14}, \qquad (4.312)$$

by (4.172) we obtain that

$$\bar{b}_3 e_8 = b_{15} + h_9 \quad \text{where } h_9 \in B \qquad (4.313)$$

and by (4.302) we obtain that

$$\bar{b}_3 s_{13} = \bar{b}_{15} + c_{16} + v_8 \quad \text{where } v_8 \in B. \tag{4.314}$$

Now (4.305) implies that $\bar{y}_{11} \in Irr(b_3 v_8)$. Therefore

$$b_3 v_8 = \bar{y}_{11} + s_{13}. \tag{4.315}$$

Now (4.174) implies that either $(\bar{v}_8, \bar{b}_3 c_{16} = 1)$ or $(\bar{v}_8, \bar{b}_3 d_5) = 1$.

Assume henceforth that $(\bar{v}_8, \bar{b}_3 d_5) = 1$ and we shall derive a contradiction. Then by (4.169) we obtain that

$$\bar{b}_3 d_5 = x_7 + \bar{v}_8 \tag{4.316}$$

and by (4.315) we obtain that

$$\bar{b}_3 v_8 = \bar{d}_5 + f_{19} \quad \text{where } f_{19} \in B.$$

Now by (4.314), (4.316) and $b_3(\bar{b}_3 v_8) = \bar{b}_3(b_3 v_8)$, we obtain that $\bar{x}_7 + b_3 f_{19} = \bar{b}_{15} + c_{16} + b_3 \bar{y}_{11}$. Hence $(f_{19}, b_3 b_{15}) = 1$, which is a contradiction to (4.172).

Now we have that $(\bar{v}_8, \bar{b}_3 c_{16}) = 1$ which implies by (4.315) that

$$\bar{b}_3 v_8 = \bar{c}_{16} + \gamma_8 \quad \text{where } \gamma_8 \in B \tag{4.317}$$

and by (4.297) we obtain that

$$\bar{b}_3 c_{16} = x_7 + b_{15} + \bar{v}_8 + \Sigma_{18} \quad \text{where } \Sigma_{18} \in \mathbb{N}B.$$

Now by (4.3), (4.5), (4.162), (4.165), (4.172), (4.313), (4.314) and $b_3(\bar{b}_3 b_{15}) = \bar{b}_3(b_3 b_{15})$ we obtain that $x_8 + b_3 t_{14} = \bar{e}_8 + \Sigma_{18} + b_{15} + h_9$, and by (4.312) we obtain that $(\bar{e}_8, b_3 t_{14}) = 1$. Therefore $(x_8, \Sigma_{18}) = 1$ which implies that

$$\bar{b}_3 c_{16} = x_7 + b_{15} + \bar{v}_8 + x_8 + h_{10} \quad \text{where } h_{10} \in B,$$

and by (4.174) and (4.299) we obtain that

$$b_3 y_{11} = x_7 + g_8 + \bar{v}_8 + h_{10}. \tag{4.318}$$

Now by (4.314), (4.315), (4.317) and $b_3(\bar{b}_3 v_8) = \bar{b}_3(b_3 v_8)$ we obtain that $c_{16} + v_8 + \bar{g}_8 = \bar{x}_8 + b_3 \gamma_8$. By (4.317) we obtain that $(v_8, b_3 \gamma_8) = 1$ which implies that $g_8 = x_8$. By (4.318) we obtain that $(y_{11}, \bar{b}_3 x_8) = (x_8, b_3 y_{11}) = 1$, and by (4.5) we obtain that $b_8 \neq x_8$. Now by (4.162) we obtain that $(b_{10}, \bar{b}_3 x_8) = 1$, which is a contradiction.

Case 3 By (4.306) we obtain that

$$b_3 e_8 = b_{10} + \beta + t \tag{4.319}$$

which implies that $(e_8, \bar{b}_3 t) = 1$, and by (4.165) we obtain that $(\bar{b}_{15}, \bar{b}_3 t) = 1$. Therefore $9 \leq |t|$ and $|\beta| \leq 5$. By (4.5) and (4.319) we obtain that $b_8 \neq e_8$. Therefore $|\beta| \neq 3$ which implies that $4 \leq |\beta| \leq 5$.

Assume henceforth that $\beta = \beta_4$ and we shall derive a contradiction. Then (4.319) implies that

$$\bar{b}_3 \beta_4 = e_8 + h_4 \quad \text{where } h_4 \in B.$$

By (4.307) we obtain that

$$\bar{b}_3 x_8 = b_{10} + \alpha_{10} + \beta_4$$

which implies that

$$b_3 \beta_4 = x_8 + g_4 \quad \text{where } g_4 \in B.$$

Now (4.319) and $b_3(\bar{b}_3 \beta_4) = \bar{b}_3(b_3 \beta_4)$ imply that $t_{10} + b_3 h_4 = \alpha_{10} + \bar{b}_3 g_4$ which implies that $t_{10} = \alpha_{10}$. By (4.165) we obtain that $(b_{15}, b_3 t_{10}) = 1$ and by (4.304) we obtain that $(\bar{c}_{16}, b_3 t_{10}) = 1$, which is a contradiction.

Now we have that $\beta = \beta_5$, and by (4.165) and (4.319) we obtain that $t = t_9$, $s = s_{18}$ and $\alpha = \alpha_9$. By (4.165), (4.172) and (4.319) we obtain that

$$\bar{b}_3 t_9 = \bar{b}_{15} + e_8 + \alpha_4 \quad \text{where } \alpha_4 \in B$$

and

$$\bar{b}_3 e_8 = \bar{b}_{15} + f_4 + f_5 \quad \text{where } f_4, f_5 \in B.$$

Now by (4.5), (4.3) (4.162), (4.165), (4.297), (4.302), (4.172) and $b_3(\bar{b}_3 b_{15}) = \bar{b}_3(b_3 b_{15})$, we obtain that $x_8 + \bar{\alpha}_4 + \bar{T}_{23} = f_5 + f_4 + H_{26}$ which implies that $\bar{f}_5 \in Irr(T_{23})$, and we have a contradiction to (4.302).

4.6 Case $(b_3 x_7, b_3 x_7) = 3$

Since $(b_3 x_7, b_3 x_7) = 3$ we obtain that

$$b_3 x_7 = c + d + e \quad \text{where } c, d, e \in B, \tag{4.320}$$

and by (4.162) we obtain that

$$\bar{b}_3 x_7 = b_{10} + w + z \quad \text{where } w, z \in B. \tag{4.321}$$

Now by (4.162) and $b_3(\bar{b}_3 x_7) = \bar{b}_3(b_3 x_7)$ we obtain that

$$\bar{b}_3 c + \bar{b}_3 d + \bar{b}_3 e = b_{15} + x_7 + x_8 + b_3 z + b_3 w. \tag{4.322}$$

Now we can assume that $(c, b_3 b_{15}) = 1$. By (4.3) we obtain that $(b_6, b_3 b_{15}) = 1$, by (4.10) we obtain that $(\bar{b}_{15}, b_3 b_{15}) = 1$ and $(b_3 b_{15}, b_3 b_{15}) = 4$. Now since $c \neq b_6, \bar{b}_{15}$ we obtain that

$$b_3 b_{15} = b_6 + \bar{b}_{15} + c + f \quad \text{where } f \in B, \tag{4.323}$$

and by (4.10) we obtain that

$$\bar{b}_3 b_{10} = b_6 + c + f. \tag{4.324}$$

By (4.320) we obtain that $|c| \le 13$ and $(\bar{b}_3 c, x_7) = 1$ and by (4.323) we obtain that $(\bar{b}_3 c, b_{15}) = 1$ which implies that

$$9 \le |c| \le 13. \tag{4.325}$$

Now (4.324) implies that

$$11 \le |f| \le 15. \tag{4.326}$$

By (4.321) we obtain that either $\bar{b}_3 x_7 = z_4 + w_7$ or $\bar{b}_3 x_7 = z_5 + w_6$.
    Assume henceforth that

$$\bar{b}_3 x_7 = b_{10} + z_4 + w_7$$

and we shall derive a contradiction. Then

$$b_3 z_4 = x_7 + g_5 \quad \text{where } g_5 \in B$$

and

$$\bar{b}_3 z_4 = \alpha + \beta \quad \text{where } \alpha, \beta \in B. \tag{4.327}$$

Now $b_3(\bar{b}_3 z_4) = \bar{b}_3(b_3 z_4)$ implies that

$$b_3 \alpha + b_3 \beta = b_{10} + z_4 + w_7 + \bar{b}_3 g_5. \tag{4.328}$$

Now we can assume that

$$(b_{10}, b_3 \alpha) = 1. \tag{4.329}$$

By (4.327) we obtain that $|\alpha| \le 9$, by (4.4) we obtain that $\alpha \ne b_6$ and by (4.326) we obtain that $f \ne \alpha$. Now we have by (4.324) that $c = \alpha$, by (4.325) we obtain that $c = c_9$ and by (4.327) we obtain that $\beta = \beta_3$. By (4.328) we obtain that $b_3 c_9 + b_3 \beta_3 = b_{10} + z_4 + w_7 + \bar{b}_3 g_5$ which implies that $(w_7, b_3 c_9) = 1$. By (4.327) we obtain that $(z_4, b_3 c_9) = 1$ and together with (4.329) we obtain that $4 \le (b_3 c_9, b_3 c_9)$. By (4.320) and (4.323) we obtain that $(b_3 c_9, b_3 c_9) \le 3$, and we have a contradiction.
    Now we have that

$$\bar{b}_3 x_7 = b_{10} + z_5 + w_6 \tag{4.330}$$

and

$$b_3 z_5 = x_7 + \Sigma_8 \quad \text{where } \Sigma_8 \in \mathbb{N} B. \tag{4.331}$$

Now $b_3(\bar{b}_3 z_5) = \bar{b}_3(b_3 z_5)$ implies that $b_3(\bar{b}_3 z_5) = b_{10} + z_5 + w_6 + \bar{b}_3 \Sigma_8$, by (4.4) we obtain that $(b_6, \bar{b}_3 z_5) = 0$ and by (4.324) we obtain that

$$\text{either} \quad (c, \bar{b}_3 z_5) = 1 \quad \text{or} \quad (f, \bar{b}_3 z_5) = 1. \tag{4.332}$$

By (4.331) we obtain that $2 \le (b_3 z_5, b_3 z_5) \le 3$.

Assume henceforth that $(b_3z_5, b_3z_5) = 3$ and we shall derive a contradiction. Then by (4.325), (4.326) and (4.332) we obtain that

$$\bar{b}_3z_5 = c_9 + m_3 + n_3 \quad \text{where } m_3, n_3 \in B.$$

Now (4.330), (4.331) and $b_3(\bar{b}_3z_5) = \bar{b}_3(b_3z_5)$ imply that $b_3c_9 + b_3m_3 + b_3n_3 = b_{10} + z_5 + w_6 + \bar{b}_3\Sigma_8$. Hence $b_3c_9 = b_{10} + z_5 + w_6 + v_6$ where $v_6 \in B$. By (4.320) and (4.323) we obtain that $(b_3c_9, b_3c_9) \leq, 3$ and we have a contradiction. Now (4.331) implies that

$$b_3z_5 = x_7 + g_8 \quad \text{where } g_8 \in B. \tag{4.333}$$

Assume henceforth that $c = c_{13}$ and we shall derive a contradiction. Then by (4.324) we obtain that $f = f_{11}$ and (4.332) implies that

$$\bar{b}_3z_5 = f_{11} + \gamma_4 \quad \text{where } \gamma_4 \in B.$$

By (4.320) we obtain that

$$b_3x_7 = c_{13} + d_4 + e_4,$$

$$\bar{b}_3d_4 = x_7 + q_5 \quad \text{where } q_5 \in B$$

and

$$\bar{b}_3e_4 = x_7 + u_5 \quad \text{where } u_5 \in B.$$

Now by (4.322) and (4.333) we obtain that $\bar{b}_3c_{13} + q_5 + u_5 = b_{15} + x_8 + g_8 + b_3w_6$ which implies that $(q_5, b_3w_6) = (u_5, b_3w_6) = 1$, and by (4.330) we obtain that $(x_7, b_3w_6) = 1$, which is a contradiction. Now we have by (4.325) that $9 \leq |c| \leq 12$ and by (4.324) we obtain that $12 \leq |f| \leq 15$.

Assume henceforth that $(f, \bar{b}_3z_5) = 1$ and we shall derive a contradiction. Then $f = f_{12}$ and

$$\bar{b}_3z_5 = f_{12} + \gamma_3 \quad \text{where } \gamma_3 \in B. \tag{4.334}$$

By (4.324) we obtain that $c = c_{12}$. By (4.1), (4.320), (4.330) and $b_3(b_3x_7) = b_3^2x_7$ we obtain that $b_3c_{12} + b_3d + b_3e = b_{10} + z_5 + w_6 + b_6x_7$. By (4.320) and since $c = c_{12}$ we obtain that $|d|, |e| \neq 3, 12$ which implies by (4.334) that $(z_5, b_3d) = (z_5, b_3e) = 0$. Hence $(z_5, b_3c_{12}) = 1$ which implies that $c_{12} = f_{12}$ and since $(b_3b_{10}, b_3b_{10}) = 3$ we have a contradiction to (4.324).

Now by (4.332) and (4.333) we obtain that

$$\bar{b}_3z_5 = c + \delta \quad \text{where } \delta \in B. \tag{4.335}$$

By (4.324) we obtain that

$$b_3c = b_{10} + z_5 + D \quad \text{where } D \in \mathbb{N}B, \tag{4.336}$$

and by (4.320) and (4.323) we obtain that

$$\bar{b}_3c = x_7 + b_{15} + C \quad \text{where } C \in \mathbb{N}B. \tag{4.337}$$

Now by (4.320), (4.323), (4.324) and $b_3(\bar{b}_3 c) = \bar{b}_3(b_3 c)$, we obtain that $d + e + \bar{b}_{15} + b_3 C = \delta + \bar{b}_3 D$. By (4.335) we obtain that $|\delta| \neq 15$ which implies that $\delta \neq \bar{b}_{15}$. Therefore $\bar{b}_{15} \in Irr(\bar{b}_3 D)$. Since $9 \leq |c| \leq 12$ we obtain by (4.5), (4.162) and (4.336) that $b_8 \notin Irr(D)$ and $\bar{b}_{10} \notin Irr(D)$. Therefore by (4.165) we can assume that $s \in Irr(D)$. Now (4.336) implies that

$$b_3 c = b_{10} + z_5 + s + E \quad \text{where } E \in \mathbb{N}B. \tag{4.338}$$

Now (4.165) implies that

$$\bar{b}_3 s = \bar{b}_{15} + c + F \quad \text{where } F \in \mathbb{N}B.$$

By (4.2), (4.165) and $(b_3 \bar{b}_3)s = b_3(\bar{b}_3 s)$ we obtain that

$$b_8 s = b_{10} + z_5 + E + b_8 + \bar{b}_{10} + s + t + b_3 F.$$

Since b_8 and s are real elements and z_5 is a non-real element, we obtain that either $\bar{z}_5 \in Irr(E)$ or $\bar{z}_5 \in Irr(b_3 F)$. If $\bar{z}_5 \in Irr(E)$ then by (4.338) we obtain that $(\bar{c}, b_3 z_5) = 1$, and since $9 \leq |c| \leq 12$ we have a contradiction to (4.333). Therefore $\bar{z}_5 \in Irr(b_3 F)$. Now by (4.333) we obtain that either $\bar{x}_7 \in Irr(F)$ or $\bar{g}_8 \in Irr(F)$. If $\bar{x}_7 \in Irr(F)$ then $(s, \bar{b}_3 x_7) = (\bar{x}_7, \bar{b}_3 s) = 1$. Since s is a real element and b_{10}, z_5 are non-real elements we obtain that $s \neq b_{10}, z_5$, so (4.330) implies that $s = w_6$, which is a contradiction to (4.165) and (4.330). Now we have that

$$\bar{b}_3 s = \bar{b}_{15} + c + \bar{g}_8 + G \quad \text{where } G \in \mathbb{N}B. \tag{4.339}$$

By (4.333) we obtain that

$$b_3 g_8 = z_5 + s + H \quad \text{where } H \in \mathbb{N}B \tag{4.340}$$

Assume henceforth that $c = c_9$ and we shall derive a contradiction. Then by (4.337) we obtain that

$$\bar{b}_3 c_9 = x_7 + b_{15} + u_5 \quad \text{where } u_5 \in B,$$

and by (4.338) we obtain that

$$b_3 c_9 = b_{10} + z_5 + s_{12}.$$

By (4.339) we obtain that

$$\bar{b}_3 s_{12} = \bar{b}_{15} + c_9 + \bar{g}_8 + u_4 \quad \text{where } u_4 \in B$$

which implies that $b_3 u_4 = s_{12}$. Now by (4.320), (4.323), (4.324), (4.335) and $b_3(\bar{b}_3 c_9) = \bar{b}_3(b_3 c_9)$ we obtain that $d + e + b_3 u_5 = \delta_6 + c_9 + \bar{g}_8 + u_4$. Hence $(b_3 u_4, b_3 u_4) \neq 1$, and we have a contradiction.

Assume henceforth that $c = c_{10}$ and we shall derive a contradiction. Then by (4.337) we obtain that

$$\bar{b}_3 c_{10} = x_7 + b_{15} + u_8 \quad \text{where } u_8 \in B$$

which implies by (4.338) that

$$b_3c_{10} = b_{10} + z_5 + s_{15}.$$

By (4.339) we obtain that

$$\bar{b}_3s_{15} = \bar{b}_{15} + c_{10} + \bar{g}_8 + G_{12}.$$

Now by (4.320), (4.323), (4.324), (4.335) and $b_3(\bar{b}_3c_{10}) = \bar{b}_3(b_3c_{10})$, we obtain that $d + e + b_3u_8 = \delta_5 + c_{10} + \bar{g}_8 + G_{12}$. By (4.320) we obtain that $|d| + |e| = 11$, and now we can assume that $d = \delta_5$, $e = e_6$ and $e_6 \in Irr(G_{12})$ which implies that

$$\bar{b}_3s_{15} = \bar{b}_{15} + c_{10} + \bar{g}_8 + e_6 + \alpha_6 \quad \text{where } \alpha_6 \in B$$

and

$$b_3e_6 = s_{15} + \alpha_3 \quad \text{where } \alpha_3 \in B.$$

By (4.320) we obtain that

$$\bar{b}_3e_6 = x_7 + \alpha_{11}.$$

Now by (4.320) and $b_3(\bar{b}_3e_6) = \bar{b}_3(b_3e_6)$ we obtain that $d_5 + b_3\alpha_{11} = \bar{b}_{15} + \bar{g}_8 + \alpha_6 + \bar{b}_3\alpha_3$ which implies that $(\bar{\alpha}_{11}, b_3b_{15}) = 1$, which is a contradiction to (4.323).

Assume henceforth that $c = c_{11}$ and we shall derive a contradiction. Then by (4.335) we obtain that $\delta = \delta_4$ and $(z_5, b_3\delta_4) = 1$ which implies that $(w_6, b_3\delta_4) = 0$. By (4.330), (4.333), (4.335), (4.340), (4.338) and $b_3(\bar{b}_3z_5) = \bar{b}_3(b_3z_5)$, we obtain that $w_6 + z_5 + H = E + b_3\delta_4$. Now we have that $w_6 \in Irr(E)$ which implies that $(w_6, b_3c_{11}) = 1$, and by (4.338) we obtain that

$$b_3c_{11} = b_{10} + z_5 + s + w_6 + Q \quad \text{where } Q \in \mathbb{N}B.$$

Hence $4 \leq (b_3c_{11}, b_3c_{11})$. Now (4.337) implies that

$$\bar{b}_3c_{11} = x_7 + b_{15} + u_5 + u_6 \quad \text{where } u_5, u_6 \in B.$$

By (4.320), (4.323), (4.324), (4.335) and $b_3(\bar{b}_3c_{11}) = \bar{b}_3(b_3c_{11})$, we obtain that $d + e + \bar{b}_{15} + b_3u_5 + b_3u_6 = \delta_4 + \bar{b}_3s + \bar{b}_3w_6$. By (4.339) we obtain that $\bar{g}_8 \in Irr(\bar{b}_3s)$, and since $(c_{11}, b_3u_5) = (c_{11}, b_3u_6) = 1$ we obtain that $(\bar{g}_8, b_3u_5) = (\bar{g}_8, b_3u_6) = 0$. Therefore we have that either $d = \bar{g}_8$ or $e = \bar{g}_8$, but by (4.169) we obtain that $|d| + |e| = 10$ and we have a contradiction.

Now we have that $c = c_{12}$ which implies by (4.335) that $\delta = \delta_3$ and $(z_5, b_3\delta_3) = 1$ which implies that $(w_6, b_3\delta_3) = 0$. By (4.330), (4.333), (4.335), (4.340), (4.338) and $b_3(\bar{b}_3z_5) = \bar{b}_3(b_3z_5)$, we obtain that $w_6 + z_5 + H = E + b_3\delta_3$. Now we have that $w_6 \in Irr(E)$ which implies that $(w_6, b_3c_{12}) = 1$, and by (4.338) we obtain that

$$b_3c_{12} = b_{10} + z_5 + s + w_6 + Q \quad \text{where } Q \in \mathbb{N}B$$

which implies that $|s| \geq 15$. If $|s| = 15$ then By (4.320), (4.323), (4.324), (4.335) and $b_3(\bar{b}_3c_{12}) = \bar{b}_3(b_3c_{12})$ we obtain that $d + e + \bar{b}_{15} + b_3C_{14} = \delta_3 + \bar{b}_3s_{15} + \bar{b}_3w_6$.

By (4.320) we obtain that $d, e \neq \delta_3$ and since $c = c_{12}$ we obtain that $(d, \bar{b}_3 s_{15}) = (e, \bar{b}_3 s_{15}) = 0$. Now we have that $(d, \bar{b}_3 w_6) = (e, \bar{b}_3 w_6) = 1$, which is a contradiction to $(c_{12}, \bar{b}_3 w_6) = 1$. If $|s| \neq 15$ then $|s| \leq 11$, which is a contradiction to (4.339).

4.7 Case $b_3 b_{10} = b_{15} + x_5 + y_5 + z_5$

Since $b_3 b_{10} = b_{15} + x_5 + y_5 + z_5$ we obtain that

$$\bar{b}_3 x_5 = b_{10} + u_5 \quad \text{where } u_5 \in B \tag{4.341}$$

and

$$\bar{b}_3 y_5 = b_{10} + v_5 \quad \text{where } v_5 \in B. \tag{4.342}$$

Now we have that $1 \leq (b_3 x_5, b_3 y_5)$ which implies that

$$b_3 x_5 = c + d \quad \text{where } c, d \in B \tag{4.343}$$

and

$$b_3 y_5 = c + e \quad \text{where } e \in B. \tag{4.344}$$

Assume henceforth that $d = e$ and we shall derive a contradiction. Then $(b_3 x_5, b_3 y_5) = 2$ and by (4.341) and (4.342), we obtain that

$$\bar{b}_3 x_5 = \bar{b}_3 y_5 = b_{10} + u_5 \tag{4.345}$$

which implies that $(x_5, b_3 u_5) = (y_5, b_3 u_5) = 1$. Now we have that

$$b_3 u_5 = x_5 + y_5 + g_5 \quad \text{where } g_5 \in B, \tag{4.346}$$

and by $b_3 (\bar{b}_3 x_5) = \bar{b}_3 (b_3 x_5)$ we obtain that

$$\bar{b}_3 c + \bar{b}_3 d = b_{15} + 2x_5 + 2y_5 + z_5 + g_5$$

which implies that

$$|c|, |d| \geq 5 \tag{4.347}$$

and we can assume that $(b_{15}, \bar{b}_3 c) = 1$. By (4.343) and (4.344) we obtain that $(x_5, \bar{b}_3 c) = (y_5, \bar{b}_3 c) = 1$ which implies that $|c| \geq 10$ and by (4.343) and (4.347) we obtain that $c = c_{10}$ and $d = d_5$. Since $(c_{10}, b_3 b_{15}) = 1$ we obtain by (4.10) that $(c_{10}, \bar{b}_3 b_{10}) = 1$ which implies by (4.4) that

$$\bar{b}_3 b_{10} = b_6 + c_{10} + \alpha + \beta \quad \text{where } \alpha, \beta \in B. \tag{4.348}$$

Now by (4.1), (4.341), (4.343) and $b_3 (\bar{b}_3 x_5) = \bar{b}_3^2 x_5$ we obtain that $\bar{b}_3 b_{10} + \bar{b}_3 u_5 = d_5 + c_{10} + \bar{b}_6 x_5$. If $(d_5, \bar{b}_3 b_{10}) = 1$ then by (4.10) we obtain that $(d_5, b_3 b_{15}) = 1$, and we have a contradiction.

Now we have that

$$(d_5, \bar{b}_3b_{10}) = 0 \tag{4.349}$$

which implies that $(d_5, \bar{b}_3u_5) = 1$ and by (4.346) we obtain that $(b_3u_5, b_3u_5) = 3$ which implies that

$$\bar{b}_3u_5 = d_5 + \gamma + \delta \quad \text{where } \gamma, \delta \in B. \tag{4.350}$$

By (4.345), (4.346) and $b_3(\bar{b}_3u_5) = \bar{b}_3(b_3u_5)$ we obtain that $b_3d_5 + b_3\gamma + b_3\delta = 2b_{10} + 2u_5 + \bar{b}_3g_5$. By (4.349) we obtain that $(b_{10}, b_3\gamma) = (b_{10}, b_3\delta) = 1$. By (4.350) and (4.4) we obtain that $\gamma, \delta \neq b_6$ and $|\gamma| + |\delta| = 10$ which implies that $\gamma, \delta \neq c_{10}$. Now (4.348) implies that $\gamma + \delta = \alpha + \beta$ and $|\alpha| + |\beta| = 14$, which is a contradiction.

Now we have that $(\bar{b}_3x_5, \bar{b}_3y_5) = 1$. In the same way, we can show that $(\bar{b}_3x_5, \bar{b}_3z_5) = (\bar{b}_3y_5, \bar{b}_3z_5) = 1$ and we obtain that

$$\bar{b}_3z_5 = b_{10} + w_5, \tag{4.351}$$

and by (4.343) and (4.344) we obtain that either $b_3z_5 = d + e$ or $b_3z_5 = c + f$ where $f \in B$. If $b_3z_5 = d + e$ then $2|c| = 15$, so

$$b_3z_5 = c + f \quad \text{where } f \in B. \tag{4.352}$$

By $b_3(\bar{b}_3x_5) = \bar{b}_3(b_3x_5)$, $b_3(\bar{b}_3y_5) = \bar{b}_3(b_3y_5)$ and $b_3(\bar{b}_3z_5) = \bar{b}_3(b_3z_5)$, we obtain that

$$\begin{aligned}
\bar{b}_3c + \bar{b}_3d &= b_{15} + x_5 + y_5 + z_5 + b_3u_5, \\
\bar{b}_3c + \bar{b}_3e &= b_{15} + x_5 + y_5 + z_5 + b_3v_5
\end{aligned} \tag{4.353}$$

and

$$\bar{b}_3c + \bar{b}_3f = b_{15} + x_5 + y_5 + z_5 + b_3w_5.$$

By $\bar{b}_3(\bar{b}_3x_5) = \bar{b}_3^2x_5$ we obtain that

$$\bar{b}_3b_{10} + \bar{b}_3u_5 = c + d + \bar{b}_6x_5. \tag{4.354}$$

Assume henceforth that $(c, b_3b_{15}) = 0$ and we shall derive a contradiction. Then $(d, b_3b_{15}) = (e, b_3b_{15}) = (f, b_3b_{15}) = 1$. In addition, we have that $|d| = |e| = |f|$ which implies by (4.4) and (4.10) that

$$\bar{b}_3b_{10} = b_6 + d_8 + e_8 + f_8.$$

Now by (4.352) we obtain that $c = c_7$ and by (4.354) we obtain that $(c_7, \bar{b}_3u_5) = 1$. By (4.4), (4.341), (4.342), (4.351) and $b_3(\bar{b}_3b_{10} = \bar{b}_3(b_3b_{10})$, we obtain that $b_8 + b_3d_8 + b_3e_8 + b_3f_8 = \bar{b}_3b_{15} + 2b_{10} + u_5 + v_5 + w_5$ which implies that $(\bar{b}_3u_5, \bar{b}_3u_5) = 2$. By (4.353) and since $(c, b_3b_{15}) = 0$ we obtain that $(b_{15}, \bar{b}_3d_8) = 1$ and together with (4.343) we obtain that

$$\bar{b}_3d_8 = b_{15} + x_5 + g_4 \quad \text{where } g_4 \in B,$$

and by (4.353) we obtain that $(g_4, b_3u_5) = 1$. Now by (4.341) and $(\bar{b}_3u_5, \bar{b}_3u_5) = 2$ we obtain that $b_3u_5 = g_4 + x_5$, which is a contradiction.

Now we have that $(b_{15}, \bar{b}_3c) = 1$ and by (4.10) we obtain that $(c, \bar{b}_3b_{10}) = 1$ which implies that

$$\bar{b}_3b_{10} = b_6 + c + \alpha + \beta \quad \text{where } \alpha, \beta \in B \tag{4.355}$$

and

$$b_3b_{15} = \bar{b}_{15} + b_6 + c + \alpha + \beta. \tag{4.356}$$

Now we have that

$$|\alpha|, |\beta| \geq 6. \tag{4.357}$$

By (4.343), (4.344) and (4.352) we obtain that $(x_5, \bar{b}_3c) = (y_5, \bar{b}_3c) = (z_5, \bar{b}_3c) = 1$, and together with $(b_{15}, \bar{b}_3c) = 1$ and $b_3x_5 = c + d$ we obtain that either $c = c_{10}$ or $c = c_{12}$. By (4.341) we obtain that $(x_5, b_3b_{10}) = (x_5, b_3u_5) = 1$ which implies that

$$(\bar{b}_3u_5, \bar{b}_3b_{10}) \geq 1. \tag{4.358}$$

Assume henceforth that $c = c_{12}$ and we shall derive a contradiction. Then by (4.356) we obtain that $\alpha = \alpha_6$, $\beta = \beta_6$ and

$$\bar{b}_3\alpha_6 = b_{15} + h_3 \quad \text{where } h_3 \in B. \tag{4.359}$$

By (4.354) and (4.356) we obtain that $b_6 + c_{12} + \alpha_6 + \beta_6 + \bar{b}_3u_5 = c_{12} + d_3 + \bar{b}_6x_5$ which implies that

$$(d_3, \bar{b}_3u_5) = 1. \tag{4.360}$$

By (4.4) we obtain that $(b_6, \bar{b}_3u_5) = 0$. By (4.355) we obtain that $(b_{10}, b_3\alpha_6) = 1$ and by (4.359) we obtain that $(b_3\alpha_6, b_3\alpha_6) = 2$ which implies that $0 = (u_5, b_3\alpha_6) = (\alpha_6, \bar{b}_3u_5)$. In the same way, we can show that $(\beta_6, \bar{b}_3u_5) = 0$ which implies by (4.358) and (4.360) that $\bar{b}_3u_5 = d_3 + c_{12}$. Now we have that $(b_3u_5, b_3u_5) = 2$ and by (4.341) we obtain that

$$b_3u_5 = x_5 + g_{10} \quad \text{where } g_{10} \in B.$$

By (4.353) we obtain that $\bar{b}_3c_{12} + \bar{b}_3d_3 = b_{15} + 2x_5 + y_5 + z_5 + g_{10}$, and we have a contradiction.

Now we have that $c = c_{10}$ which implies that

$$\bar{b}_3c_{10} = b_{15} + x_5 + y_5 + z_5 \tag{4.361}$$

and by (4.354) and (4.355) we obtain that $b_6 + \alpha + \beta + \bar{b}_3u_5 = d_5 + \bar{b}_6x_5$. By (4.357) we obtain that $\alpha, \beta \neq d_5$ which implies that

$$(d_5, \bar{b}_3u_5) = 1. \tag{4.362}$$

By (4.4) we obtain that $(b_6, \bar{b}_3 u_5) = 0$. If $(\alpha, \bar{b}_3 u_5) = 1$ then $(\bar{b}_6 x_5, \bar{b}_6 x_5) \geq 7$. By (4.2) and (4.343) we obtain that $(b_8, x_5 \bar{x}_5) = 1$ in addition $(b_8, b_6 \bar{b}_6) = 1$. Now we have that

$$x_5 \bar{x}_5 = 1 + b_8 + \Sigma_{16} \quad \text{where } \Sigma_{16} \in \mathbb{N}B$$

and

$$b_6 \bar{b}_6 = 1 + b_8 + \Sigma_{27} \quad \text{where } \Sigma_{27} \in \mathbb{N}B.$$

By (4.12) we obtain that $b_3 \bar{b}_{15} = b_8 + \bar{b}_{10} + \Sigma_{27}$ and since $(b_3 b_{15}, b_3 b_{15}) = 5$ we have a contradiction. Now (4.358) implies that $(u_5, b_3 c_{10}) = 1$. In the same way, we can show that $(v_5, b_3 c_{10}) = (w_5, b_3 c_{10}) = 1$ and by (4.355) we obtain that $(b_{10}, b_3 c_{10}) = 1$ which implies that $(b_3 c_{10}, b_3 c_{10}) \neq 4$, and we have a contradiction to (4.361).

Chapter 5
Finishing the Proofs of the Main Results

5.1 Introduction

In Chap. 5 we will freely use the definitions and notation of Chap. 2. The Main Theorem 3 of Chap. 4 left two unsolved problems for the complete classification of the NITA (A, B) generated by a faithful non-real element of degree 3 with $L_1(B) = 1$ and $L_2(B) = \emptyset$.

In Chap. 4, the Main Theorem 3 Case (2) states that there exist $b_6, b_{10}, b_{15} \in B$, where b_6 is non-real, such that $b_3^2 = \bar{b}_3 + b_6$, $\bar{b}_3 b_6 = b_3 + b_{15}$, $b_3 b_6 = b_8 + b_{10}$ and $(b_3 b_8, b_3 b_8) = 3$. Moreover, if b_{10} is real, then $(A, B) \cong_x (CH(3 \cdot A_6), Irr(3 \cdot A_6))$ of dimension 17. If b_{10} is non-real, then $(b_3 b_{10}, b_3 b_{10}) = 2$ and b_{15} is a non-real element.

In this chapter, we will almost completely solve and classify the open Case (2) of Chap. 4 when b_{10} is a non-real element, $(b_3 b_{10}, b_3 b_{10}) = 2$ and b_{15} is a non-real element.

In Chap. 4 we proved under the assumption of the open Case (2) of the Main Theorem 3 that

$$b_3 b_8 = b_3 + \bar{b}_6 + b_{15},$$

$$b_3 b_{10} = b_{15} + d_{15} \quad \text{where } d_{15} \in B \text{ where } b_{15} \neq d_{15},$$

and

$$\bar{b}_3 b_{10} = b_6 + b_{24} \quad \text{where } b_{24} \in B.$$

A Table Algebra (A, B) is an algebra of a finite dimension. In the following definition, we define a Countable Table Algebra and the dimension of that algebra is not necessarily finite.

Definition 5.1 Let $B := \{b_1 = 1, b_2, \ldots, b_l, \ldots\}$ be a distinguished basis of a countable dimensional, associative and commutative algebra A over the complex field $\mathbb{C}$ with identity element 1_A. If $a \in A$ and $b_i \in B$, let $k(b_i, a)$ be the coefficient of b_i in a. Then (A, B) is a Countable Table Algebra (B is a table basis and $|B|$ is the dimension of the Table Algebra (A, B)) if the following hold:

Z. Arad et al., *On Normalized Integral Table Algebras (Fusion Rings)*,
Algebra and Applications 16, DOI 10.1007/978-0-85729-850-8_5,
© Springer-Verlag London Limited 2011

TA1. For all $i, j, m \in \mathbb{N}$ $b_i b_j = \sum_{m=1}^{k} b_{ijm} b_m$ where $k \in \mathbb{N}$ with b_{ijm} a nonnegative real number.

TA2. There is an algebra automorphism (denoted by $-$) of A whose order divides 2, such that $b_i \in B$ implies that $\bar{b}_i \in B$ (then $\bar{i}$ is defined by $b_{\bar{i}} = \bar{b}_i$ and $b_i \in B$ is called real if $b_{\bar{i}} = b_i$).

TA3. There is a function $g : B \times B \longrightarrow \mathbb{R}^+$ such that $b_{ijm} = g(b_i, b_m) b_{\bar{j}mi}$ where $b_{ijm}, b_{\bar{j}mi}$ are defined in TA1 and TA2 for all i, j, m.

TA4. There exists a unique algebra homomorphism $| \; | : A \longrightarrow \mathbb{C}$ such that $|b| = |\bar{b}| > 0$, $b \in B$. We call it the degree homomorphism. The positive real numbers $\{|b|\}_{b \in B}$ are called the degrees of (A, B).

If for all $b_i \in B$ $b_{i\bar{i}1} = 1$, then (A, B) is a Countable Normalized Table Algebra (denoted by C-NTA) and if in addition for all $b \in B$ $|b| \in \mathbb{N}$ and for all i, j, m $b_{ijm} \in \mathbb{N} \cup \{0\}$ then (A, B) is a Countable Normalized Integral Table Algebra (denoted by C-NITA).

Note 5.1 If the dimension of a Countable Table Algebra is finite then by [AB, Lemma 2.9], axiom TA4 is a consequence of axioms TA1–TA3.

In Chaps. 2–4, we assumed that Table Algebras are NITA of finite dimension. If we assume instead that the Table Algebras are C-NITA, the same proofs of Chaps. 2–4 give us C-NITA of finite dimension, i.e., Table Algebras.

Definition 5.2 Let $D = \{(k, n) \mid n \in \mathbb{N}, k \text{ is an odd number with} -1 \leq k \leq 2n+1\}$. Define $f : D \to \mathbb{N} \cup \{0\}$ by

$$f(k, n) = \frac{(n + 1)(k + 1)(2n - k + 1)}{8}.$$

By definition $f(-1, n) = 0$, $f(2n + 1, n) = 0$, $f(1, 1) = 1$ and $f(1, 2) = 3$.

Definition 5.3 Let (A, B) be a C-NITA with a faithful non-real element of degree b_3 such that $b_3^2 = \bar{b}_3 + b_6$ where $b_6 \in B$. For $t \in \mathbb{N}$ with $t \geq 2$, or $t = \infty$, define an *array* C_t as

$$C_t = \{b_{(2j+1,i)}\}_{i=1; j=-1}^{t+1; i},$$

where $b_{(-1,i)} = b_{(2i+1,i)} = 0$ for $1 \leq i \leq t + 1$, $b_{(k,i)} \in B$ for all $1 \leq i \leq t + 1$ and odd k with $1 \leq k \leq 2i - 1$, and the following properties hold:

1. $b_{(1,1)} = 1$, $b_{(1,2)} = b_3$ and $b_{(k,i)} = \bar{b}_{(2i-k,i)}$ for all $1 \leq i \leq t + 1$ and odd k with $1 \leq k \leq 2i - 1$.
2. $b_3 b_{(k,i)} = b_{(k-2,i-1)} + b_{(k+2,i)} + b_{(k,i+1)}$ for all $2 \leq i \leq t$ and odd k with $1 \leq k \leq 2i - 1$.

Remark If an array C_t exists for $t > 2$, then properties 1 and 2 imply that C_{t-1} exists, with $C_{t-1} \subseteq C_t$ in the obvious sense. It is a straightforward consequence of properties 1 and 2 and induction on i that

$$|b_{(k,i)}| = f(k,i), \quad \text{for all } 1 \le i \le t+1 \text{ and odd } k \text{ with } 1 \le k \le 2i+1.$$

Definition 5.4 Let (A, B) be a C-NITA as in Definition 5.3. If for some $n \in \mathbb{N}$ there exists an array C_n for (A, B), but there does not exist an array C_{n+1}, then n is called the *stopping number* of (A, B). All such Table Algebras with stopping number $n \in \mathbb{N}$ are denoted (A, B, C_n). If there are arrays C_n for all $n \in \mathbb{N}$, $n \ge 2$, then the array C_∞ exists. In this case, we say that the stopping number for (A, B) is ∞, and denote such a Table Algebra as (A, B, C_∞). In either case, we say that (A, B) is a C_n-Table Algebra.

We shall show in Theorem 5.1 that there exists no C-NITA (A, B, C_n) such that $\infty > n \ge 43$. The classification of C-NITA (A, B, C_n) such that $4 \le n \le 42$ is still open. If there exists a C-NITA (A, B, C_n), it could be an infinite dimension C-NITA.

We shall show in Theorem 5.2 that (A, B, C_∞) is a uniquely determined C-NITA. Note that (A, B, C_∞) is an infinite dimensional C-NITA as defined in Definition 5.1.

In Theorem 2.1 of Chap. 2 we showed that the Table Algebra $(CH\,PSL(2, 7), Irr\,PSL(2, 7))$ is strictly isomorphic to (A, B, C_2) and $(CH(3 \cdot A_6), Irr(3 \cdot A_6))$ is strictly isomorphic to (A, B, C_3). Moreover, the Main Theorem 3 shows that the Table Algebras above are unique.

In the following two examples, we will give an example of a C-NITA (A, B, C_2) and an example of a C-NITA (A, B, C_3).

Example 5.1 We list part of the structure of the C-NITA (A, B, C_2). The full list appears in [AC]. It has a basis $\{1, b_3, \bar{b}_3, b_6, b_8, b_7\}$ of dimension 6 and by Definition 5.3 we obtain that $C_2 = \{0, 1, b_3, \bar{b}_3, b_6, b_8\}$ and the following hold:

$$b_3^2 = \bar{b}_3 + b_6,$$
$$b_3\bar{b}_3 = 1 + b_8,$$
$$b_3 b_6 = \bar{b}_3 + b_7 + b_8.$$

Note that in the above example, $b_{(1,3)} = b_{(5,3)}$. Since $b_{(1,4)} \notin Irr(b_3 b_{(1,3)})$, we obtain that all the terms of Definitions 5.3 and 5.4 hold.

Example 5.2 We list part of the structure of the C-NITA (A, B, C_3). The full list appears in Chap. 2. It has a basis $\{1, b_3, \bar{b}_3, b_6, \bar{b}_6, b_8, b_{10}, b_{15}, \bar{b}_{15}, c_3, \bar{c}_3, b_5, c_5, c_8, b_9, c_9, \bar{c}_9\}$ of dimension 17 and by Definition 5.3 we obtain that $C_3 = \{0, 1, b_3, \bar{b}_3, b_6, \bar{b}_6, c_8, b_{10}, b_{15}, \bar{b}_{15}\}$ and the following hold:

$$b_3^2 = \bar{b}_3 + b_6,$$
$$b_3\bar{b}_3 = 1 + c_8,$$
$$b_3b_6 = c_8 + b_{10},$$
$$b_3\bar{b}_6 = \bar{b}_3 + \bar{b}_{15},$$
$$b_3c_8 = b_3 + \bar{b}_6 + b_{15},$$
$$b_3b_{15} = b_6 + 2\bar{b}_{15} + c_9.$$

Note that in the above example $b_{(1,4)} = b_{(7,4)}$. Since $b_{(3,5)} \notin Irr(b_3b_{(3,4)})$, we obtain that all the terms of Definitions 5.3 and 5.4 hold.

Example 5.3   The classification of a Countable Table Algebra (A, B, C_4) is still open. If there exists a C-NITA (A, B, C_4) then $C_4 = \{0, 1, b_3, \bar{b}_3, b_6, \bar{b}_6, b_8, b_{10}, \bar{b}_{10}, b_{15}, \bar{b}_{15}, c_{15}, \bar{c}_{15}, b_{24}, \bar{b}_{24}, b_{27}\}$ and the following hold:

$$b_3^2 = \bar{b}_3 + b_6,$$
$$b_3\bar{b}_3 = 1 + b_8,$$
$$b_3b_6 = b_8 + b_{10},$$
$$b_3\bar{b}_6 = \bar{b}_3 + \bar{b}_{15},$$
$$b_3b_8 = b_3 + \bar{b}_6 + b_{15},$$
$$b_3b_{10} = b_{15} + c_{15},$$
$$b_3\bar{b}_{10} = \bar{b}_6 + \bar{b}_{24},$$
$$b_3b_{15} = b_6 + \bar{b}_{15} + b_{24},$$
$$b_3\bar{b}_{15} = b_8 + \bar{b}_{10} + b_{27}.$$

We can show that if $b_3c_{15} = b_{24} + b_{21}$ then the stopping number of (A, B) is not 4 which implies that if there exists such a C-NITA then it is either a C_n-Table Algebra where $5 \leq n$ or a C_∞-Table Algebra. Now, we have that $b_{21} \notin B$ and $b_3c_{15} = b_{24} + \Sigma_{21}$ where $\Sigma_{21} \in \mathbb{N}B \setminus B$. By the Main Theorem 3 of Chap. 4 we show that c_{15} is a non-real element but we still do not know if b_{24} is a real element.

Note 5.2   From the above example we can see that if there exists a C-NITA (A, B) that satisfies the assumptions of the open case (2) of the Main Theorem 3, then either $(A, B) \cong_x (A, B, C_n)$ where $4 \leq n$ or $(A, B) \cong_x (A, B, C_\infty)$.

Theorem 5.1   *There exists no C-NITA (A, B, C_n) such that $43 \leq n$.*

Theorem 5.2   *There exists a unique C_∞-Table Algebra which is strictly isomorphic to a C-NITA generated by the non-real polynomial representation of dimension 3 of $SL(3, \mathbb{C})$.*

Now, by the Main Theorem 3 of Chap. 4, Theorems 5.2 and 5.1 of this chapter, we state the Main Theorem 4 of this chapter as follows.

Main Theorem 4 *Let (A, B) be a C-NITA generated by a non-real element $b_3 \in B$ of degree 3 and without non-identity basis element of degree 1 and 2. Then $b_3 \bar{b}_3 = 1 + b_8$, $b_8 \in B$, and (A, B) is of one of the following types:*

(1) $(A, B) \cong_x (A, B, C_2)$ *which is the unique C_2-Table Algebra and is strictly isomorphic to $(CH\,PSL(2, 7), Irr\,PSL(2, 7))$ of dimension 6.*
(2) $(A, B) \cong_x (A, B, C_3)$ *which is the unique C_3-Table Algebra and is strictly isomorphic to $(CH(3 \cdot A_6), Irr(3 \cdot A_6))$ of dimension 17.*
(3) *There exists a unique C_∞-Table Algebra which is strictly isomorphic to a C-NITA generated by the non-real polynomial representation of dimension 3 of $SL(3, \mathbb{C})$.*
(4) *There exists no C-NITA (A, B, C_n) such that $43 \le n$ and the classification of the Table Algebra (A, B, C_n) such that $4 \le n \le 42$ is still open.*
(5) *There exist $c_3, b_6 \in B$, $c_3 \neq b_3$ or $\bar{b}_3$, such that $b_3^2 = c_3 + b_6$ and either $(b_3 b_8, b_3 b_8) = 3$ or 4. If $(b_3 b_8, b_3 b_8) = 3$ and c_3 is non-real, then $(A, B) \cong_x (A(3 \cdot A_6 \cdot 2), B_{32})$ of dimension 32. (See Theorem 2.9 of Chap. 2 for the definition of this specific NITA.) If $(b_3 b_8, b_3 b_8) = 3$ and c_3 is real, then $(A, B) \cong_x (A(7 \cdot 5 \cdot 10), B_{22})$ of dimension 22. (See Theorem 2.10 of Chap. 2 for the definition of this specific NITA.) The Case $(b_3 b_8, b_3 b_8) = 4$ is still open.*

By the Main Theorem 4 we obtain that if (A, B) is a C-NITA generated by a non-real element $b_3 \in B$ of degree 3 and without non-identity basis element of degree 1 and 2, then $b_3^2 = \bar{c}_3 + b_6$ where $c_3, b_6 \in B$. If (A, B) is of one of (1)–(4) types of Main Theorem 4, then $c_3 = \bar{b}_3$ and if (A, B) is of type (5) of the Main Theorem 4, then $c_3 \neq b_3$ or $\bar{b}_3$. The rest of this chapter is devoted to prove Theorems 5.1 and 5.2.

5.2 Proof of Theorem 5.1

Throughout this section, (A, B) denotes a C-NITA. We will use some lemmas in order to prove Theorem 5.1.

Lemma 5.1 *If $a, b \in B$ where $|a|, |b| \neq 1$ and $c, d \in Irr(ab)$, then $\frac{|d|}{|c|} \le \min\{|a|^2 - 1, |b|^2 - 1\}$.*

Proof Assume, without loss of generality, that $|a| \le |b|$. If $c = d$ then $\frac{|d|}{|c|} = 1$, and since $|a|, |b| \neq 1$ the Lemma holds. If $c \neq d$ then $|c| + |d| \le |a| \cdot |b|$. Since $c \in Irr(ab)$ we obtain that $b \in Irr(\bar{a}c)$ which implies that $|b| \le |a| \cdot |c|$. Now we have that $|c| + |d| \le |a|^2 \cdot |c|$ which implies that $\frac{|d|}{|c|} \le |a|^2 - 1$, as required. □

Lemma 5.2 *If (A, B) is a C_n-Table Algebra then*

$$\bar{b}_3 b_{(k,i)} = b_{(k,i-1)} + b_{(k-2,i)} + b_{(k+2,i+1)}$$

for all $2 \le i \le n$ and odd k with $1 \le k \le 2 - 1$.

Proof Let $i \in \mathbb{N}$ where $2 \leq i \leq n$ and let k be an odd number where $1 \leq k \leq 2i - 1$. By Definition 5.3 we obtain that

$$b_3 \bar{b}_{(k,i)} = b_3 b_{(2i-k,i)} = b_{(2i-k-2,i-1)} + b_{(2i-k+2,i)} + b_{(2i-k,i+1)}$$
$$= \bar{b}_{(k,i-1)} + \bar{b}_{(k-2,i)} + \bar{b}_{(k+2,i+1)},$$

as required. $\square$

Lemma 5.3 *If (A, B) is a C_n-Table Algebra and $i, j, k, l \in \mathbb{N}$ then $b_{(k,i)} \neq b_{(l,j)}$ where $1 \leq i \leq n$, $1 \leq j \leq n + 1$, $1 \leq k \leq 2i - 1$ is an odd number, $1 \leq l \leq 2j - 1$ is an odd number and either $k \neq l$ or $i \neq j$.*

Proof We will prove the lemma by induction on n. For $n = 1$ we obtain that $C_1 = \{b_{((1,1)}, b_{(1,2)}, b_{(3,2)}\}$. Now by Definition 5.2 $f(1, 1) \neq f(1, 2)$ and $f(1, 1) \neq f(3, 2)$ which imply that $b_{(1,1)} \neq b_{(1,2)}$ and $b_{(1,1)} \neq b_{(3,2)}$ as required. Example 5.1 implies that the lemma holds for $n = 2$. Now we assume that the lemma holds for $n = t$ where $2 \leq t$. We will prove the lemma for $n = t + 1$. By Definition 5.3 we obtain that $C_t \subset C_{t+1}$, so by the induction hypothesis we obtain that $b_{(k,i)} \neq b_{(l,j)}$ where $1 \leq i \leq t$, $1 \leq j \leq t + 1$, $1 \leq k \leq 2i - 1$ is an odd number, $1 \leq l \leq 2j - 1$ is an odd number and either $k \neq l$ or $i \neq j$. We will prove that

$$b_{(k,t+1)} \neq b_{(l,t+1)} \quad \text{where } 1 \leq k \neq l \leq 2t + 1 \tag{5.1}$$

and

$$b_{(k,i)} \neq b_{(l,t+2)}$$

where $1 \leq i \leq t + 1$, $1 \leq k \leq 2i - 1$, $1 \leq l \leq 2t + 3$ and k, l are odd numbers.

Assume henceforth that $b_{(k,t+1)} = b_{(l,t+1)}$ where $1 \leq k \neq l \leq 2t + 1$ are odd numbers, and we will derive a contradiction. So we can assume, without loss of generality, that $k < l$. By Definition 5.3 we obtain that

$$b_3 b_{(k,t+1)} = b_{(k-2,t)} + b_{(k+2,t+1)} + b_{(k,t+2)}$$

and

$$b_3 b_{(l,t+1)} = b_{(l-2,t)} + b_{(l+2,t+1)} + b_{(l,t+2)}.$$

Now we have that $b_{(k-2,t)} + b_{(k+2,t+1)} + b_{(k,t+2)} = b_{(l-2,t)} + b_{(l+2,t+1)} + b_{(l,t+2)}$, and by the induction hypothesis we obtain that $b_{(k-2,t)} \neq b_{(l-2,t)}, b_{(l+2,t+1)}$ and $b_{(l-2,t)} \neq b_{(k+2,t+1)}$ which implies that $b_{(k-2,t)} = b_{(l,t+2)}$ and $b_{(l-2,t)} = b_{(k,t+2)}$. If $k = 1$ we obtain that $b_{(-1,t)} = b_{(l,t+2)}$ where $1 \leq l \leq 2t + 1$.

By Definition 5.2 we obtain that $f(-1, t) = 0$ and $f(l, t + 2) \neq 0$, which is a contradiction. If $k \neq 1$ we obtain that $b_{(k+2,t+1)} = b_{(l+2,t+1)}$. We repeat that process $\frac{2t+3-l}{2}$ times and we obtain that $b_{(k-l+2t+3,t+1)} = b_{(2t+3,t+1)}$. Since $f(2t + 3, t + 1) = 0$ and $f(k - l + 2t + 3, t + 1) \neq 0$, we have a contradiction.

Assume henceforth that $b_{(k,i)} = b_{(l,t+2)}$ where $1 \leq i \leq t + 1$, $1 \leq k \leq 2i - 1$, $1 \leq l \leq 2t + 3$ and k, l are odd numbers, and we will derive a contradiction. Then we have the following four cases:

Case 1: $l \neq 1, 3, t + 3$.
Case 2: $l = 1$.
Case 3: $l = 3$.
Case 4: $l = t + 3$.

In the rest of this section, we derive a contradiction for the above cases.

Case 1 By Definition 5.3, Lemma 5.2 and $b_3(\bar{b}_3 b_{(l-2,t+1)}) = \bar{b}_3(b_3 b_{(l-2,t+1)})$, we obtain that

$$b_{(l-4,t+2)} + b_3 b_{(l,t+2)} = b_{(l+2,t+2)} + \bar{b}_3 b_{(l-2,t+2)}. \tag{5.2}$$

If $b_{(l-4,t+2)} = b_{(l+2,t+2)}$ then $f(l, t+2) = f(l-2, t+2)$, and by Definition 5.2 we obtain that $(l+1)(2t - l + 5) = (l-1)(2t - l + 7)$ which implies that $l = t + 3$. Since $l \neq t + 3$ we obtain that

$$f(l - 4, t + 2) \neq f(l + 2, t + 2) \tag{5.3}$$

which implies that $b_{(l-4,t+2)} \neq b_{(l+2,t+2)}$ and (5.2) implies that

$$b_{(l+2,t+2)} \in Irr(b_3 b_{(l,t+2)}).$$

By Lemma 5.2 we obtain that

$$\bar{b}_3 b_{(l-2,t+1)} = b_{(l-2,t)} + b_{(l-4,t+1)} + b_{(l,t+2)}$$

which implies that

$$b_{(l-2,t+1)} \in Irr(b_3 b_{(l,t+2)}). \tag{5.4}$$

Now we have the following two cases:

Case A. $b_{(l+2,t+2)} = b_{(l-2,t+1)}$.
Case B. $b_{(l+2,t+2)} \neq b_{(l-2,t+1)}$.

By deriving a contradiction for Cases A and B, we shall show that Case 1 is impossible.

Case A Since $b_{(l+2,t+2)} = b_{(l-2,t+1)}$ we obtain that $f(l+2, t+2) = f(l-2, t+1)$, and by (5.3) we obtain that $f(l-4, t+2) \neq f(l+2, t+2)$ which implies that $f(l-4, t+2) \neq f(l-2, t+1)$. Now we have that $b_{(l-4,t+2)} \neq b_{(l-2,t+1)}$. By Definition 5.3 we obtain that

$$b_3 b_{(l-2,t+1)} = b_{(l-4,t)} + b_{(l,t+1)} + b_{(l-2,t+2)}$$

which implies that

$$b_{(l-2,t+1)} \in Irr(\bar{b}_3 b_{(l-2,t+2)}).$$

Now by (5.2) and since $3f(l, t+2) = f(l-2, t+1) + f(l+2, t+2) + f(l, t+3)$, we obtain that

$$b_3 b_{(l,t+2)} = 2b_{(l+2,t+2)} + \Sigma_{f(l,t+3)} \quad \text{where } \Sigma_{f(l,t+3)} \in \mathbb{N}B.$$

By Definition 5.3 we obtain that

$$b_3 b_{(k,i)} = b_{(k-2,i-1)} + b_{(k+2,i)} + b_{(k,i+1)}$$

which implies that $2b_{(l+2,t+2)} + \Sigma_{f(l,t+3)} = b_{(k-2,i-1)} + b_{(k+2,i)} + b_{(k,i+1)}$, so $\Sigma_{f(l,t+3)} \in B$ which implies that the stopping number is not $t + 1$, and we have a contradiction.

Case B By (5.2) and (5.4) and since $b_{(l+2,t+2)} \neq b_{(l-2,t+1)}$ we obtain that

$$b_3 b_{(l,t+2)} = b_{(l-2,t+1)} + b_{(l+2,t+2)} + \Sigma_{f(l,t+3)} \quad \text{where } \Sigma_{f(l,t+3)} \in \mathbb{N}B.$$

By Definition 5.3 we obtain that

$$b_3 b_{(k,i)} = b_{(k-2,i-1)} + b_{(k+2,i)} + b_{(k,i+1)}$$

which implies that $b_{(l-2,t+1)} + b_{(l+2,t+2)} + \Sigma_{f(l,t+3)} = b_{(k-2,i-1)} + b_{(k+2,i)} + b_{(k,i+1)}$. So $\Sigma_{f(l,t+3)} \in B$ which implies that the stopping number is not $t + 1$, and we have a contradiction.

Case 2 Since $l = 1$ we obtain that $b_{(k,i)} = b_{(1,t+2)}$ where $1 \leq i \leq t + 1$. By Definition 5.3 we obtain that

$$b_3 b_{(1,t+1)} = b_{(3,t)} + b_{(1,t+2)}$$

which implies that

$$b_{(1,t+1)} \in Irr(\bar{b}_3 b_{(1,t+2)}).$$

By Lemma 5.2 we obtain that

$$\bar{b}_3 b_{(k,i)} = b_{(k,i-1)} + b_{(k-2,i)} + b_{(k+2,i+1)}.$$

Since $i \leq t + 1$ we obtain by the induction hypothesis that $b_{(k,i-1)} \neq b_{(1,t+1)}$. Now we have that either $b_{(k-2,i)} = b_{(1,t+1)}$ or $b_{(k+2,i+1)} = b_{(1,t+1)}$.

If $b_{(k-2,i)} = b_{(1,t+1)}$ then (5.1) and the induction hypothesis imply that $k = 3$ and $i = t + 1$. Now we have that $b_{(3,t+1)} = b_{(1,t+2)}$. Thus $f(3, t + 1) = f(1, t + 2)$ and by Definition 5.2 we obtain that $(t - 3)(t + 2) = 0$. Since $t \neq -2$ we have that $t = 3$. Now we have that $b_{(3,4)} = b_{(1,5)}$, which is a contradiction to Main Theorem 3 of Chap. 4.

If $b_{(k+2,i+1)} = b_{(1,t+1)}$ then (5.1) and the induction hypothesis imply that $i = t + 1$, so $b_{(k+2,t+2)} = b_{(1,t+1)}$ and by Case 1 we obtain that either $k = 1$ or $k = t + 1$.

If $k = 1$ then $b_{(1,t+1)} = b_{(1,t+2)}$ and $f(1, t + 1) = f(1, t + 2)$. Now by Definition 5.2 we obtain that $(t + 1)(t + 2) = (t + 2)(t + 3)$, and we have a contradiction.

If $k = t + 1$ then $b_{(t+3,t+2)} = b_{(1,t+1)}$, and by Definition 5.3 we obtain that $b_{(t+1,t+2)} = b_{(2t+1,t+1)}$. Since $f(3,4) \neq f(5,3)$ we have that $t \neq 2$. Now we have a contradiction to Case 1.

Case 3 Since $l = 3$ we obtain that $b_{(k,i)} = b_{(3,t+2)}$. By Definition 5.3, Lemma 5.2 and $b_3(\bar{b}_3 b_{(1,t+1)}) = \bar{b}_3(b_3 b_{(1,t+1)})$, we obtain that

$$b_3 b_{(3,t+2)} = b_{(5,t+2)} + \bar{b}_3 b_{(1,t+2)}$$

which implies that

$$b_{(5,t+2)} \in Irr(b_3 b_{(3,t+2)}).$$

By Lemma 5.2 we obtain that

$$\bar{b}_3 b_{(1,t+1)} = b_{(1,t)} + b_{(3,t+2)}$$

which implies that

$$b_{(1,t+1)} \in Irr(b_3 b_{(3,t+2)}).$$

By Case 1 we obtain that $b_{(1,t+1)} \neq b_{(5,t+2)}$. Now we have that

$$b_3 b_{(3,t+2)} = b_{(1,t+1)} + b_{(5,t+2)} + \Sigma_{f(3,t+3)} \quad \text{where } \Sigma_{f(3,t+3)} \in \mathbb{N}B,$$

and by Definition 5.3 we obtain that

$$b_3 b_{f(k,i)} = b_{f(k-2,i-1)} + b_{f(k+2,i)} + b_{f(k,i+1)}$$

which implies that $b_{(1,t+1)} + b_{(5,t+2)} + \Sigma_{f(3,t+3)} = b_{(k-2,i-1)} + b_{(k+2,i)} + b_{(k,i+1)}$. So $\Sigma_{f(3,t+3)} \in B$ which implies that the stopping number is not $t + 1$, and we have a contradiction.

Case 4 Since $l = t + 3$ we obtain that $b_{(k,i)} = b_{(t+3,t+2)}$ which implies by property no. 2 of Definition 5.3 that $b_{(2i-k,i)} = b_{(t+1,t+2)}$. If $t = 2$ we have a contradiction to Case 3, and if $t \neq 2$ we have a contradiction to Case 1. $\qquad\square$

Lemma 5.4 *If (A, B) is a C_n-Table Algebra where $2 \leq n$, then for $1 \leq k \leq 2n + 1$ where k is an odd number*

$$b_3 b_{(k,n+1)} = b_{(k-2,n)} + b_{(k+2,n+1)} + \Sigma_{f(k,n+2)} \quad \text{where } \Sigma_{f(k,n+2)} \in \mathbb{N}B.$$

Proof Let (A, B) be a C_n-Table Algebra. If $n = 2$ then by Example 5.1 the lemma holds, and if $n = 3$ then by Example 5.2 the lemma holds. Now we can assume that $4 \leq n$. By Lemma 5.2 we obtain that

$$\bar{b}_3 b_{(k-2,n)} = b_{(k-2,n-1)} + b_{(k-4,n)} + b_{(k,n+1)}$$

where k is an odd number and $3 \leq k \leq 2n + 1$. Now we have that

$$b_{(k-2,n)} \in Irr(b_3 b_{(k,n+1)}) \tag{5.5}$$

where k is an odd number and $3 \leq k \leq 2n + 1$. By Definition 5.3, Lemma 5.2 and $b_3(\bar{b}_3 b_{(k-2,n)}) = \bar{b}_3(b_3 b_{(k-2,n)})$, we obtain that

$$b_{(k-4,n+1)} + b_3 b_{(k,n+1)} = b_{(k+2,n+1)} + \bar{b}_3 b_{(k-2,n+1)}.$$

If $b_{(k-4,n+1)} = b_{(k+2,n+1)}$ then $f(k, n+1) = f(k-2, n+1)$, and by Definition 5.2 we obtain that $k = n + 2$. Now we have by Lemma 5.3 that if $k \neq n + 2$, then

$$b_3 b_{(k,n+1)} = b_{(k-2,n)} + b_{(k+2,n+1)} + \Sigma_{f(k,n+2)} \quad \text{where } \Sigma_{f(k,n+2)} \in \mathbb{N}B. \quad (5.6)$$

Now we show that the lemma holds for $k = n + 2$. By (5.6) and since $n \neq 3$ we obtain that

$$b_3 b_{(n-2,n+1)} = b_{(n-4,n)} + b_{(n,n+1)} + \Sigma_{f(n-2,n+2)}$$

which implies by property no. 2 of Definition 5.3 that

$$\bar{b}_3 b_{(n+4,n+1)} = b_{(n+4,n)} + b_{(n+2,n+1)} + \bar{\Sigma}_{f(n-2,n+2)}.$$

Now we have that

$$b_{(n+4,n+1)} \in Irr(b_3 b_{(n+2,n+1)}),$$

and by Lemma 5.3 and (5.5) we obtain that

$$b_3 b_{(n+2,n+1)} = b_{(n,n)} + b_{(n+4,n+1)} + \Sigma_{f(n+2,n+2)} \quad \text{where } \Sigma_{f(n+2,n+2)} \in \mathbb{N}B.$$

Now we have that

$$b_3 b_{(k,n+1)} = b_{(k-2,n)} + b_{(k+2,n+1)} + \Sigma_{f(k,n+2)}$$

where k is an odd number and $3 \leq k \leq 2n + 1$. In particular,

$$b_3 b_{(2n-1,n+1)} = b_{(2n-3,n)} + b_{(2n+1,n+1)} + \Sigma_{f(2n-1,n+2)} \quad \text{where } \Sigma_{f(k,n+2)} \in \mathbb{N}B$$

which implies by property no. 2 of Definition 5.3 that

$$\bar{b}_3 b_{(3,n+1)} = b_{(3,n)} + b_{(1,n+1)} + \bar{\Sigma}_{f(2n-1,n+2)}.$$

Now we have that

$$b_3 b_{(1,n+1)} = b_{(-1,n)} + b_{(3,n+1)} + \Sigma_{f(1,n+2)} \quad \text{where } \Sigma_{f(1,n+2)} \in \mathbb{N}B$$

and the lemma holds for $k = 1$. $\qquad\square$

Lemma 5.5 *If (A, B) is a C_n-Table Algebra where $2 \leq n$, then for $1 \leq k \leq 2n + 1$ where k is an odd number*

$$b_3 b_{(k,n+1)} = b_{(k-2,n)} + b_{(k+2,n+1)} + \Sigma_{f(k,n+2)} \quad \text{where } \Sigma_{f(k,n+2)} \in \mathbb{N}B \setminus B.$$

Proof Since the stopping number is n, Lemma 5.4 implies that there exists an odd number $1 \leq l \leq 2n + 1$ such that

$$b_3 b_{(l,n+1)} = b_{(l-2,n)} + b_{(l+2,n+1)} + \Sigma_{f(l,n+2)} \quad \text{where } \Sigma_{f(l,n+2)} \in \mathbb{N}B \setminus B. \quad (5.7)$$

Now by property no. 2 of Definition 5.3 and Lemma 5.4, we obtain that $b_3 \bar{b}_{(l,n+1)} = b_3 b_{(2n+2-l,n+1)} = b_{(2n-l,n)} + b_{(2n+4-l,n+1)} + \Sigma$ where $\Sigma \in \mathbb{N}B$. Hence $b_3 \bar{b}_{(l,n+1)} = \bar{b}(l,n) + \bar{b}_{(l-2,n+1)} + \Sigma$. Since $3f(l, n + 1) = f(l, n) + f(l - 2, n + 1) + f(l + 2, n + 2)$, we obtain that

$$\bar{b}_3 b_{(l,n+1)} = b_{(l,n)} + b_{(l-2,n+1)} + \Sigma_{f(l+2,n+2)}. \quad (5.8)$$

If $l \neq 1$ then $(b_3 b_{(l,n+1)}, b_3 b_{(l,n+1)}) = (b_3 \bar{b}_{(l,n+1)}, b_3 \bar{b}_{(l,n+1)})$, (5.7) and Lemma 5.3 imply that either $\Sigma_{f(l+2,n+2)} \notin B$ or $\Sigma_{f(l+2,n+2)} = b_{(l-2,n+1)}$.

By property no. 3 of Definition 5.3, Lemma 5.2 and $b_3(\bar{b}_3 b_{(l,n)}) = \bar{b}_3(b_3 b_{(l,n)})$, we obtain that $b_3 b_{(l+2,n+1)} = b_{(l,n)} + b_{(l+4,n+1)} + \Sigma_{f(l+2,n+2)}$. Now if $\Sigma_{f(l,n+2)} = b_{(l-2,n+1)}$ we obtain that $b_{(l+4,n+1)} = \Sigma_{f(l+2,n+2)}$ which implies that $f(l + 4, n + 1) = f(l + 2, n + 2) = f(l - 2, n + 1)$, and this is impossible.

So we have that

$$b_3 b_{(l+2,n+1)} = b_{(l,n)} + b_{(l+4,n+1)} + \Sigma_{f(l+2,n+2)} \quad \text{where } \Sigma_{f(l+2,n+2)} \in \mathbb{N}B \setminus B. \quad (5.9)$$

If $l = 1$ then Lemma 5.3 and (5.8) imply that $\bar{b}_3 b_{(1,n+1)} = b_{(1,n)} + \Sigma_{f(3,n+2)}$ where $\Sigma_{f(3,n+2)} \in \mathbb{N}B \setminus B$. Now by property no. 3 of Definition 5.3, Lemma 5.2 and $\bar{b}_3(b_3 b_{(1,n+1)}) = b_3(\bar{b}_3 b_{(1,n+1)})$, we obtain that

$$b_3 b_{(3,n+1)} = b_{(1,n)} + b_{(5,n+1)} + \Sigma_{f(3,n+2)} \quad \text{where } \Sigma_{f(3,n+2)} \in \mathbb{N}B \setminus B. \quad (5.10)$$

By (5.7), (5.9) and (5.10) we obtain that if

$$b_3 b_{(l,n+1)} = b_{(l-2,n)} + b_{(l+2,n+1)} + \Sigma_{f(l,n+2)} \quad \text{where } \Sigma_{f(l,n+2)} \in \mathbb{N}B \setminus B,$$

then

$$b_3 b_{(l+2,n+1)} = b_{(l,n)} + b_{(l+4,n+1)} + \Sigma_{f(l+2,n+2)} \quad \text{where } \Sigma_{f(l+2,n+2)} \in \mathbb{N}B \setminus B$$

where $1 \leq l \leq 2n - 1$.

The stopping number is n, which implies that $b_3 b_{(2n+1,n+1)} = b_{(2n-1,n)} + \Sigma_{f(2n+1,n+2)}$ where $\Sigma_{f(l+2,n+2)} \in \mathbb{N}B \setminus B$ which implies that $b_3 \bar{b}_{(1,n+1)} = \bar{b}_{(1,n)} + \Sigma_{f(2n+1,n+2)}$. Now $(b_3 b_{(1,n+1)}, b_3 b_{(1,n+1)}) = (b_3 \bar{b}_{(1,n+1)}, b_3 \bar{b}_{(1,n+1)})$ and Lemma 5.4 imply that $b_3 b_{(1,n+1)} = b_{(3,n+1)} + \Sigma_{f(1,n+2)}$ and either $\Sigma_{f(1,n+2)} \notin B$ or $b_{(3,n+1)} = \Sigma_{f(1,n+2)}$. If $b_{(3,n+1)} = \Sigma_{f(1,n+2)}$ then $f(3, n + 1) = f(1, n + 2)$ which implies that $n = 3$ which is impossible by Example 5.2 and the lemma holds. □

Lemma 5.6 *If (A, B) is a C_n-Table Algebra, then*

$$b_{(1,l)} b_{(1,m)} = \sum_{i=1}^{l} b_{(2i-1,m+l-i)}$$

and

$$b_{(3,l)}b_{(1,m)} = \sum_{i=2}^{l} b_{(2i-3,m+l-i-1)} + \sum_{i=1}^{l-1} b_{(2i+1,m+l-i)}$$

where $l, m \in \mathbb{N}$, $2 \leq m$, $l \leq m$ and $l + m \leq n + 2$.

Proof In the following proof we assume that $2 \leq m$. Let (A, B) be a C_n-Table Algebra. We will prove that

$$b_{(1,l)}b_{(1,m)} = \sum_{i=1}^{l} b_{(2i-1,m+l-i)} \tag{5.11}$$

where $l, m \in \mathbb{N}$, $l \leq m$ and $l + m \leq n + 2$ by induction on l. Since

$$b_{(1,1)}b_{(1,m)} = b_{(1,m)}$$

we obtain that (5.11) holds for $l = 1$. Since

$$b_{(1,2)}b_{(1,m)} = b_3 b_{(1,m)} = b_{(3,m)} + b_{(1,m+1)},$$

we obtain that (5.11) holds for $l = 2$.

By Definition 5.3 and $b_3(b_3 b_{(1,m)}) = b_3^2 b_{(1,m)}$ we obtain that

$$b_{(1,3)}b_{(1,m)} = b_{(1,m+2)} + b_{(3,m+1)} + b_{(5,m)}$$

and (5.11) holds for $l = 3$. Assume that (5.11) holds for $3 \leq l \leq t \leq m - 1$ and we will prove (5.11) for $l = t + 1$, i.e., we will prove that

$$b_{(1,t+1)}b_{(1,m)} = \sum_{i=1}^{t+1} b_{(2i-1,m+t+1-i)}.$$

By the induction hypothesis and by Definition 5.3, we obtain that

$$b_3(b_{(1,t)}b_{(1,m)}) = \sum_{i=1}^{t} b_3 b_{(2i-1,m+t-i)}$$

$$= \sum_{i=1}^{t} (b_{(2i-3,m+t-i-1)} + b_{(2i+1,m+t-i)} + b_{(2i-1,m+t-i+1)})$$

and $(b_3 b_{(1,t)})b_{(1,m)} = b_{(3,t)}b_{(1,m)} + b_{(1,t+1)}b_{(1,m)}$. Thus

$$b_{(3,t)}b_{(1,m)} + b_{(1,t+1)}b_{(1,m)}$$

$$= \sum_{i=1}^{t} (b_{(2i-3,m+t-i-1)} + b_{(2i+1,m+t-i)} + b_{(2i-1,m+t-i+1)}). \tag{5.12}$$

By the induction hypothesis, Definition 5.3, Lemma 5.2 and since $3 \leq t$, we obtain that

$$\bar{b}_3(b_{(1,t-1)}b_{(1,m)}) = \sum_{i=1}^{t-1} \bar{b}_3 b_{(2i-1,m+t-i-1)}$$

$$= \sum_{i=1}^{t-1} (b_{(2i-1,m+t-i-2)} + b_{(2i-3,m+t-i-1)} + b_{(2i+1,m+t-i)})$$

and

$$(\bar{b}_3 b_{(1,t-1)})b_{(1,m)} = b_{(1,t-2)}b_{(1,m)} + b_{(3,t)}b_{(1,m)}$$

$$= \sum_{i=1}^{t-2} b_{(2i-1,m+t-i-2)} + b_{(3,t)}b_{(1,m)}.$$

Thus

$$b_{(3,t)}b_{(1,m)} = \sum_{i=2}^{t} b_{(2i-3,m+t-i-1)} + \sum_{i=1}^{t-1} b_{(2i+1,m+t-i)}.$$

Now by (5.12) we obtain that

$$b_{(1,t+1)}b_{(1,m)} = \sum_{i=1}^{t+1} b_{(2i-1,m+t+1-i)},$$

as required. $\square$

Lemma 5.7   *If (A, B) is a C_n-Table Algebra, then*

$$\bar{b}_{(1,l)}b_{(1,m)} = \sum_{i=1}^{l} b_{(2i-1,m-l-1+2i)}$$

and

$$\bar{b}_{(3,l)}b_{(1,m)} = \sum_{i=2}^{l} b_{(2i-3,m-l+2i-1)} + \sum_{i=1}^{l-1} b_{(2i+1,m-l+2i)}$$

where $l, m \in \mathbb{N}$, $2 \leq m$, $l \leq m$ and $l + m \leq n + 2$.

Proof Let (A, B) be a C_n-Table Algebra. We will prove that

$$\bar{b}_{(1,l)}b_{(1,m)} = \sum_{i=1}^{l} b_{(2i-1,m-l-1+2i)} \tag{5.13}$$

where $l, m \in \mathbb{N}$, $l \leq m$ and $l + m \leq n + 2$ by induction on l.

Since

$$\bar{b}_{(1,1)}b_{(1,m)} = b_{(1,m)},$$

we obtain that (5.13) holds for $l = 1$. Since

$$\bar{b}_{(1,2)}b_{(1,m)} = \bar{b}_3 b_{(1,m)} = b_{(1,m-1)} + b_{(3,m+1)},$$

we obtain that (5.13) holds for $l = 2$. By Definition 5.3, Lemma 5.2 and $\bar{b}_3(\bar{b}_3 b_{(1,m)}) = \bar{b}_3^2 b_{(1,m)}$, we obtain that

$$b_{(1,3)}b_{(1,m)} = b_{(1,m-2)} + b_{(3,m)} + b_{(5,m+2)}$$

and (5.13) holds for $l = 3$.

Assume that (5.13) holds for $3 \le l \le t \le m - 1$ and we will prove (5.13) for $l = t + 1$, i.e., we will prove that

$$\bar{b}_{(1,t+1)}b_{(1,m)} = \sum_{i=1}^{t+1} b_{(2i-1,m-t-2+2i)}.$$

By the induction hypothesis, Definition 5.3 and Lemma 5.2, we obtain that

$$\bar{b}_3(\bar{b}_{(1,t)}b_{(1,m)}) = \sum_{i=1}^{t} \bar{b}_3 b_{(2i-1,m-t-1+2i)}$$

$$= \sum_{i=1}^{t} (b_{(2i-1,m-t-2+2i)} + b_{(2i-3,m-t-1+2i)} + b_{(2i+1,m-t+2i)})$$

and $(\bar{b}_3\bar{b}_{(1,t)})b_{(1,m)} = \bar{b}_{(3,t)}b_{(1,m)} + \bar{b}_{(1,t+1)}b_{(1,m)}$. Thus

$$\bar{b}_{(3,t)}b_{(1,m)} + \bar{b}_{(1,t+1)}b_{(1,m)}$$

$$= \sum_{i=1}^{t} (b_{(2i-1,m-t-2+2i)} + b_{(2i-3,m-t-1+2i)} + b_{(2i+1,m-t+2i)}). \qquad (5.14)$$

By the induction hypothesis, Definition 5.3, Lemma 5.2 and since $3 \le t$, we obtain that

$$b_3(\bar{b}_{(1,t-1)}b_{(1,m)}) = \sum_{i=1}^{t-1} b_3 b_{(2i-1,m-t+2i)}$$

$$= \sum_{i=1}^{t-1} (b_{(2i-3,m-t+2i-1)} + b_{(2i+1,m-t+2i)} + b_{(2i-1,m-t+2i+1)})$$

and

$$(b_3 \bar{b}_{(1,t-1)}) b_{(1,m)} = \bar{b}_{(1,t-2)} b_{(1,m)} + \bar{b}_{(3,t)} b_{(1,m)}$$

$$= \sum_{i=1}^{t-2} b_{(2i-1, m-t+1+2i)} + \bar{b}_{(3,t)} b_{(1,m)}.$$

Thus

$$\bar{b}_{(3,t)} b_{(1,m)} = \sum_{i=2}^{t} b_{(2i-3, m-t+2i-1)} + \sum i = 1^{t-1} b_{(2i+1, m-t+2i)}.$$

Now by (5.14) we obtain that

$$\bar{b}_{(1,t+1)} b_{(1,m)} = \sum_{i=1}^{t+1} b_{(2i-1, m-t+2i-2)}$$

as required.   □

Lemma 5.8   *If* (A, B) *is a* C_{2n-1}-*Table Algebra then*

$$n + 1 \le (\bar{b}_{(1,n+1)} b_{(1,n+2)}, \bar{b}_{(1,n)} b_{(1,n+1)}).$$

Proof By Lemma 5.6 we obtain that

$$b_{(1,n)} b_{(1,n+1)} = \sum_{i=1}^{n} b_{(2i-1, 2n+1-i)}.$$

Now by Definition 5.3 and Lemma 5.5 we obtain that

$$b_3 (b_{(1,n)} b_{(1,n+1)}) = \sum_{i=1}^{n} b_3 b_{(2i-1, 2n+1-i)}$$

$$= \sum_{i=2}^{n} (b_{(2i-3, 2n-i)} + b_{(2i-1, 2n+2-i)}) + \sum_{i=1}^{n} b_{(2i+1, 2n+1-i)}$$

$$+ \Sigma_{f(1,2n+1)}$$

where $\Sigma_{f(1,2n+1)} \in \mathbb{N} B \setminus B$. By Definition 5.3 and Lemma 5.6, we obtain that

$$(b_3 b_{(1,n)}) b_{(1,n+1)} = b_{(3,n)} b_{(1,n+1)} + b_{(1,n+1)}^2$$

$$= \sum_{i=2}^{n} b_{(2i-3, 2n-i)} + \sum_{i=1}^{n-1} b_{(2i+1, 2n+1-i)} + b_{(1,n+1)}^2.$$

Thus

$$b_{(1,n+1)}^2 = \sum_{i=2}^{n+1} b_{(2i-1, 2n+2-i)} + \Sigma_{f(1,2n+1)}. \tag{5.15}$$

By Lemma 5.6 we obtain that

$$b_{(1,n-1)}b_{(1,n+2)} = \sum_{i=1}^{n} b_{(2i-1,2n+1-i)}.$$

Now by Definition 5.3 and Lemma 5.5 we obtain that

$$b_3(b_{(1,n-1)}b_{(1,n+2)}) = \sum_{i=1}^{n-1} b_3 b_{(2i-1,2n+1-i)}$$

$$= \sum_{i=2}^{n-1}(b_{(2i-3,2n-i)} + b_{(2i-1,2n+2-i)}) + \sum_{i=1}^{n-1} b_{(2i+1,2n+1-i)}$$

$$+ \Sigma_{f(1,2n+1)}.$$

By Definition 5.3 and Lemma 5.6, we obtain that

$$(b_3 b_{(1,n-1)})b_{(1,n+2)} = b_{(3,n-1)}b_{(1,n+2)} + b_{(1,n)}b_{(1,n+2)}$$

$$= \sum_{i=2}^{n-1} b_{(2i-3,2n-i)} + \sum_{i=1}^{n-2} b_{(2i+1,2n+1-i)} + b_{(1,n)}b_{(1,n+2)}.$$

Thus

$$b_{(1,n)}b_{(1,n+2)} = \sum_{i=2}^{n} b_{(2i-1,2n+2-i)} + \Sigma_{f(1,2n+1)}.$$

Now by (5.15) we obtain that

$$n + 1 \le (b_{(1,n)}b_{(1,n+2)}, b_{(1,n+1)}^2)$$

$$= (\bar{b}_{(1,n+1)}b_{(1,n+2)}, \bar{b}_{(1,n)}b_{(1,n+1)}),$$

as required.   □

Lemma 5.9 *If (A, B) is a C_{2n-1}-Table Algebra, then*

$$\bar{b}_{(1,n+1)}b_{(1,n+2)} = \sum_{i=1}^{n-1} b_{(2i-1,2i)} + 2b_{(2n-1,2n)} + \Sigma_{f(2n+1,2n+2)-f(2n-1,2n)}$$

where $\Sigma_{f(2n+1,2n+2)-f(2n-1,2n)} \in \mathbb{N}B$.

Proof If (A, B) is a C_{2n-1}-Table Algebra then by Lemma 5.6 we obtain that

$$b_{(1,j)}b_{(1,n+2)} = \sum_{i=1}^{j} b_{(2i-1,j+n+2-i)}$$

and

$$b_{f(1,j+1)}b_{f(1,n+1)} = \sum_{i=1}^{j+1} b_{(2i-1,j+n+2-i)}$$

where $1 \le j \le n-1$. Now by Lemma 5.3 we obtain that for all $1 \le j \le n-1$

$$j = (b_{(1,j)}b_{(1,n+2)}, b_{(1,j+1)}b_{(1,n+1)}) = (\bar{b}_{(1,j)}b_{(1,j+1)}, \bar{b}_{(1,n+1)}b_{(1,n+2)}).$$

By Lemma 5.7 we obtain that

$$\bar{b}_{(1,j)}b_{(1,j+1)} = \sum_{i=1}^{j} b_{(2i-1,2i)} \quad \text{where } 1 \le j \le n.$$

Now Lemma 5.8 implies that

$$\bar{b}_{(1,n+1)}b_{(1,n+2)} = \sum_{i=1}^{n-1} b_{(2i-1,2i)} + 2b_{(2n-1,2n)} + \Sigma_{f(2n+1,2n+2)-f(2n-1,2n)}$$

where $\Sigma_{f(2n+1,2n+2)-f(2n-1,2n)} \in \mathbb{N}B$, as required. $\qquad\square$

Lemma 5.10 *If (A, B) is a C_{2n-1}-Table Algebra, then*

$$b_{(1,n+2)}\bar{b}_{(1,n+1)} = \sum_{i=2}^{n} b_{(2i-3,2i-2)} + b_3\Sigma_{f(2n+1,2n+1)} - \bar{\Sigma}_{f(2n-1,2n+1)}$$

where $\Sigma_{f(2n+1,2n+1)} \in \mathbb{N}B \setminus B$ and $\Sigma_{f(2n-1,2n+1)} \in \mathbb{N}B$.

Proof If (A, B) is a C_{2n-1}-Table Algebra then by Definition 5.3, Lemmas 5.5 and 5.7, we obtain that

$$\bar{b}_3(\bar{b}_{(1,n)}b_{(1,n+1)}) = \sum_{i=1}^{n} \bar{b}_3 b_{(2i-1,2i)}$$

$$= \sum_{i=1}^{n} b_{(2i-1,2i-1)} + \sum_{i=2}^{n} b_{(2i-3,2i)} + \sum_{i=1}^{n-1} b_{(2i+1,2i+1)} + \Sigma_{f(2n+1)}$$

where $\Sigma_{f(2n+1,2n+1)} \in \mathbb{N}B \setminus B$. By Definition 5.3 and Lemma 5.7 we obtain that

$$(\bar{b}_3\bar{b}_{(1,n)})b_{(1,n+1)} = \bar{b}_{(3,n)}b_{(1,n+1)} + \bar{b}_{(1,n+1)}b_{(1,n+1)}$$

$$= \sum_{i=2}^{n} b_{f(2i-3,2i)} + \sum_{i=1}^{n-1} b_{(2i+1,2i+1)} + \bar{b}_{(1,n+1)}b_{(1,n+1)}.$$

Thus

$$b_{(1,n+1)}\bar{b}_{(1,n+1)} = \sum_{i=1}^{n} b_{(2i-1,2i-1)} + \Sigma_{f(2n+1,2n+1)}.$$

By Lemmas 5.2 and 5.7, we obtain that

$$(\bar{b}_3 b_{(1,n)})\bar{b}_{(1,n+1)} = b_{(1,n-1)}\bar{b}_{(1,n+1)} + b_{f(,n+1)}\bar{b}_{(1,n+1)}$$

$$= \sum_{i=1}^{n-1} \bar{b}_{(2i-1,2i+1)} + b_{(3,n+1)}\bar{b}_{(1,n+1)}.$$

By Definition 5.3, Lemmas 5.5 and 5.7, we obtain that

$$\bar{b}_3(b_{(1,n)}\bar{b}_{(1,n+1)}) = \sum_{i=1}^{n} \bar{b}_3 \bar{b}_{(2i-1,2i)}$$

$$= \sum_{i=2}^{n} \bar{b}_{(2i-3,2i-1)} + \sum_{i=1}^{n} \bar{b}_{(2i+1,2i)} + \sum_{i=1}^{n-1} \bar{b}_{(2i-1,2i+1)}$$

$$+ \bar{\Sigma}_{f(2n-1,2n+1)}$$

where $\Sigma_{f(2n-1,2n+1)} \in \mathbb{N}B \setminus B$. Thus

$$b_{(3,n+1)}\bar{b}_{(1,n+1)} = \sum_{i=2}^{n} \bar{b}_{(2i-3,2i-1)} + \sum_{i=1}^{n} \bar{b}_{(2i+1,2i)} + \bar{\Sigma}_{f(2n-1,2n+1)}.$$

Now by Definition 5.3 we obtain that

$$b_3(b_{(1,n+1)}\bar{b}_{(1,n+1)}) = \sum_{i=1}^{n} b_3 b_{(2i-1,2i-1)} + b_3 \Sigma_{f(2n+1,2n+1)}$$

$$= \sum_{i=2}^{n} b_{(2i-3,2i-2)} + \sum_{i=1}^{n} (b_{(2i+1,2i-1)} + b_{(2i-1,2i)})$$

$$+ b_3 \Sigma_{f(2n+1,2n+1)}.$$

By Definition 5.3 we obtain that

$$(b_3 b_{(1,n+1)})\bar{b}_{(1,n+1)} = b_{(3,n+1)}\bar{b}_{(1,n+1)} + b_{(1,n+2)}\bar{b}_{(1,n+1)}$$

$$= \sum_{i=2}^{n} \bar{b}_{(2i-3,2i-1)} + \sum_{i=1}^{n} \bar{b}_{(2i+1,2i)} + \bar{\Sigma}_{f(2n-1,2n+1)}$$

$$+ b_{(1,n+2)}\bar{b}_{(1,n+1)}.$$

Thus

$$b_{(1,n+2)}\bar{b}_{(1,n+1)} = \sum_{i=2}^{n} b_{(2i-3,2i-2)} + b_3 \Sigma_{f(2n+1,2n+1)} - \bar{\Sigma}_{f(2n-1,2n+1)},$$

as required.   □

Corollary 5.1 *There exists no C_k-Table Algebra where k is an odd number and $43 \le k$.*

Proof By Lemmas 5.9 and 5.10, we obtain that

$$2b_{(2n-1,2n)} + \Sigma_{f(2n+1,2n+2)-f(2n-1,2n)} + \bar{\Sigma}_{f(2n-1,2n+1)} = b_3 \Sigma_{f(2n+1,2n+1)}.$$

Now if $\beta \in Irr(\Sigma_{f(2n+1,2n+2)-f(2n-1,2n)})$, then $|\beta| \le f(2n + 1, 2n + 2) - f(2n - 1, 2n)$ and there exists $\alpha \in Irr(\Sigma_{f(2n+1,2n+1)})$ such that $\beta \in Irr(b_3\alpha)$. By Lemma 5.5 we obtain that

$$\bar{b}_3 b_{(2n-1,2n)} = b_{(2n-1,2n-1)} + b_{(2n-3,2n)} + \Sigma_{f(2n+1,2n+1)}$$

which implies that

$$b_{(2n-1,2n)} \in Irr(b_3\alpha).$$

Now by Lemma 5.1 we obtain that

$$\frac{f(2n-1,2n)}{|\beta|} \le 8$$

which implies by Definition 5.2 that

$$\frac{n(2n+1)}{6n+6} \le 8,$$

so there exists no C_k-Table Algebra where k is an odd number and $49 \le k$.

Assume henceforth that (A, B) is a C_{2n-1}-Table Algebra where $43 \le 2n - 1 \le 47$, and we will derive a contradiction. If $\Sigma_{f(2n+1,2n+2)-f(2n-1,2n)} \notin B$, then there exists

$$\beta_1, \beta_2 \in Irr(\Sigma_{f(2n+1,2n+2)-f(2n-1,2n)})$$

and

$$\alpha_1, \alpha_2 \in \Sigma_{f(2n+1,2n+1)}$$

such that

$$\beta_1, b_{(2n-1,2n)} \in Irr(b_3\alpha_1)$$

and

$$\beta_2, b_{(2n-1,2n)} \in Irr(b_3\alpha_2).$$

Since

$$|\beta_1| + |\beta_2| \leq f(2n+1, 2n+2) - f(2n-1, 2n)$$

and $n \geq 22$, we obtain that either

$$\frac{f(2n-1, 2n)}{|\beta_1|} \geq 8$$

or

$$\frac{f(2n-1, 2n)}{|\beta_2|} \geq 8,$$

and we have a contradiction to Lemma 5.1. Note that if there is only one element β in $Irr(\Sigma_{f(2n+1,2n+2)-f(2n-1,2n)})$, then

$$Irr(\Sigma_{f(2n+1,2n+2)-f(2n-1,2n)}) = r\beta \quad \text{where } 2 \leq r$$

which implies that

$$2|\beta| \leq f(2n+1, 2n+2) - f(2n-1, 2n),$$

and since $n \geq 22$ we obtain that

$$\frac{f(2n-1, 2n)}{|\beta|} \geq 8.$$

Now we have that

$$\Sigma_{f(2n+1,2n+2)-f(2n-1,2n)} \in B.$$

If $n = 24$ then $b_{14,700}, \Sigma_{1,875} \in (Irrb_3\alpha)$ which implies that

$$5,525 \leq |\alpha| \leq 5,625.$$

If $|\alpha| = 5,525$ then

$$b_3\alpha_{5,525} = \Sigma_{1,875} + b_{14,700}$$

and

$$\alpha_{5,525} \in Irr(\bar{b}_3\Sigma_{1,875})$$

which implies that there exists $\gamma \in B$ such that $|\gamma| \leq 100$ and $\gamma \in Irr(\bar{b}_3\Sigma_{1,875})$, which is a contradiction.

If $|\alpha| \neq 5,525$ then there exists $\gamma \in B$ such that $\gamma \in Irr(b_3\alpha)$ and $|\gamma| \leq 300$. Since $b_{14,700} \in Irr(b_3\alpha)$, we have a contradiction to Lemma 5.1. If $n = 23$ then $b_{12,972}, \Sigma_{1,728} \in Irr(b_3\alpha)$ which implies that

$$4,900 \leq |\alpha| \leq 5,184.$$

If $|\alpha| = 4,900$ then

$$b_3\alpha_{4,900} = \Sigma_{1,728} + b_{12,972}$$

and

$$\alpha_{4,900} \in Irr(\bar{b}_3 \Sigma_{1,728})$$

which implies that there exists $\gamma \in B$ such that $|\gamma| \leq 284$ and $\gamma \in Irr(\bar{b}_3 \Sigma_{1,728})$, which is a contradiction.

If $|\alpha| \neq 4,900$ then there exists $\gamma \in B$ such that $\gamma \in Irr(b_3\alpha)$ and $|\gamma| \leq 852$, since $b_{12,972} \in Irr(b_3\alpha)$, and we have a contradiction to Lemma 5.1. If $n = 22$ then $b_{11,385}, \Sigma_{1,587} \in Irr(b_3\alpha)$ which implies that

$$4,324 \leq |\alpha| \leq 4,761.$$

If $|\alpha| = 4,324$ then

$$b_3\alpha_{4,324} = \Sigma_{1,587} + b_{11,385}$$

and

$$\alpha_{4,324} \in Irr(\bar{b}_3 \Sigma_{1,587})$$

which implies that there exists $\gamma \in B$ such that $|\gamma| \leq 437$ and $\gamma \in Irr(\bar{b}_3 \Sigma_{1,587})$, which is a contradiction.

If $|\alpha| \neq 4,324$ then there exists $\gamma \in B$ such that $\gamma \in Irr(b_3\alpha)$ and $|\gamma| \leq 1,311$, since $b_{11,385} \in Irr(b_3\alpha)$, and we have a contradiction to Lemma 5.1. So there exists no C_k-Table Algebra where k is an odd number and $43 \leq k$.   □

Lemma 5.11 *If (A, B) is a C_{2n}-Table Algebra then*

$$n + 2 \leq (\bar{b}_{(1,n+2)}b_{(1,n+2)}, \bar{b}_{(1,n+1)}b_{(1,n+1)}).$$

Proof If (A, B) is a C_{2n}-Table Algebra then by Definition 5.3, Lemmas 5.5 and 5.7, we obtain that

$$b_3(b_{(1,n+1)}\bar{b}_{(1,n+1)}) = \sum_{i=1}^{n+1} b_3 b_{(2i-1,2i-1)}$$

$$= \sum_{i=2}^{n+1} (b_{(2i-3,2i-2)} + b_{(2i+1,2i-1)}) + \sum_{i=1}^{n} b_{(2i-1,2i)}$$

$$+ \Sigma_{f(2n+1,2n+2)}$$

where $\Sigma_{f(2n+1,2n+2)} \in \mathbb{N}B \setminus B$. By Definition 5.3 and Lemma 5.7, we obtain that

$$(b_3 b_{(1,n+1)})\bar{b}_{(1,n+1)} = b_{(3,n+1)}\bar{b}_{(1,n+1)} + b_{(1,n+2)}\bar{b}_{(1,n+1)}$$

$$= \sum_{i=2}^{n+1} \bar{b}_{(2i-3,2i-1)} + \sum_{i=1}^{n} \bar{b}_{(2i+1,2i)} + b_{(1,n+2)}\bar{b}_{(1,n+1)}.$$

Thus

$$\bar{b}_{(1,n+1)}b_{(1,n+2)} = \sum_{i=1}^{n} b_{(2i-1,2i)} + \Sigma_{(2n+1,2n+2)}. \tag{5.16}$$

Now we have that

$$n + 2 \leq (\bar{b}_{(1,n+1)}b_{(1,n+2)}, \bar{b}_{(1,n+1)}b_{(1,n+2)}) = (\bar{b}_{(1,n+2)}b_{(1,n+2)}, \bar{b}_{(1,n+1)}b_{(1,n+1)}),$$

as required.   □

Lemma 5.12 *If (A, B) is a C_{2n}-Table Algebra, then*

$$\bar{b}_{(1,n+2)}b_{(1,n+2)} = \sum_{i=1}^{n} b_{(2i-1,2i-1)} + 2b_{(2n+1,2n+1)} + \Sigma_{f(2n+3,2n+3)-f(2n+1,2n+1)}$$

where $\Sigma_{f(2n+3,2n+3)-f(2n+1,2n+1)} \in \mathbb{N}B$.

Proof If (A, B) is a C_{2n}-Table Algebra, then by Lemma 5.6 we obtain that

$$b_{(1,j)}b_{(1,n+2)} = \sum_{i=1}^{j} b_{(2i-1,j+n+2-i)} \quad \text{where } 1 \leq j \leq n.$$

Now by Lemma 5.3 we obtain that for all $1 \leq j \leq n$

$$j = (b_{(1,j)}b_{(1,n+2)}, b_{(1,j)}b_{(1,n+2)}) = (\bar{b}_{(1,j)}b_{(1,j)}, \bar{b}_{(1,n+2)}b_{(1,n+2)}).$$

By Lemma 5.7 we obtain that

$$\bar{b}_{(1,j)}b_{(1,j)} = \sum_{i=1}^{j} b_{(2i-1,2i-1)} \quad \text{where } 1 \leq j \leq n+1.$$

Now Lemma 5.11 implies that

$$\bar{b}_{(1,n+2)}b_{(1,n+2)} = \sum_{i=1}^{n} b_{(2i-1,2i-1)} + 2b_{(2n+1,2n+1)} + \Sigma_{f(2n+3,2n+3)-f(2n+1,2n+1)}$$

where $\Sigma_{f(2n+3,2n+3)-f(2n+1,2n+1)} \in \mathbb{N}B$, as required.   □

Lemma 5.13 *If (A, B) is a C_{2n}-Table Algebra, then*

$$b_{(1,n+2)}\bar{b}_{(1,n+2)} = \sum_{i=1}^{n} b_{(2i-1,2i-1)} + b_3\bar{\Sigma}_{f(2n+1,2n+2)} - \bar{\Sigma}_{f(2n-1,2n+1)}$$

where $\Sigma_{f(2n+1,2n+2)} \in \mathbb{N}B \setminus B$ and $\Sigma_{f(2n-1,2n+1)} \in \mathbb{N}B$.

Proof If (A, B) is a C_{2n}-Table Algebra then by Definition 5.3, Lemmas 5.5 and 5.7, we obtain that

$$\bar{b}_3(b_{(1,n)}\bar{b}_{(1,n+2)}) = \sum_{i=1}^{n} \bar{b}_3\bar{b}_{(2i-1,2i+1)}$$

$$= \sum_{i=2}^{n} \bar{b}_{(2i-3,2i)} + \sum_{i=1}^{n} b_{(2i+1,2i+1)}$$

$$+ \sum_{i=1}^{n-1} \bar{b}_{(2i-1,2i+2)} + \bar{\Sigma}_{f(2n-1,2n+2)}$$

where $\Sigma_{f(2n-1,2n+2)} \in \mathbb{N}B \setminus B$. By Lemmas 5.2 and 5.7 we obtain that

$$(\bar{b}_3 b_{(1,n)})\bar{b}_{(1,n+2)} = b_{(1,n-1)}\bar{b}_{(1,n+2)} + b_{(3,n+1)}\bar{b}_{(1,n+2)}$$

$$= \sum_{i=1}^{n-1} \bar{b}_{(2i-1,2i+2)} + b_{(3,n+1)}\bar{b}_{(1,n+2)}.$$

Thus

$$b_{(3,n+1)}\bar{b}_{(1,n+2)} = \sum_{i=2}^{n} \bar{b}_{(2i-3,2i)} + \sum_{i=1}^{n} b_{(2i+1,2i+1)} + \bar{\Sigma}_{f(2n-1,2n+2)}.$$

Now by Definition 5.3 we obtain that

$$(b_3 b_{(1,n+1)})\bar{b}_{(1,n+2)} = b_{(3,n+1)}\bar{b}_{(1,n+2)} + b_{(1,n+2)}\bar{b}_{(1,n+2)}$$

$$= \sum_{i=2}^{n} \bar{b}_{(2i-3,2i)} + \sum_{i=1}^{n} b_{(2i+1,2i+1)} + \bar{\Sigma}_{(2n-1,2n+2)}$$

$$+ b_{(1,n+2)}\bar{b}_{(1,n+2)}.$$

By Lemma 5.2 and (5.16) of Lemma 5.11 we obtain that

$$b_3(b_{(1,n+1)}\bar{b}_{(1,n+2)}) = \sum_{i=1}^{n} b_3\bar{b}_{(2i-1,2i)} + b_3\bar{\Sigma}_{f(2n+1,2n+2)}$$

$$= \sum_{i=1}^{n} b_{(2i-1,2i-1)} + \sum_{i=2}^{n} \bar{b}_{(2i-3,2i)} + \sum_{i=1}^{n} b_{(2i+1,2i+1)}$$

$$+ b_3\bar{\Sigma}_{f(2n+1,2n+2)}.$$

Thus

$$b_{(1,n+2)}\bar{b}_{(1,n+2)} = \sum_{i=1}^{n} b_{(2i-1,2i-1)} + b_3\bar{\Sigma}_{f(2n+1,2n+2)} - \bar{\Sigma}_{f(2n-1,2n+2)},$$

as required. □

Corollary 5.2 *There exists no C_k-Table Algebra where k is even and $44 \leq k$.*

Proof By Lemmas 5.12 and 5.13 we obtain that

$$\sum_{i=1}^{n} b_{(2i-1,2i-1)} + b_3 \bar{\Sigma}_{f(2n+1,2n+2)} - \bar{\Sigma}_{f(2n-1,2n+2)}$$

$$= \sum_{i=1}^{n} b_{(2i-1,2i-1)} + 2b_{(2n+1,2n+1)} + \Sigma_{f(2n+3,2n+3)-f(2n+1,2n+1)}.$$

Hence

$$\bar{b}_3 \Sigma_{(2n+1,2n+2)} = 2b_{(2n+1,2n+1)} + \Sigma_{f(2n-1,2n+1)} + \Sigma_{f(2n+3,2n+3)-f(2n+1,2n+1)}.$$

Now if $\beta \in Irr(\Sigma_{f(2n+3,2n+3)-f(2n+1,2n+1)})$ then $|\beta| \leq f(2n+3,2n+3) - f(2n+1,2n+1)$ and there exists $\alpha \in Irr(\Sigma_{f(2n+1,2n+2)})$ such that $\beta \in Irr(\bar{b}_3\alpha)$. By Lemma 5.5 we obtain that

$$b_3 b_{(2n+1,2n+1)} = b_{(2n-1,2n)} + b_{(2n+3,2n+1)} + \Sigma_{f(2n+1,2n+2)}$$

where $\Sigma_{f(2n+1,2n+2)} \in \mathbb{N}B \setminus B$ which implies that

$$b_{(2n+1,2n+1)} \in Irr(\bar{b}_3\alpha).$$

Now by Lemma 5.1 we obtain that

$$\frac{f(2n+1,2n+1)}{|\beta|} \leq 8$$

which implies by Definition 5.2 that

$$\frac{(n+1)^3}{(n+2)^3 - (n+1)^3} \leq 8,$$

so there exists no C_k-Table Algebra where k is even and $48 \leq k$.

Assume that toward a contradiction that (A, B) C_{2n}-Table Algebra where $44 \leq 2n \leq 46$. If $\Sigma_{f(2n+3,2n+3)-f(2n+1,2n+1)} \notin B$. Then there exists

$$\beta_1, \beta_2 \in \Sigma_{f(2n+3,2n+3)-f(2n+1,2n+1)}$$

and

$$\alpha_1, \alpha_2 \in \Sigma_{f(2n+1,2n+2)}$$

such that

$$\beta_1, b_{(2n+1,2n+1)} \in Irr(\bar{b}_3\alpha_1)$$

and

$$\beta_2, b_{(2n+1,2n+1)} \in Irr(\bar{b}_3\alpha_2).$$

Since

$$|\beta_1| + |\beta_2| \le f(2n+3, 2n+3) - f(2n+1, 2n+1)$$

and $n \ge 22$ we obtain that either

$$\frac{f(2n+1, 2n+1)}{|\beta_1|} \ge 8$$

or

$$\frac{f(2n+1, 2n+1)}{|\beta_2|} \ge 8,$$

and we have a contradiction to Lemma 5.1. Note that if there is only one element β in $Irr(\Sigma_{f(2n+3,2n+3)-f(2n+1,2n+1)})$ then

$$Irr(\Sigma_{f(2n+3,2n+3)-f(2n+1,2n+1)}) = r\beta \quad \text{where } 2 \le r$$

which implies that

$$2|\beta| \le f(2n+3, 2n+3) - f(2n+1, 2n+1)$$

and since $n \ge 22$ we obtain that

$$\frac{f(2n-1, 2n)}{|\beta|} \ge 8.$$

Now we have that $\Sigma_{f(2n+2,2n+2)-f(2n+1,2n+1)} \in B$.

If $n = 23$ then $b_{13,824}, \Sigma_{1,801} \in Irr(\bar{b}_3\alpha)$ which implies that $5,209 \le |\alpha| \le 5,403$, so there exists $\gamma \in B$ such that $\gamma \in Irr(\bar{b}_3\alpha)$ and $|\gamma| \le 584$, since $b_{13,824} \in Irr(\bar{b}_3\alpha)$, and we have a contradiction to Lemma 5.1.

If $n = 22$ then $b_{12,167}, \Sigma_{1,657} \in Irr(\bar{b}_3\alpha)$ which implies that

$$4,608 \le |\alpha| \le 4,971.$$

If $|\alpha| = 4,608$ then

$$\bar{b}_3\alpha_{4,608} = \Sigma_{1,657} + b_{12,167}$$

and

$$\alpha_{4,608} \in Irr(b_3\Sigma_{1,657})$$

which implies that there exists $\gamma \in B$ such that $|\gamma| \le 363$ and $\gamma \in Irr(b_3\Sigma_{1,657})$, which is a contradiction. If $|\alpha| \ne 4,608$ then there exists $\gamma \in B$ such that $\gamma \in Irr(\bar{b}_3\alpha)$ and $|\gamma| \le 1,089$. Since $b_{12,167} \in Irr(\bar{b}_3\alpha)$, we have a contradiction to Lemma 5.1. So there exists no C_k-Table Algebra where k is even and $44 \le k$. $\square$

By Corollaries 5.1 and 5.2, the proof of Theorem 5.1 is completed. $\square$

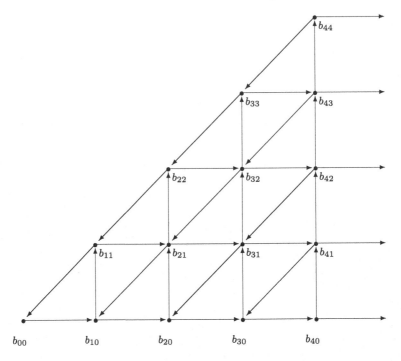

Fig. 5.1 The representation graph of $b_{1,0}$

5.3 Proof of Theorem 5.2

Let (A, B) be an Integral Normalized Table Algebra with countable basis B, the elements of which are parametrized by pairs (m, ℓ), $0 \leq \ell \leq m$, $m = 0, 1, \ldots$ in such a way that

$$b_{1,0}b_{m,\ell} = b_{m+1,\ell} + b_{m,\ell+1} + b_{m-1,\ell-1} \tag{5.17}$$

and $b_{m,\ell} = 0$ unless $0 \leq \ell \leq m$. Note that $b_{0,0} = 1$. Denote $b := b_{1,0}$. It follows from (5.17) that $bb_{1,1} = b_{0,0} + b_{2,1} = 1 + b_{2,1}$; therefore $b_{1,1} = \bar{b}$. The indexing here is different from Sect. 5.1 of this chapter, i.e., the index pair (k, n) of Sect. 5.1 of this chapter corresponds to a pair $(n - 1, (k - 1)/2)$ used in this section.

The representation graph of $b_{1,0}$ is depicted in Fig. 5.1. Note that it is isomorphic to a representation graph of a SITA generated by a non-real element of degree 3 [B].

Let $\mathcal{I}$ denote the set of pairs $\{(m, \ell) \mid 0 \leq \ell \leq m, m = 0, 1, \ldots\}$. On the set $\mathcal{I}$ we impose a lexicographic linear ordering, that is $(a, b) \leq (c, d) \iff (a < c) \vee (a = c \wedge b \leq d)$. Thus

$$(0, 0) < (1, 0) < (1, 1) < (2, 0) < (2, 1) < (2, 2) \ldots$$

In what follows we set $\mathcal{I}_{m,\ell} := \{(i, j) \in \mathcal{I} \mid (i, j) < (m, \ell)\}$.

The linear ordering of $\mathcal{I}$ induces a linear order on the set of monomials $b^i \bar{b}^j, i, j \geq 0$ as follows:

$$b^i \bar{b}^j \leq b^m \bar{b}^\ell \iff (i + j, j) \leq (m + \ell, \ell).$$

Thus we have the following linear ordering of monomials:

$$1, b, \bar{b}, b^2, b\bar{b}, \bar{b}^2, b^3, b^2\bar{b}, b\bar{b}^2, \bar{b}^3, \ldots$$

A pair (i, j) will be called a *degree* of the monomial $b^i \bar{b}^j$. A monomial of highest degree (together with its coefficient) which appears in a polynomial expression $p = \sum_{ij} \alpha_{i,j} b^i \bar{b}^j$ with non-zero coefficient will be called a *leading term* of p.

Proposition 5.1 *For each pair $(s, t), 0 \leq s, 0 \leq t$ it holds that*

$$b^s \bar{b}^t = b_{s+t,t} + \sum_{(i,j) \in \mathcal{I}_{s+t,t}} \alpha_{ij} b_{i,j}. \tag{5.18}$$

Proof Induction on $s + t$. For $s + t = 0$ and $s + t = 1$ our statement follows from $b_{0,0} = 1, b_{1,0} = b, b_{1,1} = \bar{b}$. Assume that (5.18) holds for all pairs (s', t') with $s' + t' = m \geq 2$. Pick an arbitrary pair (s, t) with $s + t = m + 1$.

Consider first the case of $s = 0$. Then $t = m + 1 \geq 3$ which implies that

$$\bar{b}^{m+1} = \bar{b} \cdot \bar{b}^m = \bar{b}\left(b_{m,m} + \sum_{(i,j) \in \mathcal{I}_{m,m}} \alpha_{ij} b_{i,j} \right)$$

$$= b_{m+1,m+1} + b_{m,m-1} + \sum_{(i,j) \in \mathcal{I}_{m,m}} \alpha_{ij}(b_{i+1,j+1} + b_{i-1,j} + b_{i,j-1}).$$

It follows from $(i, j) \in \mathcal{I}_{m,m}$ that

$$(i + 1, j + 1) < (m + 1, m + 1), \qquad (i - 1, j) < (m + 1, m + 1),$$
$$(i, j - 1) < (m + 1, m + 1).$$

Therefore

$$\bar{b}^{m+1} = b_{m+1,m+1} + \sum_{(i,j) \in \mathcal{I}_{m+1,m+1}} \beta_{ij} b_{i,j}.$$

Assume now that $s \geq 1$. Then $b^s \bar{b}^t = b \cdot b^{s-1} \bar{b}^t$ and by (5.17), we obtain

$$b \cdot b^{s-1} \bar{b}^t = b b_{s+t-1,t} + \sum_{(i,j) \in \mathcal{I}_{s+t-1,t}} \alpha_{ij} b b_{i,j}$$

$$= b_{s+t,t} + b_{s+t-1,t+1} + b_{s+t-2,t-1}$$

$$+ \sum_{(i,j) \in \mathcal{I}_{s+t-1,t}} \alpha_{ij}(b_{i+1,j} + b_{i,j+1} + b_{i-1,j-1}).$$

It follows from $(i, j) \in \mathcal{I}_{s+t-1,t}$ that

$$(i+1, j), (i, j+1), (i-1, j-1) \in \mathcal{I}_{s+t,t}.$$

Therefore

$$b^s \bar{b}^t = b_{s+t,t} + \sum_{(i,j)\in\mathcal{I}_{s+t,t}} \beta_{ij} b_{i,j}.$$

$\square$

Since the summation in (5.18) is done within a linear ordered sets $\mathcal{I}_{s+t,t}$, it is invertible which gives us

$$b_{s+t,t} = b^s \bar{b}^t + \sum_{(u,v)\in\mathcal{I}_{s+t,t}} \beta_{uv} b^{u-v} \bar{b}^v. \tag{5.19}$$

Theorem 5.3 *The Table Algebra (A, B) is isomorphic to $\mathbb{Z}[x, y]$. Its standard basis $p_{u,v}(x, y)$ is defined as follows: $p_{0,0} = 1$, $p_{1,0} = x$, $p_{1,1} = y$ and for $m \geq 1$*

$$p_{m+1,\ell} = \begin{cases} x p_{m,\ell} - p_{m,\ell+1} - p_{m-1,\ell-1}, & 0 \leq \ell \leq m \\ y p_{m,m} + p_{m,m-1}, & \ell = m \end{cases} \tag{5.20}$$

provided that $p_{i,j} = 0$ for all $(i, j) \notin \mathcal{I}$. In particular, the Table Algebra (A, B) is unique if it exists.

Proof It follows from (5.19) that each $b_{i,j}$, $(i, j) \in \mathcal{I}$ is a polynomial in b and $\bar{b}$. Therefore the mapping $x \mapsto b$, $y \mapsto \bar{b}$ has a unique extension to an algebra epimorphism from $\mathbb{Z}[x, y]$ onto A. Let us denote this epimorphism as φ. Then $\varphi(x^i y^j) = b^i \bar{b}^j$. It follows from (5.18) and (5.19) that the monomials $b^i \bar{b}^j$, $0 \leq i, j$ form a $\mathbb{Z}$-basis of A. Therefore the kernel of φ is trivial, which implies that φ is an algebra isomorphism.

Denote $p_{m,\ell}(x, y) := \varphi^{-1}(b_{m,\ell})$. Then $p_{0,0} = 1$, $p_{1,0} = x$, $p_{1,1} = y$. Now the formulae (5.20) follow directly from (5.17). $\square$

Theorem 5.3 implies that if a Table Algebra (A, B) exists, then it is exactly isomorphic to $(\mathbb{Z}[x, y], \{p_{u,v}\}_{(u,v)\in\mathcal{I}})$. In particular, this means that an existence implies uniqueness. Unfortunately, we cannot check directly that the basis $p_{u,v}$ defined by recurrence relations (5.20) is a table basis of $\mathbb{Z}[x, y]$, (that is, whether a product of two polynomials from a table basis is a linear combination of $p_{u,v}$ with non-negative integer coefficients). To show existence it is sufficient to find an example of a group whose irreducible representations satisfy (5.17).

It is well-known, [M], that irreducible polynomial representations of a linear group $\mathsf{SL}_3(\mathbb{C})$ are parameterized by pairs $(m, \ell) \in \mathcal{I}$. An irreducible character of $\mathsf{SL}_3(\mathbb{C})$ corresponding to a pair (m, ℓ), $m \geq \ell \geq 0$ will be denoted as $\chi_{m,\ell}$. The value of $\chi_{m,\ell}$ on a matrix $A \in \mathsf{SL}_3(\mathbb{C})$ is equal to $s_{(m,\ell,0)}(\alpha_1, \alpha_2, \alpha_3)$ where $s_{(m,\ell,0)} \in \Delta[x_1, x_2, x_3]$ is a Schur function and $\alpha_1, \alpha_2, \alpha_3$ are the eigenvalues of A.

Note that the irreducible characters $\chi_{m,\ell}$ are in one-to-one correspondence with Schur functions of the form $s_{(m,\ell,0)}$.

In fact, any Schur function $s_\lambda \in \Lambda[x_1, x_2, x_3]$ yields an irreducible polynomial character of $SL_3(\mathbb{C})$. Two Schur functions define the same character iff they are congruent modulo the ideal generated by the polynomial $x_1 x_2 x_3 - 1$. In particular, the characters produced by $s_{(\lambda_1,\lambda_2,\lambda_3)}$ and $s_{(\lambda_1-\lambda_3,\lambda_2-\lambda_3,0)}$ are the same.

The product of two Schur functions $s_\lambda, s_\mu \in \Lambda[x_1, x_2, x_3]$ is a linear combination of Schur functions with non-negative integer coefficients:

$$s_\lambda s_\mu = \sum_\nu c_{\lambda\mu}^\nu s_\nu.$$

The coefficients $c_{\lambda\mu}^\nu$ are defined by the Littlewood-Richardson rule [M]. It yields the following formula

$$s_{(1,0,0)}s_{(\mu_1,\mu_2,0)} = \begin{cases} s_{(\mu_1+1,\mu_2,0)} + s_{(\mu_1,\mu_2+1,0)} + s_{(\mu_1,\mu_2,1)} & \text{if } \mu_1 > \mu_2 > 0 \\ s_{(\mu_1+1,\mu_2,0)} + s_{(\mu_1,\mu_2,1)} & \text{if } \mu_1 = \mu_2 > 0 \\ s_{(\mu_1+1,0,0)} + s_{(\mu_1,1,0)} & \text{if } \mu_2 = 0 \end{cases}$$

Taking into account that $s_{(\mu_1,\mu_2,1)}$ and $s_{(\mu_1-1,\mu_2-1,0)}$ yield the same character, namely, χ_{μ_1-1,μ_2-1}, we obtain that

$$\chi_{1,0}\chi_{\mu_1,\mu_2} = \begin{cases} \chi_{\mu_1+1,\mu_2} + \chi_{\mu_1,\mu_2+1} + \chi_{\mu_1-1,\mu_2-1} & \text{if } \mu_1 > \mu_2 > 0 \\ \chi_{\mu_1+1,\mu_2} + \chi_{\mu_1-1,\mu_2-1} & \text{if } \mu_1 = \mu_2 > 0 \\ \chi_{\mu_1+1,0} + \chi_{\mu_1,1} & \text{if } \mu_2 = 0 \end{cases}$$

This gives us a NITA which satisfies (5.17). This completes the proof of Theorem 5.2 of this chapter.   □

It is worthwhile to note that polynomial irreducible representations of the group $SL_n(\mathbb{C})$ form an Infinite Dimensional Normalized Integral Table Algebra which contains a non-real faithful element of degree n.

References

[AB]   Arad, Z., Blau, H.: On table algebras and applications to finite group theory. J. Algebra **138**, 137–185 (1991)

[AC]   Arad, Z., Chen, G.: On four normalized table algebras generated by a faithful nonreal element of degree 3. J. Algebra **283**, 457–484 (2005)

[B]   Blau, H.: Table algebras, Eur. J. of Comb. **30**, 1426–1455 (2009)

[M]   Macdonald, I.G.: Symmetric Functions and Hall Polynomials, Oxford Mathematical Monographs, 2nd edn. The Clarendon Press, Oxford University Press (1995)

Index

Z. Arad et al., *On Normalized Integral Table Algebras (Fusion Rings)*,
Algebra and Applications 16, DOI 10.1007/978-0-85729-850-8,
© Springer-Verlag London Limited 2011